Electric Power Systems

Electric Power Systems

Third Edition Revised

B. M. WEEDY

Department of Electrical Engineering,
The University of Southampton

JOHN WILEY & SONS

CHICHESTER   NEW YORK   BRISBANE   SINGAPORE   TORONTO

Reprinted December 1989
Reprinted September 1990
Reprinted February 1991

Library of Congress Cataloging in Publication Data:

Weedy, B. M. (Birron Mathew)
  Electric power systems

  Includes index.
  1. Electric power systems.   2. Electric power transmission.
  I. Title.
  TK1001.W4 1987   621.319′1   87-9187
  ISBN 0 471 91659 5

British Library Cataloguing in Publication Data:

Weedy, B. M.
  Electric power systems—3rd ed., rev.
  1. Electric power transmission
  I. Title
  621.319′1′   TK3001

  ISBN 0 471 91659 5

Printed in Great Britain by Courier International Ltd, Tiptree, Essex

Preface to First Edition

In writing this book the author has been primarily concerned with the presentation of the basic essentials of power system operation and analysis to students in the final year of first degree courses at Universities and Colleges of Technology. The emphasis is on the consideration of the system as a whole rather than on the engineering details of its constituents, and the treatment presented is aimed at practical conditions and situations rather than theoretical nicety.

In recent years the contents of many undergraduate courses in electrical engineering have become more fundamental in nature with greater emphasis on electromagnetism, network analysis, and control theory. Students with this background will be familiar with much of the work on network theory and the inductance, capacitance, and resistance of lines and cables, which has in the past occupied large parts of textbooks on power supply. In this book these matters have been largely omitted resulting in what is hoped is a concise account of the operation and analysis of electric power systems. It is the author's intention to present the power system as a system of interconnected elements which may be represented by models, either mathematically or by equivalent electrical circuits. The simplest models will be used consistent with acceptable accuracy and it is hoped that this will result in the wood being seen as well as the trees. In an introductory text such as this no apology is made for the absence of sophisticated models of plant (synchronous machines in particular) and involved mathematical treatments as these are well catered for in more advanced texts to which reference is made.

The book is divided into four main parts, as follows:

(a) Introduction, including the establishment of equivalent circuits of the components of the system, the performance of which, when inter-connected, forms the main theme.
(b) Operation, the manner in which the system is operated and controlled to give secure and economic power supplies.
(c) Analysis, the calculation of voltage, power, and reactive power in the system under normal and abnormal conditions. The use of computers is emphasized when dealing with large networks.

(d) Limitations of transmittable power owing to the stability of the synchronous machine, voltage stability of loads, and the temperature rises of plant.

It is hoped that the final chapter will form a useful introduction to direct current transmission which promises to play a more and more important role in electricity supply.

The author would like to express his thanks to colleagues and friends for their helpful criticism and advice. To Mr. J. P. Perkins for reading the complete draft, to Mr. B. A. Carre on digital methods for load flow analysis, and to Mr. A. M. Parker on direct current transmission. Finally, thanks are due to past students who over several years have freely expressed their difficulties in this subject.

Southampton 1967   B. M. WEEDY

Preface to Third Edition

Since the appearance of the second edition the overall energy situation has changed considerably and this has created great interest in alternative sources of energy and energy conservation. Although this does not effect the basic theory and operation of power systems it does influence policies which have considerable impact on electric power supply. Chapter 1 has been enlarged and now includes a critical summary of new energy sources and conservation measures, and in particular their possible impact on the electricity supply industry. In addition, the influences of environmental constraints are included in the discussion of generation and transmission.

One object of the second edition was to provide a text, mainly at under-graduate level, which would cover a wide range of power-system engineering, not merely network analysis. In furthering this aim a new section on overhead line design is now combined with the previous material on underground cables.

A further major change is the bringing together of introductory network material into a new chapter called Basic Concepts. This includes a summary of three-phase theory which it is hoped will ease the transition of students into the practical world of power systems. All chapters have been revised to bring the material up to date and to improve clarity.

Southampton 1978   B. M. WEEDY

Preface to Third, Revised Edition

Overall the third edition has remained up to date since its publication. However, in certain areas significant changes have occurred and to include these a new edition has been prepared. These changes are small in number and comprise; decoupled load-flows, a digital method for calculating system transients and an introduction to state estimation security analysis.

With the new material it is hoped that the book now includes a comprehensive account of power-system engineering at the senior undergraduate level.

Southampton 1987   B. M. WEEDY

Contents

Symbols Used

$\mathbf{A, B, C, D}$ = Generalized circuit constants

a–b–c = Phase rotation (alternatively R–Y–B)

a = Operator $1\underline{/120°}$

C = Capacitance (farad)

c_p = Specific heat at constant pressure (J/gm per °C)

D = Diameter

$\mathbf{E}$ = e.m.f. generated

F = Cost function (units of money per hour)

f = Frequency (Hz)

G = Rating of machine

g = Thermal resistivity (deg C m/w)

H = Inertia constant (seconds)

h = Heat transfer coefficient (watts/m^2 per deg C)

$\mathbf{I}$ = Current (A)

I_d = In-phase current

I_q = Quadrature current

j = $1\underline{/90°}$ operator

K = Stiffness coefficient of a system (MW/Hz)

k = Thermal conductivity (W/(m°C))

L = Inductance (H)

1n = Natural logarithm

M = Angular momentum (J-s per rad or MJ-s per electrical degree)

N = Rotational speed (rev/min, rev/s, rad/s)

$\mathbf{P}$ = Propagation constant $(\alpha + j\beta)$

P = Power (W)

$\dfrac{\mathrm{d}P}{\mathrm{d}\delta}$ = Synchronizing power coefficient

p.f. = Power factor

p = Iteration number

Q = Reactive power (VAr)

q = Loss dissipated as heat (watts)

R = Resistance (ohms) also thermal resistance (deg C/w)

R–Y–B = Phase rotation (British practice)

$\mathbf{S}$ = Complex power = $P \pm jQ$

s = Slip

SCR = Short-circuit ratio

T = Absolute temperature (K)

t = Time

Δt = Interval of time

t = Off-nominal transformer tap ratio
U = Velocity
$\mathbf{V}$ = Voltage; ΔV scalar voltage difference
W = Volumetric flow of coolant (m^3/s)
X' = Transient reactance of a synchronous machine
X'' = Subtransient reactance of a synchronous machine
X_d = Direct axis synchronous reactance of a synchronous machine
X_q = Quadrature axis reactance of a synchronous machine
X_s = Synchronous reactance of a synchronous machine
$\mathbf{Y}$ = Admittance (p.u. or ohms)
$\mathbf{Z}$ = Impedance (p.u. or ohms)
$\mathbf{Z}_0$ = Characteristics or surge impedance (ohms)
α = Delay angle in rectifiers and inverters—d.c. transmission
α = Attenuation constant of line
α = Reflexion coefficient
β = Phase-shift constant of line
$\beta = (180 - \alpha)$—in inverters
β = Refraction coefficient $(1 + \alpha)$
γ = Commutation time in converters
δ = Load angle of synchronous machine or transmission angle across a system—electrical degrees
δ_0 = Recovery angle of valve
ε = Permittivity
η = Viscosity (gm/(cm-s))
θ = Temperature rise (degC) above reference or ambient
λ = Lagrange multiplier
ρ = Electrical resistivity (Ω-m)
ρ = Density (kg/m^3)
τ = Time constant
ϕ = Angle between voltage and current phasors (power factor angle)
ω = Angular frequency (pulsatance) rad/s

Subscripts 1, 2, and 0 = Refer to positive, negative, and zero symmetrical components respectively

I

Introduction

1.1 Historical

In 1882 Edison inaugurated the first central generating station in the United States. This had a load of 400 lamps each consuming 83 W. At about the same time the Holborn Viaduct Generating Station in London was the first in Britain to cater for consumers generally, as opposed to specialized loads. This scheme comprised a 60 kW generator driven by a horizontal steam engine; the voltage of generation was 100 V direct current.

The first major alternating current station in Great Britain was at Deptford where power was generated by machines of 10,000 h.p. and transmitted at 10 kV to consumers in London. During this period the battle between the advocates of alternating current and direct current was at its most intense and acrimonious level. During this same period similar developments were taking place in the U.S.A. and elsewhere. Owing mainly to the invention of the transformer the advocates of alternating current prevailed and a steady development of local electricity generating stations commenced, each large town or load centre operating its own station.

In the U.S.A. development over the years has resulted in many electric utilities, some serving their own localities, others very large areas. These range from investor-owned, municipal, federal, and state organizations, to cooperatives. Investor-owned utilities produce by far the greatest quantity of electricity, followed by federal and municipal authorities. The cooperatives have been largely concerned with making electricity available in rural areas. The growth of installed generating capacity in the U.S. is shown in Figure 1.1 Tremendous capital investment is required continuously to meet increased demands. Much of this is acquired from the public (an estimated $12 billion in 1974) in investor-owned utilities in the form of bonds and stock issues. Only 10 per cent of construction capital is obtained from retained earnings and the problem of financing new schemes is of critical concern in investor-owned utilities.

The variation of rate of growth of electricity usage in the U.S. is shown in Figure 1.2 with the projected value to 1990.[23] Also shown in Figure 1.2 is the percentage change in the U.S. Gross National Product (GNP) indicating the close relationship between this quantity and the use of electricity.

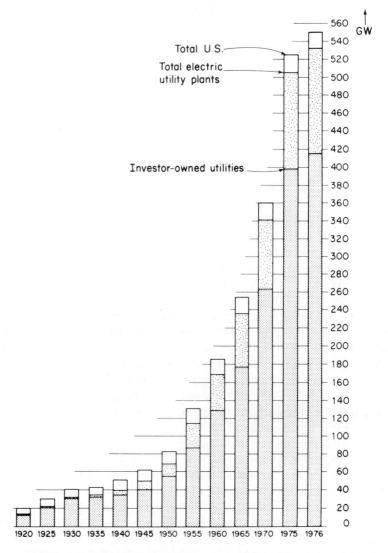

Figure 1.1 Generating capacity by year—U.S.A. (*Permission of Edison Electric Institute.*)

The U.S.A. comprises a mixed variety of organizations producing and distributing electricity, with the investor-owned sector predominating. In most other countries electricity supply is now essentially an agency of central government. In the early days the industry consisted mainly of municipally run local companies, but with the use of large generating stations and bulk transmission systems national control has become the norm.

In 1926 in Britain an Act of Parliament set up the Central Electricity Board with the object of interconnecting the best of the 500 stations then in operation

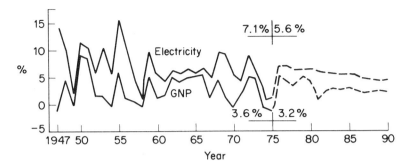

Figure 1.2 U.S.A.—yearly variation of percentage electricity growth and gross national product. (*Permission of General Electric*)

with a high-voltage network known as the *grid*. In 1948 the British supply industry was nationalized and two organizations set up: (a) the *area boards* which are mainly concerned with distribution and consumer service; and (b) the *generating boards* responsible for generation and the operation of the high-voltage transmission network or grid.

The growth of electricity consumption for the world and the U.S. are shown in Figure 1.3 During the 1970s dramatic changes in the general situation occurred, mainly as a result of large increases in oil prices. In many countries this has reduced the expected increases in the consumption of electricity and other forms of energy. In the light of the world shortage of oil and natural gas which may occur during the careers of most readers of this book, these events could be looked on as an effective warning that serious thought to alternatives must be given at once. In this situation it is difficult realistically to see any alternative to the increasing role of electricity as the major energy form. Furthermore, the generation of this electricity will be based on coal and nuclear fission for many years. There is no short-term shortage of coal, but the nuclear issue is clouded by safety and environmental concerns, especially regarding the breeder reactor. However, a realistic look at alternatives for large-scale generation, e.g. solar, ocean, geothermal, must point to an overwhelming reliance on coal and nuclear power into the twenty-first century. Hopefully, in the not too distant future, fusion will solve the problem of energy resources. There is, therefore, not an energy shortage but a need to utilize effectively the forms of energy available and also to ensure that new forms of energy sources are developed in time to replace depleted fossil fuels.

1.2 Characteristics Influencing Generation and Transmission

There are three main characteristics of electricity supply which, however obvious, have a profound effect on the manner in which it is engineered. They are as follows:

(a) Electricity, unlike gas and water, cannot be stored and the supplier has small control over the load at any time. The control engineers endeavour to

4

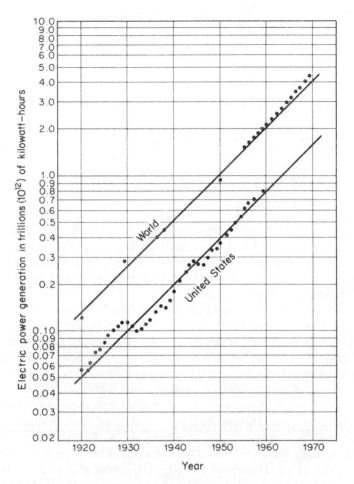

Figure 1.3   World growth rate for electricity. (*Permission of Edison Electric Institute, Historic Studies of the Electric Industry.*)

keep the output from the generators equal to the connected load at the specified voltage and frequency; the difficulty of this task will be apparent from a study of the daily load curves in Figure 1.4. It will be seen that the load consists basically of a steady component known as the 'base load', plus peaks depending on the time of day, popular television programmes, and other factors. The effect of an unusual television programme is shown in Figure 1.5.

(b) There is a continuous increase in the demand for power as indicated in Figure 1.3. Although in many industrialized countries the rate of increase has declined in recent years, even the modest rate entails massive additions to the existing systems. A large and continuous process of adding to the system thus exists. Networks are evolved over the years rather than planned in a clear-cut manner and then left untouched.

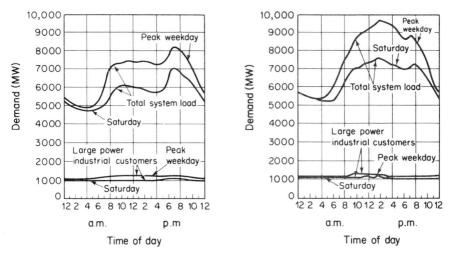

Figure 1.4 Daily load curves (U.S. utility) showing industrial component. (a) Winter. (b) Summer. (*Copyright © 1976 by the Institute of Electrical and Electronics Engineers, Inc. Reprinted by permission from I.E.E.E. Spectrum, Vol. 13, No. 9 (Sept. 1976), pp. 50–53.*)

(c) The distribution and nature of the *fuel* available. This aspect is of great interest as coal is mined in areas not necessarily the main load centres; hydroelectric power is usually remote from the large load centres. The problem of station siting and transmission distances is an involved exercise in economics. The greater use of nuclear energy will tend to modify the existing pattern of supply.

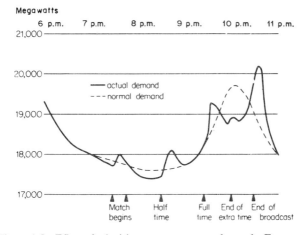

Figure 1.5 Effect of television programmes on demand—European Cup soccer match televised in Britain on 29 May 1968. Peaks caused by connexion of kettles, etc. in intervals and at end. (*Permission of Central Electricity Generating Board.*)

6

(d) In recent years environmental considerations have assumed major importance and influence the siting, construction cost, and operation of generating plants. Planning is also affected because of delays in making a start to projects because of legal proceedings, etc. Of particular importance at the present time is the question of the environmental impact of nuclear plants, especially the proposed fast breeder reactor.

1.3   Energy Conversion Employing Steam

The combustion of coal or oil in boilers produces steam at high temperatures and pressures which is passed to steam turbines. Oil has economic advantages when it can be pumped from the refinery through pipelines direct to the boilers of the generating station. The use of energy resulting from nuclear fission is being progressively extended in electricity generation; here also the basic energy is used to produce steam for turbines. The axial-flow type of turbine is in common use with several cylinders on the same shaft.

The steam power-station operates on the Rankine cycle, modified to include superheating, feed-water heating, and steam reheating. Increased thermal efficiency results from the use of steam at the highest possible pressure and temperature. Also, for turbines to be economically constructed the larger the size the less the capital cost. As a result turbogenerator sets of 500 MW and over are now being used. With steam turbines of 100 MW capacity and over the efficiency is increased by reheating the steam after it has been partially expanded, by an external heater. The reheated steam is then returned to the turbine where it is expanded through the final stages of blading. A schematic diagram of a coal-fired station is shown in Figure 1.6. In Figure 1.7 the flow of energy in a modern steam station is shown. Despite continual advances in the design of boilers and in the development of improved materials, the nature of the steam cycle is such that efficiencies are comparatively low and vast

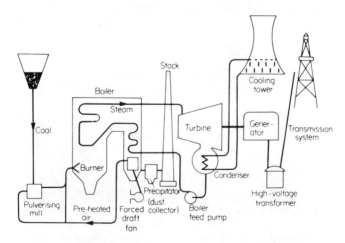

Figure 1.6   Schematic view of coal-fired generating station.

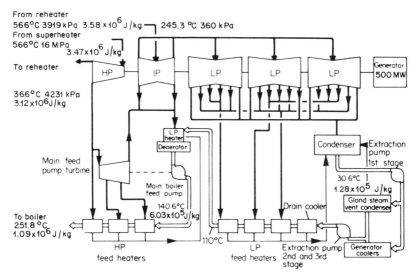

Figure 1.7 Energy flow diagram for a 500 MW turbo-generator.
(*Permission of Electrical Review.*)

quantities of heat are lost in the condensate. However, the great advances in design and materials in the last few years have increased the thermal efficiencies of coal stations to about 40 per cent.

Progress in improving efficiencies by increased steam pressures and temperatures is shown in Figures 1.8 and 1.9. Efficiency is improved and capital cost reduced by the use of large turbine/generator units. The continued increase in the size of units is shown in Figure 1.10 and the improvement in coal weight per kilowatt-hour of electricity produced over the years is shown in Figure 1.11.

In coal-fired stations coal is conveyed to a mill and crushed into fine powder i.e. pulverised. The pulverized fuel is blown into the boiler where it mixes with a

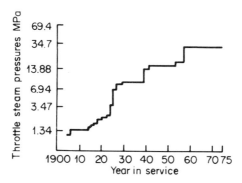

Figure 1.8 Maximum steam pressures with year (1 p.s.i. $= 6.89 \times 10^3$ N/m^2). (*Permission of General Electric.*)

8

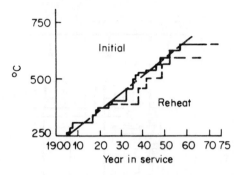

Figure 1.9 Maximum steam temperatures with year. (*Permission of General Electric.*)

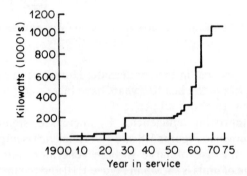

Figure 1.10 Largest generator installed with year. (*Permission of General Electric.*)

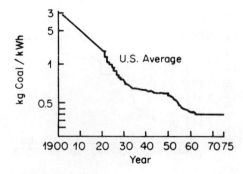

Figure 1.11 Coal used per kilowatt-hour of electricity produced—U.S. average (12,500 BTU coal). (*Permission of General Electric.*)

supply of air for combustion. The exhaust from the LP turbine is cooled to form condensate by the passage through the condenser of large quantities of sea- or river-water. Where this is not possible cooling towers are used (see section on thermal pollution).

Fluidized-bed boilers For typical coals, combustion gases contain 0.2–0.3 per cent sulphur dioxide by volume. If the gas flow-rate through the granular bed of a grate-type boiler is increased the gravity pull is balanced by the upward gas force and the fuel-bed takes on the character of a fluid. In a travelling grate this increases the heat output and temperature. The ash formed conglomerates and sinks into the grate and is carried to the ash pit. The bed is limited to the ash-sintering temperature of 1050–1200°C. Secondary combustion occurs above the bed where CO burns to CO_2 and H_2S to SO_2. This type of boiler is undergoing extensive development and is attractive because of the lower pollutant level and better efficiency.

1.4 Energy Conversion Using Water

Perhaps the oldest form of energy conversion is by the use of water power. In the hydro-electric station the energy is obtained free of cost. This attractive feature has always been somewhat offset by the very high capital cost of construction, especially of the civil engineering works. Today, however, the capital cost per kilowatt of hydroelectric stations is becoming comparable with that of steam stations. Unfortunately, the geographical conditions necessary for hydro-generation are not commonly found, especially in Britain. In most highly developed countries hydroelectric resources are used to the utmost. There still exists great hydroelectric potential in many under-developed countries and this will doubtless be utilized as their load grows.

An alternative to the conventional use of water energy, pumped storage, enables water to be used in situations which would not be amenable to conventional schemes. The utilization of the energy in tidal flows in channels has long been the subject of speculation. The technical and economic difficulties are very great and few locations exist where such a scheme would be feasible. An installation using tidal flow has been constructed on the La Rance estuary in northern France where the tidal height range is 9.2 m (30 ft) and the tidal flow is estimated at 18,000 m^3/s.

Before discussing the types of turbine used, a brief comment on the general modes of operation of hydroelectric stations will be given. The vertical difference between the upper reservoir and the level of the turbines is known as the head. The water falling through this head gains kinetic energy which it then imparts to the turbine blades. There are three main types of installation as follows:

(a) *High Head* or *Stored*—the storage area or reservoir normally fills in over 400 h;

(b) *Medium Head* or *Pondage*—storage fills in 200–400 h;

(c) *Run of River*—storage fills in less than 2 h and has 3–15 m head.

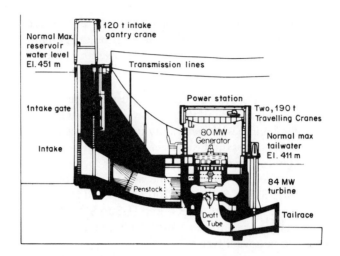

Figure 1.12 Hydroelectric scheme—Kainji, Nigeria. Section through the intake dam and power house. The scheme comprises an initial four 80 MW Kaplan turbine sets with the later installation of eight more sets. Running speed 115.4 rev/min. This is a large-flow scheme with penstocks 9 m in diameter. (*Permission of Engineering.*)

A schematic diagram for type (c) is shown in Figure 1.12

Associated with these various heights or heads of water level above the turbines are particular types of turbine. These are:

(a) *Pelton*   This is used for heads of (184–1840 m (600–6000 ft) and consists of a bucket wheel rotor with adjustable flow nozzles.
(b) *Francis*   Used for heads of 37–490 m (120–1600 ft) and is of the mixed flow type.
(c) *Kaplan*   Used for run of river and pondage stations with heads of up to 61 m (200 ft). This type has an axial-flow rotor with variable-pitch blades.

Typical efficiency curves for each type of turbine are shown in Figure 1.13. As the efficiency depends upon the head of water which is continually fluctuating, often water consumption in cubic meters per kilowatt-hour is used and is related to the head of water. Hydroelectric plant has the ability to start up quickly and the advantage that no losses are incurred when at a standstill. It has great advantages, therefore, for generation to meet peak loads at minimum cost, working in conjunction with thermal stations. By using remote control of the hydro sets, the time from the instruction to start up to the actual connexion to the power network can be as short as 2 min.

At certain periods when water availability is low or when generation is not required from hydro sets, it may be advantageous to run the electric machines as motors supplied from the power system. These then act as synchronous compensators, to be discussed in Chapter 5. To reduce the amount of power

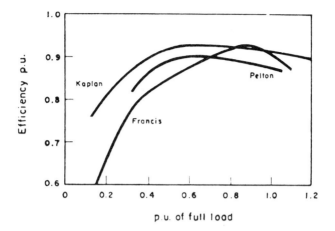

Figure 1.13 Typical efficiency curves of hydraulic turbines.

required the water is pushed below the turbine runner by compressed air. This is achieved by closing the water inlet valve and injecting compressed air which pushes the water towards the lower reservoir. The runner now rotates in air and thus requires much less motive power than in water.

The power available from a hydro scheme is given by

$$P = \rho g W H \ \text{(W)}$$

where

$W =$ flow rate (m^3/s) through the turbine,

$\rho =$ density (1000 kg/m^3),

$g = 9.81$ m/s^2,

$H =$ head, i.e. height of upper water level above the lower (m).

Substituting,

$$P = 9.81 W H \ \text{kW}$$

Tides

An effective method of utilizing the tides is to allow the incoming tide to flow into a basin, thus operating the turbine, and then at low tide to release the stored water, again operating a set of turbines. This gives continuous if varying head operation. If the tidal range from high to low water is h (m) and the area of water enclosed in the basin is A (m^2), then the energy in the full basin

$$= \rho g A \int_0^h x \, dx$$

$$= \tfrac{1}{2} \rho g h^2 A$$

The total energy for both flows is therefore twice this value, and the average power is $\rho g A h^2 / T$, where T is the period of tidal cycle, normally 12 h 44 min. The number of sites with good potential is small, typical of those which have been studied are listed below along with values of h, A, and mean power.

Passamaquoddy Bay (N. America)	5.5 m, 262 km², 1800 MW
Minas-Cohequid (N. America)	10.7, 777, 19,900
San Jose (S. America)	5.9, 750, 5870.
Severn (U.K.)	9.8, 70, 1680

1.5   Gas Turbines

The use of the gas turbine as a prime mover has certain advantages over steam plant, although with normal running it is less economical to operate. The main advantages lies in the ability to start and take up load quickly. Hence the gas turbine is coming into use as a method for dealing with the peaks of the system load. At the time of writing there are many installations, many generators being of 100 MW output. A further use for this type of machine is as a synchronous compensator to assist with maintaining voltage levels. Even on economic grounds it is probably advantageous to meet peak loads by starting up gas turbines from cold in the order of 2 min rather than running spare steam plant continuously.

The installation consists of a turbine, a combustion chamber, and a compressor driven by the turbine. The compressed air is delivered to the combustion chamber where continuous combustion of the injected fuel oil is maintained; the hot gases then drive the turbine.

1.6 . Magnetohydrodynamic (MHD) Generation

Whether the fuel used is coal, oil, or nuclear, the result is the production of steam which then drives the turbine. Attempts are being made to generate electricity without the prime mover or rotating generator. In the magneto-hydrodynamic method, gases at 2500°C are passed through a chamber in which

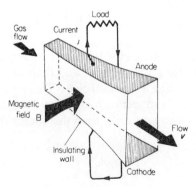

Figure 1.14   The principle of MDH power generation. (*Permission of English Electric Co. Ltd.*)

a strong magnetic field has been created (Figure 1.14). If the gas is hot enough it is electrically slightly conducting (it is seeded with potassium to improve the conductivity) and constitutes a conductor moving in the magnetic field. An e.m.f. is thus induced which can be collected at suitable electrodes. Great efforts are at present being made in several countries to construct an economical large MHD generator. In its practical form it will be used in conjunction with traditional plant and Figure 1.15 shows a schematic layout of a possible system incorporating normal steam operation plus MHD generation. There are, however, great doubts as to the economic viability of this form of generation.

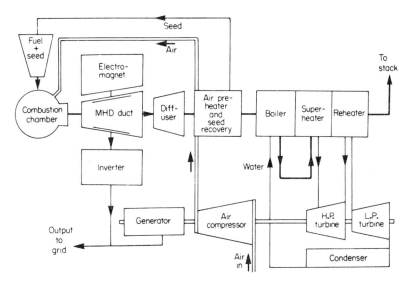

Figure 1.15   Block diagram of open cycle MHD power station.
(*Permission of English Electric Co. Ltd.*)

1.7   Nuclear Power

Fission

So far power has only been successfully obtained from the fission reaction which involves the splitting of a nucleus. Compared with chemical reactions very large amounts of energy are released per atomic event with both fusion and fission, typically in the range 10–200 MeV (15–300×10^{-13} J). Metal extracted from the base ore consists mainly of two isotopes, uranium-238 (99.3 per cent by weight) and U-235 (0.7 per cent). Only uranium-235 is fissile, i.e. when struck by slow-moving neutrons its nucleus splits into two substantial fragments plus several neutrons and 3×10^{-11} J of kinetic energy. The fast-moving fragments hit surrounding atoms producing heat before coming to rest. The neutrons travel further, hitting atoms and producing further fissions. Hence the number of neutrons increases, causing, under the correct conditions, a chain reaction. In conventional reactors the core or moderator slows down the moving neutrons to achieve more effective splitting of the nuclei.

Fuels used in reactors have some component of U-238. Natural uranium is sometimes used, and although the energy density is considerably less than for the pure isotopes it is still much better than fossil fuels. The uranium used at present comes from metal-rich ores, but these are a limited world resource (about 2×10^6 tons) and the requirement for a reactor which breeds fuel, the breeder reactor, in the long term is essential. The energy breakdown in the fission process is as follows (in MeV): kinetic energy of fission fragments 168, KE of neutrons 5, gamma radiation 5, beta and gammas emitted by fission products 7 and 6 respectively, and neutrinos 11.

When struck by the neutrons certain non-fissile materials, if placed around the core of a reactor, are transformed into fissionable material. For example, U-238 and thorium-232 are converted into plutonium-239 and uranium-233. When more fissionable material is produced than is consumed the reactor is said to breed.

The basic reactor consists of the fuel in the form of rods or pellets situated in an environment (moderator) which will slow down the neutrons and fission products and in which the heat is evolved. The moderator can be light or heavy water or graphite. Also situated in the moderator are movable rods which absorb neutrons and hence exert control over the fission process. In some reactors the cooling fluid is pumped through channels to absorb the heat which is then transferred to a secondary loop in which steam is produced for the turbine. In water reactors the moderator itself forms the heat-exchange fluid.

There are a number of versions of the reactor in use with different coolants and types of fissile fuel. In Britain the Magnox reactor has been used in which natural uranium in the form of rods is enclosed in magnesium alloy cans. The fuel cans are placed in a structure or core of pure graphite made up of bricks called the *moderator*. This graphite core slows down the neutrons to the correct range of velocities in order to provide the maximum number of collisions. The fission process is controlled by the insertion of control rods made of neutron-absorbing material; the number and position of these controls the heat output of the reactor. Heat is removed from the graphite via carbon dioxide gas pumped through vertical ducts in the core. This heat is then transferred to water to form steam via a heat exchanger. Once the steam has passed through the high-pressure turbine it is returned to the heat exchanger for reheating as in a coal- or oil-fired boiler. A schematic diagram showing the basic elements of such a reactor is shown in Figure 1.16.

A more recent reactor similar to the Magnox is the *advanced gas-cooled reactor (AGR)*. A reinforced-concrete steel-lined pressure vessel contains the reactor and heat exchanger. Enriched uranium dioxide fuel in pellet form, encased in stainless steel cans, is used; a number of cans form a cylindrical fuel element which is placed in a vertical channel in the core. Carbon dioxide gas at a higher pressure than in the Magnox type removes the heat. The control rods are made of boron steel. Spent fuel elements when removed from the core are

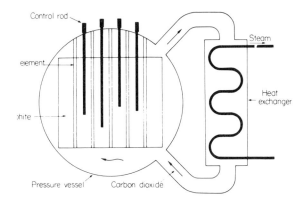

Figure 1.16   Schematic view, nuclear reactor—British Magnox type.

stored in a special chamber for about a week and then dismantled and lowered
into a pond of water where they remain until the level of radioactivity has
decreased sufficiently for them to be removed from the station.

In the U.S.A. pressurized water and boiling water reactors are used. In the
pressurized water type the water is pumped through the reactor and acts as a
coolant and moderator, the water being heated to 315°C. The steam pressure is
greater than the vapour pressure at this temperature and the water leaves the
reactor at below boiling point. The fuel is in the form of pellets of uranium
dioxide in bundles of stainless steel tubing. The *boiling water reactor* was
developed later than the above type and is now used extensively. Inside the
reactor heat is transferred to boiling water at a pressure of $690 \, \text{N/cm}^2$.
Schematic diagrams of these reactors are shown in Figures 1.17 and 1.18.

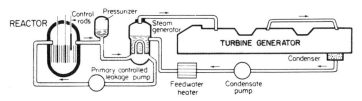

Figure 1.17   Schematic diagram of a pressurized water reactor.
(*Permission of Edison Electric Institute.*)

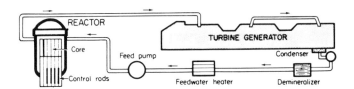

Figure 1.18   Schematic diagram of a boiling water reactor. (*Permission of Edison Electric Institute.*)

All reactors that use uranium produce plutonium in the reaction. Most of this is not utilized in the reactor. *Fast breeder reactors* which are being developed at the present time breed new fuel in considerable quantities during the reaction, as well as producing heat. In the liquid-metal fast breeder reactor, shown in Figure 1.19 liquid sodium is the coolant which leaves the reactor at 650°C at atmospheric pressure. The heat is then transferred to a secondary sodium circuit which transfers it via a heat exchanger to produce steam at 540°C.

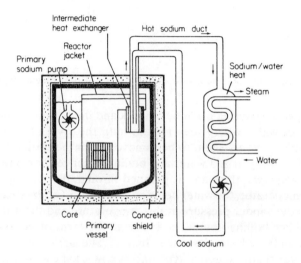

Figure 1.19   Schematic diagram of a liquid-metal fast breeder reactor. (*Permission of Edison Electric Institute.*)

Both pressurized and boiling water reactors use *light water*. The practical pressure limit for the pressurized water reactor is about 167 bar (2500 p.s.i.) which limits its efficiency to about 30 per cent. However, the design is relatively straightforward and experience has shown them to be stable and dependable. In the boiling water reactor the efficiency of heat removal is improved by use of the latent heat of evaporation. The steam produced flows directly to the turbine, causing possible problems of radioactivity in the turbine. The fuel for both water reactors is enriched to 1.44 per cent U–235. These reactors are probably the cheapest to construct, however the steam produced is saturated requiring wet-steam turbines, although the possibilities of producing superheated steam are being investigated. A further type of water reactor is the heavy-water CANDU type developed by Canada. Its operation and construction are similar to the light water variety.

In the high-temperature reactor (HTR) under development, steam is provided at the turbine at 540°C and 100 bar, improving the thermal efficiency to 35 per cent. The fuel is 13.8 per cent enriched in U-235, the moderator is graphite, and the core is gas (helium) cooled.

Fast breeder reactors In these no deliberate attempt is made to slow down the neutrons, although collisions with structural and fuel materials have this effect. The power per kilogram of fuel is high and hence the core is small. It is essential, therefore, that the coolant has good thermal properties and sodium is used. The ability to transform U-238 into plutonium-239 increases the known nuclear energy reserves very considerably and brings them to about a factor of 10 times the coal reserves. Also, the mining of lower-grade ores becomes an economic possibility. The fuel for a fast breeder consists of 20 per cent plutonium plus 80 per cent uranium oxide.

Experience is being gained in fast breeder technology. In 1959 an experimental reactor at Dounreay in Scotland commenced operation at 15 MW and after many years of operation was shut down in 1977. A 250 MW prototype fast reactor is now built at Dounreay and is giving power. Similar reactors have been built in the U.S.S.R. and France. Criticism of the fast reactor is made on the grounds that it uses plutonium as a fuel, a toxic material and also capable of use in an atomic bomb. Processing of the material would be required before the latter use is possible. Typical power densities (MW/m^3) in fission reactor cores are as follows: gas cooled 0.53, high-temperature gas cooled 7.75, heavy water 18.0, boiling water 29.0, pressurized water 54.75, fast breeder reactor 760.0.

Safety and environmental considerations The translation of energy states in a nucleus creates the emission of γ- and β-rays, α and fission fragments. The half-lives of the substances created are: $^3_1H_2$ (tritium) 12.26 years, ^{90}Sr (strontium) 28.8 years, ^{137}Cs (caesium) 30.2 years, ^{131}I (iodine) 8 days, ^{85}Kr (krypton) 10.76 years and ^{133}Xe (xenon) 5.27 days. Generally, materials with a long half-life have a lower intensity of radioactivity than those with a short half-life. Tritium is produced in small amounts and mostly retained in the fuel. Xenon and krypton escape from fuel elements which have cladding defects and remain in free form in the coolant. Because of its long life krypton-85 constitutes the greater problem. In the water-cooled cores the fission and activation products are present in the coolant. The more active of such wastes are concentrated by evaporation, mixed with concrete, and shipped for storage. The lower level wastes are eventually released to the condenser cooling-water discharge at low concentration levels.

High-level wastes, e.g. strontium-90 are produced in processing the used fuel elements. At the moment the wastes are concentrated in liquid form and stored in stainless steel containers. The storage of such wastes creates great controversy, the material still being active after centuries. Any mistakes made now will create serious problems for future generations. With future development such long-lived wastes will probably be converted to solid form (e.g. glass) and stored underground in stable geological situations such as salt domes.

Any accident involving substantial heating and rupture of the structure will involve the release of fission fragments held in the fuel rods into the atmosphere. With a breeder reactor the release of plutonium, an extremely radiotoxic material, would add to the problem. In the design and construction of reactors great care is taken to cover every contingency. Many facilities, e.g. control systems, are at least duplicated with alternative electrical supplies.

One aspect which has received much attention has been the *loss of coolant accident* (LOCA). In this situation the fission process must be reduced to the lowest possible energy level by the control rods to prevent overheating and possible melt down of the core and structure. The sequence of events in a boiling water reactor subsequent to an LOCA has been anticipated as follows:

0 s	Break in water coolant pipe.
1 s	Shut-down control rods inserted, pumps started, emergency diesel generators started.
0–7 s	Water-steam leaves reactor, some cooling still operative.
7 s	Reactor core isolated.
8–10 s	Water in lower part of reactor boils rapidly.
10–30 s	Reactor core temperature rising rapidly.
30 s	Pumps supply water to spray top of core and to cause flooding.
1–2 min	Rod fractures may occur.
1 min	Channels full of water; 3 min Core is re-covered with water.

Stringent regulations have been laid down to provide a shield or covering to retain the core material and coolant inside the reactor assembly in such an accident. Similarly, an outer barrier to contain radiation and small pressure increases is provided.

Over the past few years there has been considerable controversy regarding the safety of reactors. Experience is still relatively small and situations have not as yet been revealed in which important safety features fail. However, human error is always a possibility. However, the health controls in the atomic power industry have from the outset been much more rigorous than in any other industry.

Fusion

Energy is produced by the combination of two light nuclei to form a single heavier one. Neutron emission is not required, the reaction being sustained by the very high temperature of the reactants which maintain continual collisions. The most promising fuels are isotopes of hydrogen known as deuterium (D) (mass 2) and tritium (T) (mass 3). The product of fusion is the helium isotope (mass 3), hydrogen, neutrons, and heat. As tritium is not a naturally occurring isotope it is produced in the reactor shield by the interaction of the fusion neutrons and the lithium isotope of mass 6. The dueterium–deuterium fusion requires higher temperatures than deuterium–tritium and the latter is more likely to be used initially.

Reserves of lithium have been estimated to be roughly equal to those of fossil fuels. Deuterium, on the other hand, is contained in sea water of a concentration of about 34 parts per million. The potential energy-resource is therefore vast.

The fusion reactions are as follows:

$$_1D^2 + _1D^2 \rightarrow _2He^3 + _0n^1 + 3.27 \text{ MeV}$$

$$_1D^2 + _1D^2 \rightarrow _1T^3 + _1H^1 + 4.03 \text{ MeV}$$

$$_1D^2 + _1T^3 \rightarrow _2He^4 + _0n^1 + 17.6 \text{ MeV}$$

$$_1D^2 + _2He^3 \rightarrow _2He^4 + _1H^1 + 18.3 \text{ MeV}$$

For the successful operation of a deuterium–tritium fusion reactor the temperature must be high enough for the power extracted to exceed the power put in plus losses. Also the number of reacting nuclei must be sufficient to maintain the reaction. The temperature should be above 8×10^7 K and the nuclei density about $10^{15}/\text{cm}^3$. In order to contain these high temperatures in the plasma it must be kept away from the containing vessel walls as there is a natural tendency for the plasma to expand and extinguish. This may be achieved by transverse and coaxial magnetic fields created by coils around a toroidal vessel as in the Russian Tokamak system as shown in Figure 1.20.

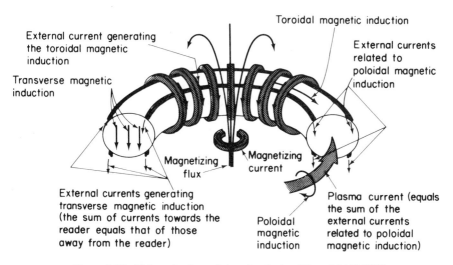

Toroidal magnetic induction

External current generating the toroidal magnetic induction

External currents related to poloidal magnetic induction

Transverse magnetic induction

Magnetizing flux

Magnetizing current

External currents generating transverse magnetic induction (the sum of currents towards the reader equals that of those away from the reader)

Poloidal magnetic induction

Plasma current (equals the sum of the external currents related to poloidal magnetic induction)

Figure 1.20   Tokamak scheme for nuclear fusion. (*Copyright © 1977 by the Institute of Electrical and Electronics Engineers, Inc. Reprinted by permission from I.E.E.E. Spectrum, Vol. 14, No. 7 July 1977), pp. 32–38.*

At these temperatures the fuel is in the form of ionized gas or plasma, i.e. the outer electrons are freed from their atoms and the gas conducts electricity. For useful amounts of energy to be obtained the plasma must be at the required temperature for more than a critical time. The product of plasma density and confinement time is all-important, tritium giving the lowest value of 10^{13}–10^{14} cm^{-3}-s. The working value to give useful energy must be higher than this. In the steady-state reactor, fuel is continuously fed into the plasma and spent fuel continuously removed. In the pulsed version an initial charge of fuel is ignited and burns for the prescribed time, after which the spent fuel is removed and a fresh charge introduced and ignited. The Tokamak concept is intermediate between these two, involving cyclic burning (20 min) and down times (1 min). In one proposed design, energy is transferred via two heat-transfer loops, supplying heat to a conventional steam system.

Based on radioactive considerations the impact of fusion reactors would be much less than with fission reactors. In fission reactors the loss of coolant accident and the 'after-heat' generated after shut-down (fission which continues after full 'shut-down' control action) may lead to vaporization and dispersal of radioactive material. With fusion there is much less power density under these conditions, possibly 1/50th of the equivalent value for a fast breeder reactor. The main source of radioactive waste from fusion reactors would be the structural material which undergoes damage due to radiation and hence requires occasional renewal. This could be recycled after a 50-year period, compared with centuries for strontium-90 and caesium-137 from fission.

About 90 per cent of fusion costs would be capital expenditure compared with 70 per cent for light water fission reactors (20 per cent fuel and 10 per cent operation and maintenance). A figure of \$2000/kW (1976 dollars) has been suggested for a Tokamak reactor, although it is too early to suggest hard figures. Certainly the generation of electricity via fusion would solve the long-term fuel shortages with a minimum of environmental problems. Intensive international research is being undertaken to demonstrate the scientific feasibility of a large-scale reactor. If this is successful the engineering problems are still formidable and the project must be viewed in the long term. Attempts are being made to demonstrate a commercial reactor by the year 2000.

1.8 Generation and Fuel

With new and more efficient generating sets being brought into operation there exists a wide range of plant available for use. As previously mentioned, the load consists of a base plus a variable element, depending on the time of day and other factors. Obviously the base load should be supplied by the more efficient plant which then runs 24 hr per day with the remaining load met by the less efficient stations. In addition to the machines supplying the load, a certain proportion of available plant is held in reserve to meet sudden contingencies. A proportion of this reserve must be capable of being brought into operation in 5 min and hence some machines must be run at, say, 75 per cent of full load to allow for this spare generating capacity.

Reserve margins are allowed in the *total* generation plant available to cope with unavailability of plant due to faults and maintenance. It is common practice to allow a planned margin of about 20 per cent over the annual peak demand. In a power system there is a mix of plants, i.e. hydro, coal, oil and nuclear, and gas turbine. The optimum mix gives the most economic operation, but this is highly dependent on fuel prices which can fluctuate with time and from region to region. Hence in the U.S. with plentiful coal resources in the west–north–central and also in certain eastern areas, coal plants would be expected to be predominant in these regions. In the south and other areas not endowed with coal, nuclear would tend to be dominant. Of course, hydro would be exploited wherever possible for cost and environmental reasons.

Plant costs are complicated by inflation and new regulatory requirements. Nuclear and coal plant costs in the U.S.A. rose approximately by 15 per cent per annum from 1971 to 1975. Anticipated costs for 1981 are given in Table 1.1.[24] Relatively small changes in fuel costs in some regions can radically effect the hypothetical mix of fuel types for generation. For a future situation, perhaps without oil, nuclear power will largely fill the gap, with gas turbines used for peak loading. With a constrained availability of capital, however, gas turbines would probably displace higher capital-cost nuclear plant. This would eventually raise electricity prices above those with no constraints on money supply.

Table 1.1

	Nuclear	Coal	Gas turbine
Plant cost $/kW in 1984	2450	1240	300
Fuel cost c/kWh 1975	0.15	0.12–0.4	0.7–0.8

Energy parks

A possible development which would effect costs and fuel types is the concept of the energy park in which very large power concentrations will exist alongside associated processes such as nuclear fuel processing. Apart from cost and legislative advantages such parks would close the fuel cycle for nuclear generation, minimizing the transport of nuclear fuel and wastes. Disadvantages include vulnerability to natural and man-made disasters, requirement of cooling water, concentration of environmental effects, and the need to move large blocks of power into the main supply network. Proposed nuclear parks include 20 1300 MW sets (heat rejection of 50 GW!) and an area of 2.73×10^8 m^2 for transmission right of ways at 765 kV (average length 280 km). A total capital cost of $1195/kW is quoted compared with a value of $1475/kW for dispersed sites (projected prices for late 1980s). For various reasons the capacity of a park should not exceed 10 per cent of the total utility capacity.

1.9 Unconventional Energy Sources

There is considerable international effort into the development of alternative energy sources to supplement fossil fuels. Many of the 'novel' sources (some of them in fact have been in use for centuries!) are in fact manifestations of solar energy, e.g. wind, sea waves, ocean thermal gradients, and photosynthesis. The average incident solar energy received on the earth's surface is about 600 W/m^2, but the actual value of course varies considerably. In the following section the potentialities of various methods of utilizing this energy will be discussed.

Solar energy—thermal conversion

There are two distinct applications, (a) space and water heating on a domestic scale, and (b) central station, large-scale heat collection, used for steam raising to generate electricity; both influence power systems. The former effects the load demands and in particular the problem utilities will face in having to provide a sufficient back-up supply to customers who normally would use solar, but in certain weather conditions would require large amounts of electricity. This involves the provision of the normal amount of utility plant but with much reduced sales of energy.

As the temperature of a solar collecting surface rises it radiates heat (infra-red). The energy distributions with wavelength of solar energy and infra-red radiation are shown in Figure 1.21. It is possible to design a selective

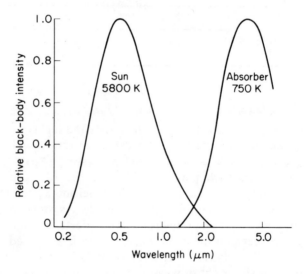

Figure 1.21 Relative black-body intensity of radiation with wavelength.

cover plate over the collecting surface such that it would pass nearly all the solar radiation and reflect all the radiated infra-red. Selective absorbers consist of a smooth metallic sheet covered with either a thin semiconducting surface or a finely divided metallic powder. The former reflects the infra-red and provides a good thermal contact between the hot absorbing layer and cooling fluid. A diagram of a simple collecting system is shown in Figure 1.22.

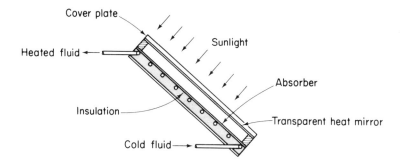

Figure.1.22   Simple solar energy panel for water heating.

The energy received by the collector per square meter (net):

$$q = I\alpha\tau - (\varepsilon_F + \varepsilon_B)\sigma(T^4 - T_0^4)$$

where ε_F and ε_B = front and back emissivities of absorber;

σ = Stefan–Boltzmann const. = $5.67 \times 10^{-8}/K^4 - m^2$;

τ = transmittance of cover plate (e.g. 0.93);

T_0 = temperature of cover plate (K);

I = incident radiation normal to surface;

$T\,(K)$ and α = temperature and absorptivity of absorbing panel.

   In large-scale (central station) installations the sun's rays may be concentrated by lenses or mirrors. Both require accurately curved surfaces and steering mechanisms to follow the motion of the sun. Concentrators may be designed to follow the sun's seasonal movement, or additionally to track the sun throughout the day (double-axis system). The former is less expensive and concentration factors of 30 have been obtained. However, in the French solar furnace in the Pyrenees two-axis mirrors are used and a concentration factor of 16,000 is achieved. A diagram of the central receiver system proposed for major generation of electricity is shown in Figure 1.23. The reflectors concentrate the rays on to a single receiver (boiler) hence raising steam. A collector area of 1 km^2 for each 100 MW (e) of output has been suggested with projected costs of 25–30 mill/kWh (1974 prices)* with capital costs of \$30/m^2 (mirrors etc.) and thermal storage costs of \$15 per kWh of electricity. A less attractive alternative to this scheme (because of the lower temperatures) is the use of many individual absorbers tracking the sun in one direction only, the thermal energy being transferred by a fluid (water or liquid sodium) to a central boiler.
   In all solar thermal schemes storage is essential because of the fluctuating nature of the sun's energy, although it has been proposed that the schemes be

* 1 mill = \$0.001.

24

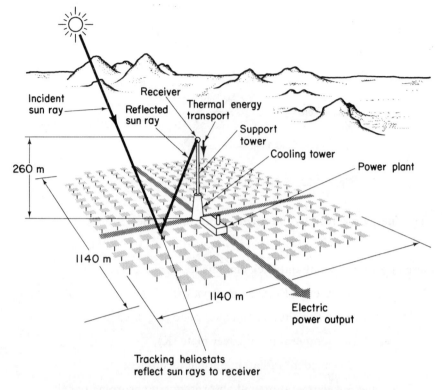

Figure 1.23   Central receiver scheme for electric power generated by
solar energy. (*Copyright © 1975 by the Institute of Electrical and
Electronics Engineers, Inc. Reprinted by permission from I.E.E.E.
Spectrum, Vol. 12, No. 12 (Dec. 1975), pp. 47–52.*)

used as pure fuel savers. This feature is common to all of the sources discussed
with the exception of geothermal and constitutes a very serious drawback.
Fluctuating sources, as well as fluctuating loads, would complicate still further
the process of electricity supply.

Solar energy—direct conversion to electricity

Photovoltaic conversion occurs in a thin layer of suitable material, e.g. silicon
when hole–electron pairs are created by incident solar photons and the
separation of these holes and electrons at a discontinuity in electrochemical
potential creates a potential difference. Whereas theoretical efficiencies are
about 25 per cent, practical values are lower. Single-crystal silicon and gallium-
arsenide cells have been constructed with efficiencies of 10 and 16 per cent
respectively. The cost of fabricating and interconnecting cells is high (used
mainly to date in spacecraft). Polycrystalline silicon films having large-area
grains (i.e. long continuous crystals) with efficiencies of over 10 per cent have
been made by techniques amenable to mass production. Although these

devices do not pollute they will, in the large-power context, occupy large areas. It has been estimated that to produce 10^{12} kWh per year (about 65 per cent of the 1970 U.S. generation value) the necessary cells would occupy about 0.1 per cent of the U.S. land area (highways occupied 1.5 per cent in 1975) assuming an efficiency of 10 per cent and a daily insolation of 4 kWh/m^2. Development of automated cell production to reduce costs to $500 per kW is in progress.

Other forms of conversion of lesser large-scale importance come under the heading of thermoelectricity. The Seebeck effect gives a potential difference between the hot and cold ends of joins of different metals, a typical value being 150 μV/K. Solar energy can heat a cathode of a diode-type tube from which electrons will be liberated by thermionic emission. These electrons drift to the anode and return through the external circuit. It is doubtful whether these devices will make any impact on the energy situation.

Windmills

The windmill is age-old and well tried. For electric generation purposes there are three scales of operation:

(a) Small, 0.5–10 kW for isolated single premises.
(b) Medium, 10–100 kW for communities.
(c) Large, say 1.5 MW for connection to power supply systems.

The best-developed rotors to date are the two- or three-bladed propeller type with horizontal axes. Energy output can be increased by a larger rotor and hence larger tower, but this must be traded against the extra capital cost, bearing in mind that wind speeds of up to 180 km/h must be withstood. The theoretical power in a wind stream is given by

$$\tfrac{1}{2}\rho A V^{3} (\text{W})$$

where

ρ = density of air (1201 gm/m^3 at NTP);

V = mean air velocity (m/s); and

A = swept area (m^2).

For a rotor of 17 m diameter and a velocity of 48 km/h the theoretical power is 265 kW. The practical values obtained are about half the theoretical. It is not economic to design windmills for the whole range of wind speed likely to be encountered and governor control is used which spills the excess wind. The output controlling mechanism is usually blade-pitch changing. To produce an optimum design the rated wind speed is assumed to be 1.5–2 times the average value and the average power is less than the rated value. A higher ratio means a large rotor for full power at low speeds and vice versa. If a 2 GW continuously operating steam plant is replaced by 2 MW wind-driven generators with mill rotor diameters of 70 m and a wind velocity of 6.7 m/s assumed (specific output of 1800 kWh per kW per year), then a total of 4800 (i.e. $2 \times 10^6 \times 8760/(1800 \times 2000)$) such machines would be required to produce the same amount of energy per year. These would occupy a very considerable area and by comparison reduce the environmental impact of overhead lines to insignificance!

Sea waves

The energy content of sea waves is very high. The Atlantic waves along the north-west coast of Britain have an average value of 80 kW/m of wave crest length. The energy is obviously very variable, ranging from greater than 1 MW/m for 1 per cent of the year to near zero for a further 1 per cent. Hence, over several hundreds of kilometres a vast source of energy is available.

A simple progressive sea wave of amplitude a, wavelength λ, and period T, in water of depth greater than $\lambda/2$, has a surface profile given by

$$y = a \sin (kx - \omega t),$$

$k = 2\pi/\lambda$ = wavenumber and $\omega = 2\pi/T$. The total energy (E) per square metre of wave surface area (kinetic and potential) $= \frac{1}{2}\rho g a^2$, where ρ = density of sea water. The power

$$P = E \times \text{velocity of propagation (group velocity)}$$

$$= \rho g^2 a^2 T/8\pi \text{ per unit width of wave front}$$

There are two kinds of waves:

(a) Ocean swell—long-wavelength, low-frequency waves generated by the wind probably thousands of kilometres distant and little attenuated over this distance. The energy and power are reasonably predicted by the above equations.

(b) Wind sea, which must be treated as the sum of many monochromatic waves of random phase, distributed both in direction and frequency. The principal direction is that of the wind. The energy and power of such waves are given in terms of statistical quantities obtained by measurement.

The sea motion can be converted into mechanical energy in a number of ways. One method is by using the Salter cam (or duck!) a cross-section of which

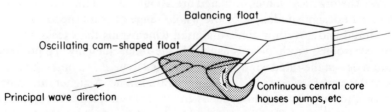

Figure 1.24 Conversion of wave energy by floating cam. (*Permission of Central Electricity Generating Board*).

is shown in Figure 1.24. The cam rotates about its axis and is shaped to minimize back-water pressure. Lengths of the cam could be moored say 80 km off the coast. Conversion of the float movement to electricity is difficult because of the slow oscillations. If reciprocating generators of some form are used, a device producing 1 MW while oscillating at 0.1 Hz with an amplitude of 15° would require a torque of 1200 t–m! Owing to the randomness of the waves the system must be rated much higher than the design average output level.

Two forms of energy conversion are being investigated. Firstly, a highly geared reciprocating alternator giving a variable-frequency output which is rectified. The combined outputs of several units would tend to smooth the d.c. waveform. Hydrogen could be produced by the varying d.c. output. Secondly, the energy could be transmitted hydraulically using an accumulator system to provide short-term storage before transfer to a turbogenerator set. The conversion equipment would be inside the cams and the electrical outputs connected to a floating substation by cable. Rectification would take place on the substation and the combined output of several cams transmitted to the shore by high-voltage d.c. cable.

The supply load factor is defined as energy available over a given period (up to a specified power level) divided by the energy if the specified power is maintained over the period. Supply load-factor curves based on measurements from 1955 to 1965 on a weather ship off the British coast are shown in Figure 1.25. Although the engineering problems associated with wave-power are formidable, the amount of energy available is vast and development work is in progress.

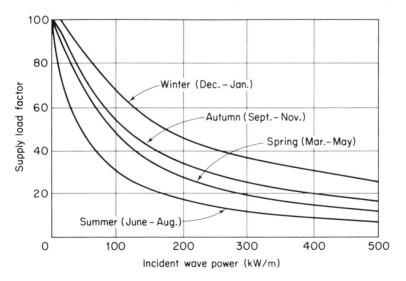

Figure 1.25 Load factor of energy supply from sea waves for various incident wave powers—recorded by a weather ship in the North Atlantic. (*Permission of Central Electricity Generating Board*).

Sea thermal-energy conversion

In 1881 D'Arsonval proposed the utilization of the temperature difference between the surface and lower layers of tropic seas. For practical purposes the layers need to be in reasonably close proximity to each other. The absorption of solar energy by the surface layers causes a thermal-syphon action, the warm surface water flowing towards the earth's poles from whence it moves back towards the tropics as cold water at a greater depth. The Gulf Stream carries, on average, about 2 km^3/min of warm water and the energy potential in this is vast. Warm and cold currents are close to each other off the Florida coast, the temperature difference range being 9–25°C. This gives a Carnot efficiency of about 3.4 per cent, and with losses an overall efficiency of say 2 per cent. Hence to produce large powers, vast quantities of water, and large process plants are required. On the other hand an Arctic Ocean scheme could utilize the temperature differential between the water (at say +2°C) and the subzero air temperature.

A proposed plant would be situated 25 km east of Miami, where the temperature difference is 17.5°C, and is shown in Figure 1.26. The best

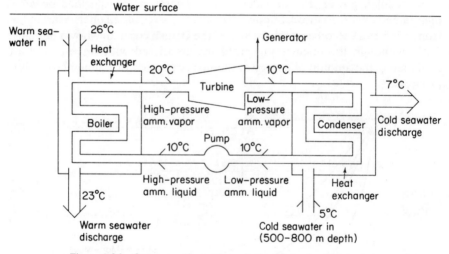

Figure 1.26 Concept of power plant using sea-temperature differences.

working fluids are ammonia (NH$_3$) or propane. A typical sea-water flow rate for a 400 MW installation is 115,000 m^3/min and the diameter of the cold-water pipe would need to be 24 m and the length 500–600 m. The plant is submerged and anchored. The warm water boils a high-pressure working fluid and the vapour expands through vapour turbines and then condenses back to a liquid in the condenser.

In 1972 the cost of producing electricity from fossil plants was 10–12 mill/kWh. Estimates for the sea thermal plant are between 14 and

37 mill/kWh. The surface temperature of the tropical oceans would be lowered by about 1°C if 60,000 GW of electricity is generated in this manner.

Geothermal energy

In most parts of the world the vast amount of heat in the earth's interior is too deep to be tapped. In some areas, however, hot springs or geysers and molten lava streams are close enough to the surface to be used. Thermal energy from hot springs has been used for producing electricity for many years, starting in 1904 in Italy. In the U.S. the major geothermal power plants are located in northern California on a natural steam field called the Geysers. Steam from a number of wells is passed through turbines. The present utilization is about 500 MW and the total estimated capacity is about 2000 MW. Because of the lower pressure and temperatures the efficiency is less than with fossil-fuelled plants, but the capital costs are less and of course the fuel is free.

The Geysers represent a dry steam field which is preferable for power generation via steam turbines. Other basic types of geothermal energy reservoirs are: hot water, hot dry rock, geopressured water, and the normal thermal gradient in the earth's crust. It is more common for wells to produce a mixture of steam and hot water, this combination being much less useful than dry steam. Electricity may be generated from hot-water wells by passing the water under pressure through a heat exchanger where it causes the vaporization of a volatile liquid such as freon. The latter expands through a turbine.

The largest potential is associated with the heat in deposits of dry rock. High-pressure water would be forced down a deep shaft and creates cracks in the rock at the bottom. Pressurized water is then forced through the cracks to extract heat. Hot rocks contracts as it cools, thereby creating fresh cracks and extending the catchment volume. At Los Alamos in New Mexico a 780 m shaft was drilled and the hydraulic creation of cracks achieved at pressures of about 100 atm. All parts of the earth's surface have heat in the rock beneath, but the temperature gradients are very modest. Although the potential amount of heat is vast the technology involved to extract this heat successfully has still to be developed.

1.10 Energy Storage

The tremendous difficulty in storing electricity in any large quantity has shaped the technology of power systems as they stand today. Various options for the large-scale storage of energy which may be converted into electricity exist, and although the basic form of generation is unchanged these methods can be used to ease operation and effect overall economies. Storage of any kind is, however, expensive and care must be taken in the economic evaluation. The options available are as follows: pumped storage, compressed air, heat, hydrogen gas, secondary batteries, and of doubtful promise, flywheels and superconducting coils.

Although gas turbines are used to meet daily load peaks, very rapid changes in load may occur (e.g. 1300 MW/min at the end of the 'Miss World' contest on

British TV!) or the outage of lines or generators. Gas turbines take 2 min to start up and a considerable amount of conventional steam plant must operate underloaded as a reserve. This is very expensive because there is a fixed heat loss for a steam turbogenerator regardless of output and also the efficiency is reduced. A significant amount of storage capable of instantaneous use would be an effective method of meeting such loadings, and by far the most important method to date is that of pumped storage.

Pumped storage

A method of obtaining the advantages of hydro plant where suitable water supplies are not available is by the use of pumped storage. This consists of an upper and a lower reservoir and turbine-generators which can be used as motor-pumps. The upper reservoir has sufficient storage usually for 4–6 h of full-load generation with a reserve of 1–2 h.

The sequence of operation is as follows. During times of peak load on the network the turbines are driven by water from the upper reservoir in the normal manner. The generators then change to synchronous-motor action and, being supplied from the general power network, drive the turbine which is now acting as a pump. During the night when only base load stations are in operation, and electricity is being produced at its cheapest, the water in the lower reservoir is pumped back into the higher one ready for the next day's peak load. Typical operating efficiencies attained are:

Motor and generator	96%
Pump and turbine	77%
Pipeline and tunnel	97%
Transmission	95%

giving an overall efficiency of 67 per cent. A further advantage is that the synchronous machines can be easily used as synchronous compensators if required.

Dams must be constructed if existing reservoirs are inadequate or non-existent. A large scheme in Britain uses six 330 MVA pump-turbine (Francis-type reversible) generator-motor units generating at 18 kV. The flow of water and hence power output is controlled by guide vanes associated with the turbine. The maximum pumping power is 1830 MW. The machines are 92.5 per cent efficient as turbines and 91.7 per cent as pumps and the speed is 500 rev/min. Such a plant can be used to provide fine frequency control for the whole British system when the machines will be expected to start and stop about 40 times a day.

Compressed air storage

Air is pumped into large receptacles (e.g. underground caverns or old mines) at night and used to drive gas turbines for peak, day loads. A German utility is, at the time of writing, installing a 290 MW scheme at a capital cost in the order of $100 per KW (1975 prices). It generates 580 MW-h of on-peak electricity and consumes 930 MW-h of fuel plus 480 MW-h of off-peak electricity. The energy stored is equal to the product of the air pressure and volume. The compressed air allows fuel to be burnt in the gas turbines at twice the normal efficiency. The general scheme is illustrated in Figure 1.27.

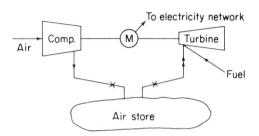

Figure 1.27   Storage using compressed air in conjunction with gas-turbine generator.

One disadvantage of the above scheme is that much of the input energy to the compressed air manifests itself as heat and is wasted. Heat could be retained after compression, but there would be possible complications with the store walls due to the temperature of 450°C at 20 bar pressure. A solution would be to have a separate heat store which could comprise stacks of stones or pebbles which store heat cheaply and effectively. This would enable more air to be stored because it is now cool. At 100 bar pressure, in the order of 30 m^3 of air is stored per megawatt-hour output.

Heat storage

No large-scale storage involving heat has yet evolved. Water has many advantages as a heat-storage medium, including good specific and latent heats. Liquid sodium has also been suggested and also various salts. In a steam generating plant, if the load is low, boilers may be kept at full output if the unwanted steam is bled to the feed-water heaters. During full-load output the steam produced goes wholly to generation, whilst the hot water is drawn from the store to give heated feed water. A 500 MW(e) boiler requires 405 kg/s of feed water and would require a storage of 17,000 t of water.

Secondary batteries

Large-scale battery use is unlikely and the two areas where the use of secondary batteries will have impact are in the electric car and local fluctuating energy sources such as windmills or solar. The popular lead–acid cell, although reasonable in price has a low energy density (15 W-h/kg). Nickel-cadmium cells are better (40 W-h/kg) but much more expensive. Under intensive development is the sodium–sulphur battery (200 W-h/kg) which has a solid electrolyte and liquid electrodes and operates at a temperature of 300°C. Other combinations of materials are under active development in attempts to increase output and storage per unit weight.

Fuel cells

A fuel cell converts chemical energy to electrical energy by electrochemical reactions. Fuel is continuously supplied to one electrode and an oxidant (usually oxygen) to the other electrode. A simple hydrogen–oxygen fuel cell is shown in Figure 1.28 in which hydrogen gas diffuses through a porous metal

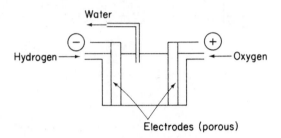

Figure 1.28   Hydrogen–oxygen fuel cell.

electrode (nickel). A catalyst in the electrode allows the absorption of H_2 on the electrode surface as hydrogen ions which react with the hydroxyl ions in the electrolyte to form water $(2H_2 + O_2 \rightarrow 2H_2O)$. A theoretical e.m.f. of 1.2 V at 25°C is obtained. Other fuels for use with oxygen are, carbon monoxide (1.33 V at 25°C), methanol (1.21 V at 25°C) and methane (1.05 V at 25°C). In practical cells conversion efficiencies of 80 per cent have been attained. A major use of the fuel cell could be in conjunction with an hydrogen energy system. At the present time the cost of fuel cells is very high and their future use will depend on a substantial reduction in price.

Hydrogen energy systems

Much publicity has been given in recent years to the use of hydrogen as a medium for energy transmission and storage. Transmission of natural gas via a network of pipes is well established. The transmission capacity of a pipe carrying natural gas is high compared with electrical links and the transmission efficiency is much higher.

   For hydrogen gas the calorific value is 12×10^6 J m^{-3} (ATP).

   The power transmitted = volumetric flow rate × volumetric calorific value (at working pressure)

For long transmission distances the pressure drop is compensated by booster compressor stations. A typical gas system uses a pipe of internal diameter 0.914 m, and with natural gas a power transfer of 12 GW is possible at a pressure of 68 atm and a velocity of 7 m/s. A 1 m diameter pipe carrying hydrogen gas can transmit 8 GW of power.

   The major advantage of hydrogen is of course that it can be stored, the major disadvantage that it must be produced from water via electrolysis. Alternative methods of production are under laboratory development, e.g. use of heat from nuclear stations to 'crack' water and so release hydrogen; however, temperatures of 3000°C are required. Very large electrolysers can attain efficiencies of about 60 per cent. This, coupled with the efficiency of electricity production from a nuclear plant, gives an overall efficiency of hydrogen production of about 21 per cent. However, it would involve nuclear plant at off-peak periods and this combined with the facility for storage could be attractive.

A schematic of such a system is shown in Figure 1.29. As fossil fuel prices increase and allowing for industry's considerable needs for hydrogen the idea will become more attractive, even if on a limited scale. Large-scale fuel cell installations are under active development (e.g. 30 MW) for the conversion of hydrogen energy back to electrical. Also the use of hydrogen as fuel for aircraft and automobiles could give impetus to its large-scale production, storage, and distribution.

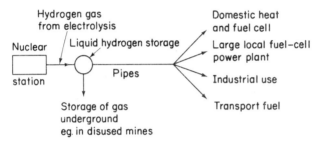

Figure 1.29   Hydrogen storage and transmission scheme.

1.11   Environmental Considerations

Atmospheric

Whereas a few years ago economics and security of supply were governing factors, over the last decade or so the impact on the environment of generation and transmission has become of great importance in planning and operation. In the U.S.A., legislation, e.g. the Clean Air Act of 1970, has had a marked effect on power generation. At certain times load is distributed amongst the generating stations of a power pool on the basis of the minimization of sulphur oxide emissions into the atmosphere rather than minimization of costs. The effects of the 1970 Clean Air Act have been marked. For example between 1970 and 1975 sulphur dioxide concentrations over the whole U.S.A. were reduced by 25 per cent, whilst particulate matter (dust, smoke, soot) reduced by 14 per cent between 1970 and 1973. By 1975 82 per cent of stationary sources of air pollution were complying with emission regulations or meeting an abatement schedule.

The emissions associated with power plants are mainly sulphur oxides, particular matter, and nitrogen oxides. Of the former, sulphur dioxide accounts for about 95 per cent and is a by-product of the combustion of coal or oil. The sulphur content of coal varies from 0.3 to 5 per cent and for generation purposes is specified by some U.S. state laws to be below a certain percentage. This has led to the widespread use of coal from western states in the eastern U.S. because of its lower sulphur content. Sulphur dioxide forms H_2SO_4 in the air which causes damage to buildings and vegetation. Sulphate concentrations of 9–10 μg/m^3 of air aggravate asthma and lung and heart disease. This level has been frequently exceeded in the past, a notorious episode being the London fog of 1952 (caused by domestic coal burning and not by the electricity industry). It should be noted that although sulphur does not accumulate in the air it does in the soil.

Sulphur oxide emission can be controlled by:

(a) the use of fuel with, say, less than 1% sulphur;
(b) the use of chemical reaction to remove the sulphur in the form of sulphuric acid from the combustion products, e.g. limestone scrubbers;
(c) removing the sulphur from the coal by gasification or flotation processes.

The provision of sulphur-control measures such as scrubbers significantly increases the cost of generation plant.

Particulate matter refers to particles in the air. These in sufficient concentrations are injurious to the respiratory system and by weakening resistance to infection may well effect the whole body. Apart from settling on the ground or buildings to produce dirt, a further effect is the reduction of the solar radiation entering the polluted area. Reported densities (particulate mass in one cubic metre of air) are, 10 μg/m^3 in rural areas to 2000 μg/m^3 in polluted areas. The average value in U.S. cities is about 100 μg/m^3.

About one-half of the oxides of nitrogen in the air in populated areas are due to power plants and originate in high-temperature combustion processes. At levels of 25–100 parts per million they can cause acute bronchitis and pneumonia.

A 1000 MW(e) coal plant burns approximately 9000 T of coal per day. If this has a sulphur content of 3 per cent the amount of SO_2 emitted per year is 2×10^5 t. Such a plant produces the following pollutants per hour (in kg); CO_2 8.5×10^5, CO 0.12×10^5, sulphur oxides 0.15×10^5, nitrogen oxides 3.4×10^3 and ash.

Precipitators The concentration of pollutants can be reduced by dispersal over a wider area by the use of high stacks. If in the stack a vertical wire is held at a high negative potential relative to the wall, the expelled electrons from the wire are captured by the gas molecules moving up the stack. Negative ions are formed which accelerate to the wall, collecting particles on the way. When a particle hits the wall the charge is neutralized and the particle drops down the stack and is collected. Precipitators have particle-removing (by weight) efficiencies of up to 99 per cent, but this is misleading as performance is poor for small particles of, say, less than 0.1 μm in diameter. The efficiency based on number of particles removed is therefore less. Disposal of the resulting fly-ash is expensive, but the ash can be used for industrial purposes, e.g. building blocks. Unfortunately the efficiency of precipitators is enhanced by a reasonable sulphur content in the gases. For a given collecting area the efficiency decreases from 99 per cent with 3 per cent sulphur to 83 per cent with 0.5 per cent at 150°C. This results in much larger and expensive precipitator units with the low-sulphur coal.

Thermal pollution

Steam from the low-pressure turbine is liquefied in the condenser at the lowest possible temperatures to maximize the steam-cycle efficiency. Where copious supplies of water exist the condenser is cooled by 'once-through' circulation of sea or river water. Where water is more restricted in availability, e.g. away from the coasts, the condensate is circulated in cooling towers in which it is sprayed in nozzles into a rising volume of air. Some of the water is evaporated, providing cooling. The latent heat of water is 2×10^6 J/kg compared with a sensible heat of 4200 J/kg per deg C in 'once-through' cooling. A disadvantage of such towers is the increase in humidity produced in the local atmosphere.

Dry cooling towers in which the water flows through enclosed channels (similar to a car radiator) past which air is blown avoids local humidity problems, but at a much higher cost than 'wet towers'. Cooling towers emit evaporated water to the atmosphere in the order of 75,000 litres/min for a 1000 MW(e) plant. A crucial aspect of once-through cooling in which the water flows directly into the sea or river is the increased temperature of the latter due to the large volume per minute (typically $360 \text{ m}^3/\text{s}$ for a coolant rise of 2.4 degC for a 2.4 GW nuclear station) of heated coolant. Because of their lower thermal efficiency nuclear power stations require more cooling water than fossil-fuelled plants.

The chemical reaction rate doubles for each 10 deg C rise in temperature, causing an increased demand for oxygen, but the ability of the water to dissolve oxygen is less at the higher temperatures. Hence extreme care must be taken to safeguard marine life, although the higher temperatures can be used effectively for marine farming if conditions can be controlled. Affecting as it does the siting of generating plant, thermal pollution poses a problem of increasing difficulty for power supply companies as the amount of installed capacity increases.

Transmission lines

The presence of overhead lines constitutes an environmental problem (perhaps the most obvious one) on several counts.

(a) Space is used which could be used for other purposes. The land allocated for the line is known as the right of way (or wayleave in Britain). The area used for this purpose is already very appreciable.

(b) Lines are considered by many to mar the landscape. This is of course a subjective matter, but it cannot be denied that several lines converging

on a substation or plant, especially from different directions, is offensive
to the eye.
(c) Radio interference (RI), audible noise (AI), and safety considerations.

Although most of the above objections could be overcome by the use of
underground cables these are not free of drawbacks. The limitation to cable
transmitting current because of temperature-rise considerations coupled with
high manufacture and installation costs results in the ratio of the cost of
transmitting energy underground to that for overhead being between 10 and
20, this increasing with operating voltage. With novel cables under development it is hoped to reduce this disadvantage. Estimates made of the increased
costs resulting from the undergrounding of whole or large parts of a power
system indicate that an intolerable financial burden would result on the utilities
and hence the consumer. However, it may be expected that in future a larger
proportion of circuits will be carried underground, especially in suburban
areas. In large urban areas circuits are invariably underground, this posing
increasing problems as load intensities increase.

1.12 Loads

The major consumption groups are industrial, residential (domestic), and
commercial. Industrial consumption accounts for up to 40 per cent of the total
in many industrialized countries and a significant item is the induction motor.
The percentage of electricity in the total industrial use of energy is expected to
continue to increase due to greater mechanization and the growth of energy-
intensive industries such as chemicals and aluminium. In the U.S. the following
six industries account for over 70 per cent of the industrial electricity
consumption: metals (25 per cent), chemicals (20 per cent), paper and products
(10 per cent), foods (6 per cent), petroleum products (5 per cent), transportation equipment (5 per cent). Over the past 25 years the amount of
electricity per unit of industrial output increased annually by 1.8 per cent, but
due to increased fuel costs this may decline.

Residential growth in electricity consumption is attributable more to
increased use per consumer than to increased number of consumers. The
breakdown of consumption in 1975 was as follows (in the U.S.): space heating
14 per cent, water heating 17 per cent, air conditioning 18 per cent, ranges
(ovens) 7 per cent, freezers 6 per cent, refrigerators 16 per cent, dryers 5 per
cent, lighting, and other 17 per cent. Over the period 1975–90 an annual
growth rate of 5.4 per cent has been forecast, resulting in a total consumption of
1285 billion kWh in 1990.

The commercial sector comprises offices, shops, schools etc. The consumption here is related to personal consumption for services, traditionally a
relatively high-growth quantity. In this area, however, conservation of energy
measures may be particularly effective and so modify the growth rate. For all
sectors in the U.S.A. a total electric utility sales (shown in Figure 1.30) are
projected to increase by an annual rate in the order of 5 per cent to 1990.

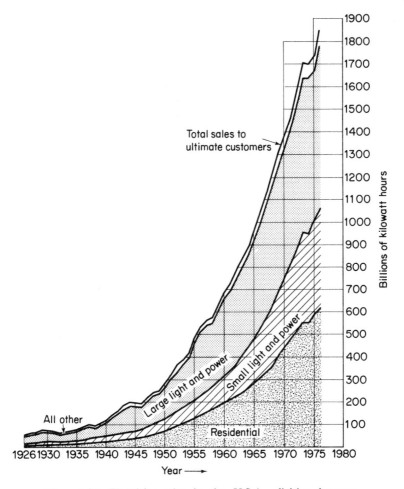

Figure 1.30 Electricity sales in the U.S.A.—division between consumer types. (*Permission of Edison Electric Institute.*)

Quantities used in measurement of loads are defined as follows:

Maximum load   The average load over the half hour of maximum output.

Load factor   The units of electricity exported by the generators in a given period divided by the product of the maximum load in this period and the length of the period in hours. The load factor should be high; if it is unity all the plant is being used over all of the period. It varies with the type of load, being poor for lighting ,(about 12 per cent) and high for industrial loads, e.g. 100 per cent for pumping stations.

Diversity factor   This is defined as the sum of individual maximum demands of the consumers, divided by the maximum load on the system. This factor measures the diversification of the load and is concerned with the installation of sufficient generating and transmission plant. If all the demands occurred simultaneously, i.e. unity diversity factor, many more generators would have to be installed. Fortunately, the factor is much higher than unity, especially for domestic loads.

A high diversity factor could be obtained with four consumers by compelling them to take load as shown in Figure 1.31. Although compulsion obviously cannot be used, encouragement can be provided in the form of tariffs. An example is the two-part tariff in which the consumer has to pay an amount dependent on the maximum demand he requires, plus a charge for each unit of energy consumed. Sometimes the charge is based on kilovolt-amperes instead of power to penalize loads of low power factor.

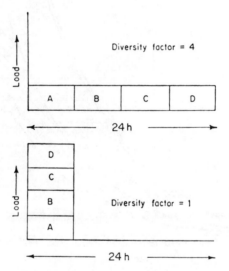

Figure 1.31   Two extremes of diversity factor in a system with four consumers.

Load loss factor

The average losses in the system, e.g. transmission line losses, depend upon the load curves for any given period. The *load loss factor* is defined as the ratio of the actual losses over a period to the losses obtaining if the maximum load is maintained. A detailed study of load curves peculiar to Great Britain has shown that the load loss factor is related to the load factor by the following expression:

$$\text{load loss factor} = 0.2(\text{load factor}) + 0.8\,(\text{load factor})^2$$

Tariffs

The cost of generation and distribution is governed approximately by the expression $(a + b \times kW + c \times kWh)$ per annum, where a is a fixed charge for the central organization and similar activities and is independent of the power output; b depends on the maximum demand on the system and hence on the interest and depreciation on the plant installed; and c depends on the units generated and hence on the fuel costs and the wages of personnel directly engaged in the production of electricity.

The manner in which customers pay for electricity varies according to the nature of their requirements. Prices should be such as to cover total costs and a reasonable return on assets. Tariff structures may be arranged to influence the load curve and to improve the load factor. A successful example has been the introduction in Britain of cheap, night-rate electricity for domestic space (storage) heating.

Kilowatt-hour rates are based on costs of fossil fuel and usually contain a fuel-cost adjustment clause to cope with rapid changes. Industrial tariffs contain a maximum demand or kilovolt-ampere component. In some industrial tariffs an 'interruptible' clause is included, which allows for compulsory load shedding. Load factors (per cent) for representative U.S. utilities over various periods are given in Table 1.2. The effect of load factor on the economics of one

Table 1.2 Average period load factors in per cent for representative U.S. electric systems

	System	Annual	Period load factor (%)		
			Seasonal	Weekly	Daily
Summer	A	48	59 ± 9.9	75 ± 4.4	83 ± 2.7
Peaking	B	60	66 ± 5.1	75 ± 2.2	80 ± 2.4
Systems	C	68	70 ± 2.0	76 ± 2.9	85 ± 1.2
Winter	A'	55	65 ± 3.6	75 ± 2.9	81 ± 1.2
Peaking	B'	63	69 ± 1.7	76 ± 1.4	82 ± 1.5
Systems	C'	78	80 ± 1.7	84 ± 3.2	89 ± 1.3

Permission of I.E.E.E.

fuel versus another is very dependent on load factor, as is illustrated in the following example.

Example 1.1   Compare the overall cost per unit generated by a conventional coal station with that from a certain type of nuclear station, assuming the load factor $\left(\dfrac{\text{average load}}{\text{installed capacity}}\right)$ of the coal station to be 50 per cent and of the nuclear station 60 per cent. The following data applies:

Capital cost ($/kW) 700 (coal), 800 (nuclear)
Interest and depreciation per annum, 10 per cent (coal), 10 per cent (nuclear)
Running cost per unit (c./kWh) 0.341 (coal), 0.153 (nuclear)

Solution   It is convenient to work on the basis of 1 kW output on annual basis. The average loads are therefore 0.5 kW and 0.6 kW over the year, for coal and nuclear, respectively.

The capital costs on an annual basis require,

$$\$\left(700 \times \frac{10}{100}\right) \text{ for coal} \quad \text{and} \quad \$\left(800 \times \frac{10}{100}\right) \text{ for nuclear}$$

The energy (kWh) generated per annum

$$= 0.5 \times 365 \times 24 = 4380 \text{ for the coal station}$$

and

$$0.6 \times 365 \times 24 = 5250 \text{ for the nuclear station}$$

$$\text{Cost of these units} = 0.341 \times 4380 = 1493 \text{ c. (coal)}$$

$$0.153 \times 5250 = 803 \text{ c. (nuclear)}$$

The total costs per annum are:

$$\text{coal, } 70 + 14.93 = \$84.93$$

$$\text{nuclear, } 80 + 8.03 = \$88.03$$

The corresponding costs per unit are:

$$\text{coal, } \frac{84.93}{4380} = \$0.0194$$

and

$$\text{nuclear, } \frac{88.03}{5250} = \$0.0168$$

With a load factor of 60 per cent for both fuels the costs per unit become, $0.0167 (coal) and $0.0168 (nuclear).

Load management

Attempts to modify the shape of the load curve to produce economy of operation have already been mentioned. These have included tariffs, pumped storage, and the use of seasonal or daily diversity between interconnected systems. A more direct method would be the control of the load either through tariff structure or direct electrical control of appliances, the latter, say, in the form of remote on/off control of electric water heaters where inconvenience to the consumer is least. For many years this has been achieved with domestic time switches, but recent schemes use switches radio-controlled from the utility to give greater flexibility. This permits load reductions almost instantaeneously and defers the hot-water load until after system peaks. Attempts to extend control to air-conditioning are more limited because these appliances can only be turned off for a few minutes before personal discomfort. Preliminary test results indicate that perhaps 25 per cent of the air-conditioning load could be 'shaved off' by remote control.

Systems for load management are varied. Ripple control (use of a ripple frequency over supply cables) has been tried in Europe. Other power-line carrier systems are available or under development in the U.S.A. The control signal frequency should be low enough to avoid high attenuation in the network. On the other hand at low frequencies the network noise level goes up as the reciprocal of frequency. One system uses a 6 kHz carrier of signal power 400 W whilst another uses a frequency of several tens of kilohertz. Effort is also going into remote kilowatt-hour-meter reading by carrier systems.

Power systems have been traditionally designed to optimize in terms of generation mix and network design a system related to given load curves. Changes in the load curve will obviously have repercussions on the existing system and these must be thoroughly investigated before large-scale load management is pursued. The yearly load factor for U.S. systems varies between 50 and 80 per cent, whereas the daily factor varies from 75 to 90 per cent. Hence, on a daily basis there is less requirement to shift load than the annual load factors would suggest. Most of the potential for load control lies in the domestic area, especially water heating.

Load forecasting

It is evident that load forecasting is a crucial activity in electricity supply. Forecasts are based on the previous year's loading for the period in question updated by factors such as general load increases, major new loads, and weather trends. Both power demand and energy (kWh) forecasts are used, the latter often being the more readily obtained. From energy forecasts demand values may be determined. Energy trends tend to be less erratic than peak power demands and are considered better growth indicators; however, load factors are also erratic in nature.

As weather has a much greater influence on residential than on industrial demands it may be preferable to assemble the load forecast in constituent parts to obtain the total. In many cases the seasonal variations in peak demand are caused by weather-sensitive domestic appliances, e.g. heaters and air conditioning. A knowledge of the increasing use of such appliances is therefore essential. Several techniques are available for forecasting. These range from simple curve fitting and extrapolation, to stochastic modelling and are given in detail in reference 8. The many physical factors affecting loads, e.g. weather, national economic health, popular TV programmes, public holidays, etc. make forecasting a complex process demanding experience and high analytical ability.

1.13 Representation of Power Systems

Modern electricity supply systems are invariably three-phase. The design of distribution networks is such that normal operation is reasonably close to balanced three-phase working, and often a study of the electrical conditions in

42

one phase is sufficient to give a complete analysis. Equal loading on all three phases of a network is ensured by allotting as far as possible equal domestic loads to each phase of the low-voltage distribution feeders; industrial loads usually take three-phase supplies.

A very useful and simple way of graphically representing a network is the schematic or line diagram in which three-phase circuits are represented by single lines. Certain conventions for representing items of plant are used and are shown in Figure 1.32. A typical line or schematic diagram of a part of a

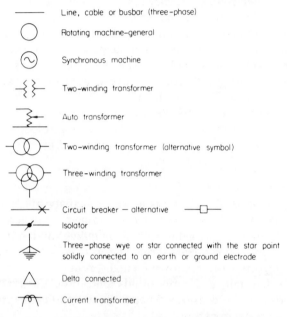

Figure 1.32   Symbols for representing the components of a three-phase power system.

power system is shown in Figure 1.33. In this, the generator is star connected, with the star point connected to earth through a resistance. The nature of the connexion of the star point of rotating machines and transformers to earth is of vital importance when considering faults which produce electrical imbalance in

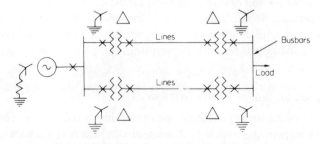

Figure 1.33   Line diagram of a simple system.

the three phases. The generator feeds two three-phase lines (overhead or underground). The line voltage is increased from that at the generator terminals by transformers connected as shown. At the end of the lines the voltage is reduced for the secondary distribution of power. Two lines are provided to improve the *security* of the supply, i.e. if one line develops a fault and has to be switched out the remaining one still delivers power to the receiving end. It is not necessary in straightforward current and voltage calculations to indicate the presence of switches, etc. on the diagrams, but in some cases such as stability calculations the location of switches, current transformers, and protection is very useful.

Although the use of jargon is avoided as much as possible, certain terms will be constantly recurring which may be rather vague to the newcomer to the subject. A short list follows of such terms with explanations.

Systems This is used to describe the complete electrical network, generators, loads, and prime movers.

Load This may be used in a number of ways: to indicate a device or collection of devices which consume electricity; to indicate the power required from a given supply circuit; the power or current being passed through a line or machine.

Busbar An electrical connexion of zero impedance joining several items such as lines, loads, etc. Often this takes the form of actual busbars of copper or aluminium.

Earthing (*Grounding*) The connexion of a conductor or frame of a device to the main body of the earth. Thus must be done in such a manner that the resistance between the item and the earth is below prescribed limits. This often entails the burying of large assemblies of conducting rods in the earth and the use of connectors of large cross-sectional area.

Fault This is a malfunctioning of the network, usually due to the short-circuiting of two conductors or live conductors connecting to earth.

Security of supply Provision must be made to ensure continuity of supply to consumers even with certain items of plant out of action. Usually two circuits in parallel are used and a system is said to be secure when continuity is assured. This is obviously the item of first priority in design and operation.

1.14 Nature of Transmission and Distribution Systems

By transmission, is normally implied, the bulk transfer of power by high-voltage links between main load centres. Distribution, on the other hand, is mainly concerned with the conveyance of this power to consumers by means of lower voltage networks.

The machines usually generate voltage in the range 11–25 kV which is increased by transformers to the main transmission voltage. At substations the connexions between the various components of the system, such as lines and transformers, are made and the switching of these components carried out. Large amounts of power are transmitted from the generating stations to the load-centre substations at 400 kV and 275 kV in Britain, and 345 kV, 765 kV and 500 kV in the U.S.A. The network formed by these very high-voltage lines is sometimes referred to as the *supergrid*. Most of the large and efficient stations feed through transformers directly into this network. This grid in turn feeds a subtransmission network operating at 132 kV in Britain and 115 kV in the U.S.A. Some of the older and less efficient stations feed into this system which in turn supplies networks which are concerned with distribution to consumers in a given area., In Britain these networks operate at 33 kV, 11 kV, or 6.6 kV and supply the final consumer feeders at 415 kV three-phase, giving 240 V per phase. Other voltages exist in isolation in various places, e.g. the 66 kV London cable system. A typical part of a supply network is shown schematically in

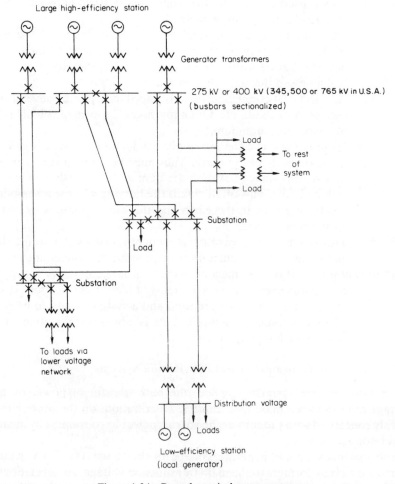

Figure 1.34   Part of a typical power system.

Figure 1.34. The power system is thus made up of networks at various voltages. There exist in effect voltage tiers as represented in Figure 1.35. Figure 1.36 shows a schematic of the transmission network in England and Wales.

Summarizing, transmission networks deliver to wholesale outlets at 132 kV and above; subtransmission networks deliver to retail outlets at voltages from 115 kV or 132 kV, and distribution networks deliver to final step-down transformers at voltages below 132 kV.

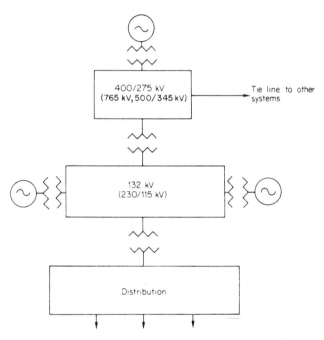

Figure 1.35 Schematic diagram of the constituent networks of a supply system. U.S.A. voltages in parentheses.

Reasons for interconnexion

(a) Generating sets are becoming increasingly large and 2000 MW capacity stations are now the norm. Not only is the initial cost per kilovolt-ampere of stations of very large capacity lower than that of smaller ones but their efficiencies are substantially higher. Hence, regardless of geographical position, it is more economical to use these efficient stations to full capacity 24 h a day and transmit energy considerable distances than to use less efficient more local stations. The main base load therefore is met by these large stations which must be interconnected so that they feed into the general system and not into a particular load.

(b) In order to meet sudden increases in load a certain amount of generating capacity known as the 'spinning reserve' is required. This consists of generators running at normal speed and ready to supply power instantaneously. If the

46

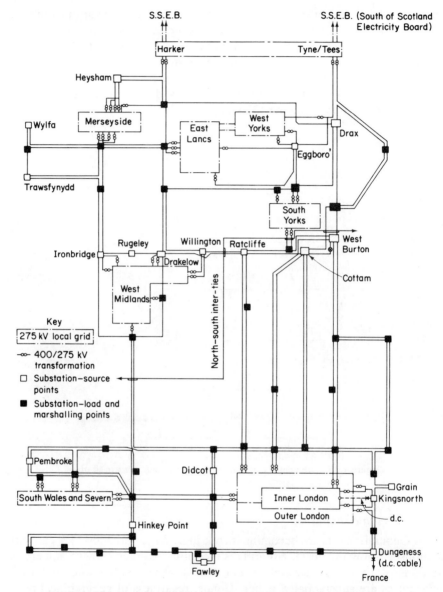

Figure 1.36 Main 400 kV transmission system in England and Wales. (*Permission of Institution of Electrical Engineers.*)

machines are stationary a reasonable time is required (especially for steam turbo-alternators) to run up to speed; this can approach 1 h. It is more economical to have certain stations only serving this function than to have each station carrying its own spinning reserve.

(c) The electricity supplies over the entire country are synchronized and a common frequency exists.

(d) In an interconnected network consisting essentially of loops, continuity of supply is maintained, since substations can be fed from either direction.

1.15 Statistics of Systems

Throughout the world the general form of power systems follows the same pattern. Magnitudes of voltage vary from country to country, the differences originating mainly from geographical and historical reasons.

Several frequencies have existed, although now only two values 50 Hz and 60 Hz remain: 60 Hz is used on the American continent, whilst most of the rest of the world uses 50 Hz. The value of frequency is a compromise between higher generator speeds (and hence higher output per unit machine volume) and the disadvantage of high system reactance at higher frequencies. Historically the lower limit was set by the need to avoid visual discomfort from electric lamps.

The topology of the power system and the voltage magnitudes used are greatly influenced by geography. Very long lines are to be found in North and South American nations and the U.S.S.R. This has resulted in higher transmission voltages, e.g. 765 kV with the possibility of voltages in the range 1000–1500 kV being actively considered. In some South American countries, e.g. Brazil, vast amounts of hydroelectric resources are in the process of being developed, resulting in very long transmission links. In highly developed countries the available hydro resources have been developed and a considerable proportion of new generation is nuclear. In geographically smaller countries as exist in Europe, the degree of interconnexion is much tighter with shorter transmission distances, the upper voltage being about 420 kV.

Systems are universally a.c. with the use of high-voltage d.c. links used for specialist purposes, e.g. very long circuits and submarine cable connexions. The use of d.c. has been limited by the high cost of the conversion equipment. This requires overhead line lengths of a few hundred kilometres in order for the reduced line costs to offset the conversion costs. Below are presented summaries of the systems of some countries.

U.S.A.

The organization of the U.S. supply system has already been described. The following statistics give some idea of the scope and nature of the industry. The whole of the country is synchronously connected with the exception of the State of Texas. The nature of the load in North America differs seasonally from one part of the country to another, and also load diversity occurs due to time zones. Generally, the summer and winter loads are much more comparable than in Europe due to the extensive use of air-conditioning in the summer. The ratio (summer peak)/(peakload in the following December) is used to assess this characteristic, and in several parts is greater than unity. The basic consumer voltage is 120 V single-phase with preferred nominal distribution voltages below 69 kV of 46, 34.5, 23, 14.4, 13.2, 7.2, 4.8, and 2.4 kV. The domestic consumers are supplied via a 240 V single-phase transformer secondary winding with a centre tap. This gives 240 V across the whole winding for major appliances and 120 V across half the winding for lighting, etc.

The circuit kilometres of overhead line by popular voltage classes in the U.S.A. in 1976 were as follows: 22 to 60 kV—258,600; 66 to 75 kV— 146,355; 110 to 120 kV—128,211; 132 to 140 kV—93,654; 220 to 240 kV—84,462; 330 to 345 kV—45,100; 500 kV—22,417; 765 kV and 800 kV—4077. The average length (km) of lines at various voltage levels are as follows (for 1970): 115 kV—34, 230 kV—74, 345 kV—108, 500 kV—108, 765 kV—118. In 1965 there were 160 GW-km (circuit length times line capacity or rating) of transmission for each gigawatt of generation. By 1970 this had risen to 225 GW-km/GW of generation due mainly to the increase of voltages and the pooling of systems.

The growth of generating capacity was shown in Figure 1.1. The energy generated (in 1976) was comprised as follows (in 10^9 kWh): hydro 283.7, conventional steam 1556.6, nuclear 191.1, internal combustion 5.08. Sales of electric energy are divided between main consumer groups as shown in Figure 1.30. Large light and power corresponds to industrial consumers.

The average kilowatt-hours used and the average annual bill per consumer (1976) were as follows: residential—8360 and $288.42; commercial (small light and power)—50,733 and $1755.0; total (all consumers)—22,361 and $646.23. The average revenues per kilowatt-hour (1976) were; residential 3.45 c, commercial 3.46 c., overall 2.89 c. In 1976 the ratio (kWh generated/kW of capacity) was 3917.

In Figure 1.37 the 765 kV system of the American Electric Power Company is shown on a map of the area covered by this organization. It is expected that by 1980 the load will be 20,000 MW and by 1990, 34,000 MW. A number of factors influence the nature of the 765 kV network and these include system load, size of generators, internal transmission capability for handling power transfers, and the strength of the inter-connexions to outside systems. The system has 1150 miles of 765 kV line, 14,700 MVA of transformer capacity, and 27 circuit breakers. There are seven 765/345 kV, two 765/500 kV, and two 765/138 kV substations. In addition there are shunt reactors connected directly to the 765 kV line terminals of a total of 4200 MVA. In Figure 1.38 the relationship between power transmittable over 765 kV and 345 kV circuits and length of circuit is shown. The corresponding relationship between the cost of transmission and power transmitted is shown in Figure 1.39, showing the cross-over powers for 345 and 765 kV for various distances.

European

England and Wales The following statistics refer to the year 1981/82. The installed capacity (GW) was 55.185 divided as follows: conventional steam 47.006, nuclear 4.467, gas turbine 3.240, hydro 0.112, pumped storage 0.36, diesel 0.01. The electricity produced annually is shown in Figure 1.40. The overall thermal efficiency was 34.14 per cent and the highest value for an

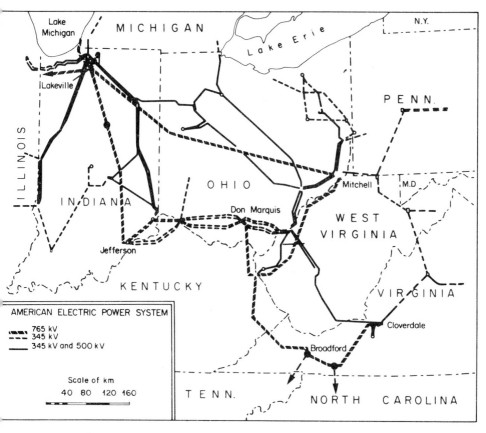

Figure 1.37 American Electric Power Co. system. (*Permission of I.E.E.E.*)

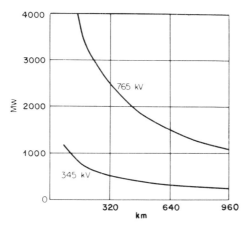

Figure 1.38 Relationships between power transmission capacity and circuit length for 765 and 345 kV circuits.

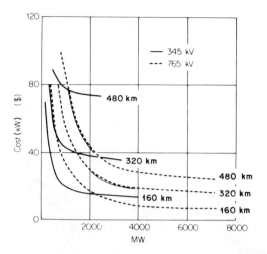

Figure 1.39   Relationship between transmission costs and power transmitted for 345 and 765 kV circuits of various lengths.

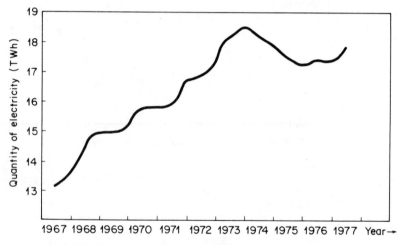

Figure 1.40   Electricity supplied by the Central Electricity Generating Board system to England and Wales. (*Permission of the Electricity Council.*)

individual station 37.87 per cent. The system load factor was 57.4 per cent, the kilowatt-hours consumed per head of population 4000, and the kilowatt-hours consumed per domestic consumer 3870. The Central Electricity Generating Board transmission system's circuit—kilometres of line were: 400 kV—9435 (88 underground); 275 kV—3896 (532 underground); 132 kV—695; 66 kV—116.

Circuit-kilometres of lines for distribution were 132 kV—18474, 66 kV—4642, 33 kV—35110, 11 kV—217,901. The London network comprises the

	132 kV	66 kV	33 kV	11 kV
Underground	406	520	829	5980
Overhead	41	13	—	—

Of the £8134 million turnover of the Generating Board, fuel accounted for 47 per cent, salaries and wages 16 per cent and capital charges 22 per cent. The average costs of generation (1981–82) are 2.18 p/kWh (fossil fuelled) and 1.98 p/kWh (nuclear).

The installed capacities (GW) of the larger of the European systems are as follows (for 1980):

	German Federal Republic	France	Italy
Conventional steam	53.256	21.973	22.747
Hydro	3.847	18.445	14.371
Pumped storage	2.616	0.84	1.445
Nuclear	8.468	14.388	1.113

West Berlin In contrast to the large systems discussed above, West Berlin, being effectively an island in the German Democratic Republic, possesses its own self-contained system with no interconnection to the outside network. At

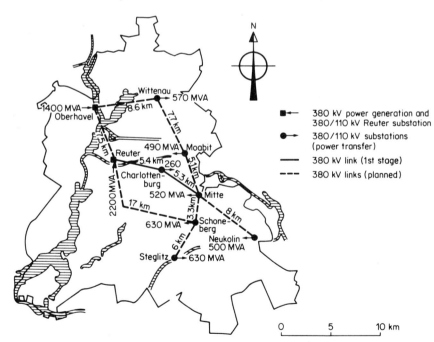

Figure 1.41 Transmission system of West Berlin. (*Permission of South African Institute of Electrical Engineers. Reprinted from Transactions (April 1977), p. 93.*)

present it has about 2 GW installed capacity and two completely isolated 110 kV networks, and because of high fault levels it will in the future be necessary to subdivide the network even more. The requirement of fuel storage and cooling water means that future generation will be situated in the north-west fringe and will total about 3.6 GW. The present 110 kV network will no longer be able to transmit this economically over the 15–25 km to the load centres. Also, the currents would exceed the ratings of available 110 kV switchgear. For these reasons a 380 kV grid is to be constructed with distribution of 110, 10, and 0.4 kV. The system as planned is shown in Figure 1.41.

Japan   In 1977 the power consumed in Japan was 401×10^9 kWh and is expected to be 518×10^9 kWh in 1981. The peak load (August) was 83.4 GW (load factor 58.8) in 1977 and is predicted to be 111 GW in 1981 with a load factor of 57.1. In 1976 there were 1862 km of 500 kV overhead lines and 9495 km of lines at 187–275 kV. There were 161 km of underground cable at 187–275 kV in 1976, the projected figure for 1986 being 990 km.

1.16   Distribution Systems

Although in this book there is a bias towards transmission networks and the associated problems, the general area of distribution should not be considered as of secondary importance. Distribution networks differ from transmission networks in several ways, quite apart from voltage magnitude. The number of branches and sources is much higher in distribution networks and the general structure or topology is different. A typical system consists of a step-down (e.g. 132/11 kV) on-load tap-changing transformer at a bulk supply point feeding a number of lines which can vary in length from a few hundred metres to several kilometres. A series of step-down three-phase transformers, e.g. 11 kV/415 V in Britain or 4.16 kV/240 V in the U.S.A., are spaced along the route and from these are supplied the consumer three-phase, four-wire networks which give 240 V or, in the U.S.A. 120 V, single-phase to houses and similar loads.

The structure of the network varies with location. In rural areas radial feeders are often used (usually overhead line), whereas in urban areas a well-defined low-voltage area or block is fed from the higher voltage network. For security reasons such areas are connected through fuses to neighbouring and similar areas which are fed from different feeders. In such systems the network has essentially a topology of a loop nature. As most faults tend to be of a transient nature, auto-reclose circuit breakers are widely used. These open on a fault and then reclose after a short period. If the fault persists they reopen and the sequence continues three times. Typical arrangements for supplying the distribution network are shown in Figure 1.42. A typical system for rural areas is shown in Figure 1.43. Recent developments in single-phase high-voltage (12 kV) supplies to transformers are described in 11.13. The standard distribution voltage in Europe is 20 kV.

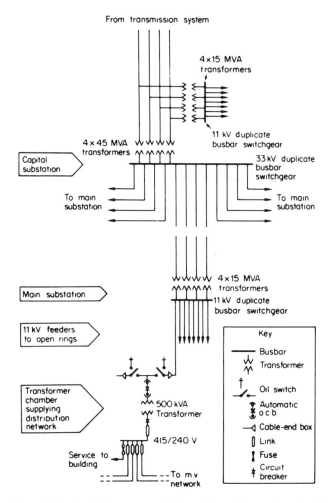

Figure 1.42   Typical arrangement of supply to medium voltage (m.v.) network—British practice. (*Permission of Institution of Electrical Engineers.*)

1.17   Use of Digital Computers

The most dramatic change in power-system analysis in the last decade has been the widespread use of computers. Their use has enabled very large systems to be analysed and controlled much more effectively and economically. The use of computers in power systems takes two main forms, off-line and on-line applications.

Off-line applications include research, routine calculation of system performance and data assimilation and retrieval.

On-line applications include data-logging and the monitoring of the system state. It is an essential part of overall system control in the steady state

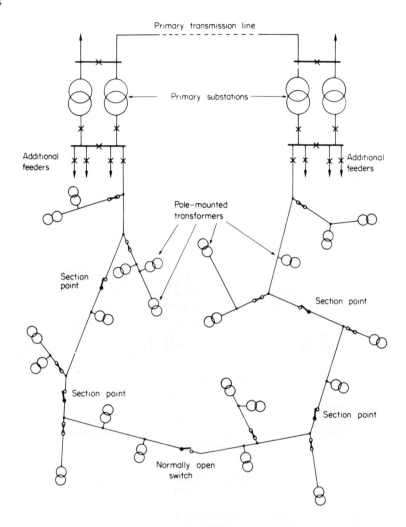

Figure 1.43 Typical rural distribution system. (*Permission of the Electricity Council.*)

including switching, safe interlocking, loading of plant, post-fault control, and load shedding. A recent application is its application to protective gear in which the high-speed measurement of system parameters is used to compare relevant quantities and so replace slower, more conventional, devices.

The use of a large central computer for economic and secure control of a large system at the moment presents difficulties, and an alternative is to subdivide it into smaller networks with a central coordinating facility. A typical subnetwork could control say, 100 generators and about 600 lines with 1200 circuit breakers. The functions of the on-line computer control would be as follows:

(a) Prediction of demand at demand points.
(b) Security of generation and transmission.
(c) Economic operation of generating plant and the network system.
(d) Maintaining system voltages and reactive power flows within specified limits.
(e) Maintaining a file of current plant capability and constraints.
(f) Costing proposed operations.
(g) Despatching load magnitudes to generating units and ensuring that instructions are implemented.
(h) Monitoring station performance relative to targets and costing and deviation.
(i) Revising hot stand-by plant.
(j) Revising network configuration.

In addition to the system aspects above, micro-processors and computers installed in generating stations control various processes such as the 'running up' from the cold state of turbine generators. These computers are completely local in their application.

References

BOOKS

1. *Electrical Transmission and Distribution Reference Book*, Westinghouse Electric Copr., East Pittsburgh, Pennsylvania.
2. Waddicor, H., *Principles of Electric Power Transmission*, Chapman and Hall, London, 5th edn., 1964.
3. Guthrie Brown, I. (Ed.), *Hydro-Electric Engineering Practice*, Blackie, London, Vols. 1–3, 1970.
4. Langford, T. E., *Electricity Generation and the Ecology of Natural Waters*, Liverpool University Press, 1983.
5. Sporn, P., *Research in Electric Power* Pergamon, London, 1966.
6. Central Electricity Generating Board, *Modern Power Station Practice*, Pergamon, 2nd edn., 1976.
7. Zerban, A. H., and E. P. Nye, 'Power Plants', International Textbook Company, Serantan, 2nd edn., 1956.
8. Sullivan, R. L., *Power System Planning*, McGraw-Hill, New York, 1977.
9. Sutton, G. W., *Direct Energy Conversion*, McGraw-Hill, New York, 1966.
10. Golding, E. W., *The Generation of Electricity by Wind Power*, Chapman & Hall, London, 1976.
11. McMullan J. T. *et al.*, *Energy Resources and Supply*, Wiley, London, 1976.
12. *Energy and Power, A Scientific American Book*, W. H. Freeman, San Francisco, 1974.
13. Fisher, J. C., *Energy Crisis in Perspective*, Wiley, New York, 1974.
14. Glasstone, S., and A. Sesank, *Nuclear Reactor Engineering*, Van Nostrand, Reinhold, New York, 1967.
15. Salmon A., *The Nuclear Reactor*, Longman, New York, 1964.
16. Bennett, D. J., *The Elements of Nuclear Power*, Longman, 1972.
17. Liebhofsky, H. A., and E. J. Cairns, *Fuel Cells and Fuel Batteries*, Wiley, 1968.

56

18. Berkowitz, D. A., *Power Generation and Environmental Change*, M.I.T. Press, Cambridge, Mass., 1972.
19. Energy Policy Staff, Office of Science and Technology, *Electric Power and the Environment*, U.S. Government Printing Office, 1970.
20. Edison Electric Institute, *E.H.V. Transmission Line Reference Book*, 1968.
21. *Statistical Year Book of the Electric Utility Industry for 1976*, Edison Electric Institute (U.S.A.), Oct. 1977.
22. *Handbook of Electricity Supply Statistics, 1982*, The Electricity Council London (British).

PAPERS

23. Bensky, M. H. and E. B. Hutchins, 'Making electricity for America—how much?', *Electric Forum*, **2**, No 1 (1976), 3–7, published by General Electric (U.S.A.).
24. Folger, S. R. *et al.*, 'Making electricity for America—what mix?', *Electric Forum*, **2** (1), 1976, 11–15.
25. Lieberman, I. A., 'Making electricity for America—Where?', *Electric Forum*, **2** (1), 1976, 17–23.
26. I.E.E.E. Power Engineering Society, 'Bibliography on corporate-financial models', *I.E.E.E. Trans.*, **PAS-101** (1982), 3084.
27. Schmidt, R. W. *et al.*, 'An economic analysis of energy over the next 75 years', *I.E.E.E.*, **PAS-96** (1977), 1353–1361.
28. Rose, D. J., 'Controlled nuclear fusion, status and outlook', *Science* **172** No. 3985 (1971), 797–808.
29. Steiner, D., 'Nuclear fusion: focus on Tokamak', *I.E.E.E.*, *Spectrum*, **1972**, 32–38.
30. Goodenough, J. B., 'The options for using the sun', *Technology Review*, Oct/Nov 1976, 63–71.
31. Workshop on Alternative Energy Strategies, *Energy—Global Prospect*, 1985–2000, McGraw-Hill, 1977.
32. Rippen, S. 'Plutonium', *I.E.E. Electronics and Power*, **July 1977**, 552–555.
33. Sebo, S. A., 'Ocean power', *Maritime Studies and Management*, **2** (1975), 202–214.
34. Kaplan, G., 'For solar power—sunny days ahead?', *I.E.E.E. Spectrum*, **Dec. 1975**, 47–52.
35. Sporn, P., 'Our environment—options on the way into the future', *I.E.E.E. Spectrum* **May 1977**, 49–58.
36. *Power Plants and Clean Air*, I.E.E.E. Power Engineering Society Special Publication 246 (1973).
37. Hernen, B. A., 'The Dinorwic pumped storage scheme,' *I.E.E. Electronics and Power*, **23**, No. 10, 828–832.
38. Nennia, D., Batteries: today and tomorrow', *I.E.E.E., Spectrum*, **March 1976**, 36–42.
39. 'Energy management—no magical solution', *Power Engineering*, **Aug. 1976**, 42–50.
40. Kaplan, G., 'Two-way communication for load management', *I.E.E.E. Spectrum*, **Aug. 1977**, 47–50.
41. Sachdev, M. S., 'Load forcasting—bibliography', *I.E.E.E., Trans.*, **PAS-96** (1977), 697–700.
42. Comerford, R. B. *et al.* 'The application of classical load forecasting techniques to load management', *I.E.E.E. Trans.*, **PAS-101** (1982), 4656.
43. Milne, A. G., and J. H. Maltby, 'An integrated system of metropolitan electricity supply', *Proc. I.E.E.*, **114**, (1967), 745.
44. Papers describing the American Electric Power 765 kV System, *I.E.E.E.*, **PAS-88**, (1969), 1313.

45. Booth, E. S. *et al.*, 'The 400 kV grid system for England and Wales', *Proc. I.E.E.*, **9**, Pt. A (1962).
46. Kinloch, D. H. *et al.*, 'Impacts of solar heating options upon electric power systems', *I.E.E.E. Trans.*, **PAS-101** (1982), 1271.
47. Ramakumar, R. *et al.*, 'Solar energy conversion and storage systems of the future', *I.E.E.E. Trans.*, **PAS-94** (1975), 1926–1934.

Problems

1.1 In the U.S.A. in 1971 the total area of right of ways for HV overhead lines was 16,000 km^2. Assuming a growth rate for the supply of electricity of 7 per cent per annum calculate what year the whole of the U.S. will be covered with transmission systems (assume area to approximate 4800×1600 km). Justify any assumptions made and discuss critically why the result is meaningless.
(Answer: 91.25 years)

1.2. The calorific value of hydrogen gas at atmospheric pressure and temperature to 12×10^6 J/m^3. Calculate the energy transfer in a pipe of 1 m diameter with gas at 60 atm (guage) flowing at 5 m/s. If *liquid* hydrogen was transferred at the same velocity calculate the energy transfer if the calorific value of liquid hydrogen is 650 times the value for gas at 1 atmosphere.
(Answer: 2.8 GW, 30.6 GW).

1.3. (a) An electric car has a steady *output* of 10 kW over its range of 100 km when running at a steady 40 km/h. The efficiency of the car (including batteries) is 65 per cent. At the end of the car's range the batteries are recharged over a period of 10 h. Calculate the average charging power if the efficiency of the battery charger is 90 per cent.

(b) The calorific value of gasoline is roughly 16,500 kJ/gallon. By assuming an average filling rate at a pump estimate the rate of energy transfer on filling a gasoline-driven car.
(Answer: (a) 4.3 kW; (b) 1.4 MW)

1.4. Discuss in a critical manner the possible impact of solar energy on the elctrical supply industry

(a) as a means of central generation;
(b) as a domestic energy source.

A solar panel (5 m × 2 m) is used for heating domestic water. The absorbing panel has an absorbivity of 0.93, a top-plate emissivity of 0.9, and a back-plate emissivity of 0.05. The transmittance through the glass top cover is 0.85.

With a peak radiation of 700 W/m^2 the absorber temperature is 100°C and the top cover 60°C. If all the absorber heat is transferred to the water which flows through ducts in the absorber with negligible heat transfer resistance, determine the water mass flow rate (kg/s) necessary to raise the water temperature from 15°C (inlet) to 60°C at peak radiation. State clearly any assumptions made. (Stefan–Boltzmann constant = 5.7×10^{-8} W/m^2 K^4).
(Answer: 0.0092 kg/s)

1.5. The variation of load (P) with time (t) in a power supply system is given by the expression,

$$P(\text{kW}) = 4000 + 8t - 0.00091t^2$$

where t is in hours over a total period of one year.

This load is supplied by three 10 MW generators and it is advantageous to fully load a machine before connecting the others. Determine:

(a) the load factor on the system as a whole;
(b) the total magnitude of installed load if the diversity factor is equal to 3;

58

(c) the minimum number of hours each machine is in operation;
(d) the approximate peak magnitude of installed load capacity to be cut off to enable only two generators to be used.

(Answer: (a) 0.74, (b) 65 MW, (c) 8760, 7300, 2500 h, (d) 5.4 MW)
1.6. A 1000 MW(e) plant operates at 33.3 per cent efficiency and uses river water to cool the condensate directly (i.e. without a cooling tower). If the river-water temperature is not to gain more than 10°C as it passes through the condenser calculate the water flow (m^3/s) required.

If this station were to use a cooling tower (evaporative) calculate the volume of water used in evaporative cooling. Assume all heat is absorbed by evaporation of water.
Data. Water:

$$C_P = 4200 \text{ J/kg K heat of vaporization} = 2260 \times 10^3 \text{ J/kg}$$

(Answer: 48 m^3/s; 0.89 m^3/s)
1.7. Compare the energies available, before processing to electricity, from strip mining and solar energy, assuming the same area of land. Assume depth of coal strip to be 3 m and the density of coal to be 1440 kg/m^3 and take calorific values and solar radiation levels appropriate to your geographical region.
1.8. A lake of area 500 km^2 is fed from a drainage area of 6000 km^2 (including the lake). The level of the water in the lake is 500 m at the beginning of the month (720 h) and 500.5 m at the end. Over this period the total rainfall is 10 cm with a 40 per cent loss due to evaporation. The only outlet from the lake is a river which supplies a hydro-station, the head above the turbines being 50 m. The power loss due to friction is 3 per cent of the total in the river. If the overall efficiency of the turbine-generators is 80 per cent calculate the average power output.
(Answer: 16.8 MW)
1.9. A compressed air system comprises separate cool-air and heat-storage facilities and operates in conjunction with gas-turbine-driven generators. During off-peak power system conditions the air is compressed to 100 atmospheres and stored over 16 h. Over this period the heat evolved is stored at a rate of 85 MW. The separate cool air store has a volume of 0.7 × 10^6 m^3.

At peak system operation the above energy, together with fuel of energy content 100 × 10^5 MJ, is fed into the gas turbines over a period of 8 h. Calculate the total energy input over the 8 h and estimate the electrical energy produced. Assume:

(a) a 20 per cent heat loss from the heat-store over the 16 h of storage;
(b) the pressure in the air store should never be below 20 atm.

Discuss the relative merits of separate heat and air stores and the use of only one store for the compressed air. (1 atm = 1.01 × 10^5 N m^{-2}).
(Answer: Total energy 19.5 × 10^6 MJ)
1.10 Outline the general concept of an energy system based on the production and distribution of hydrogen gas.

The calorific value of hydrogen gas at atmospheric pressure and temperature is 12 × 10^6 Jm^{-3}. Determine the rate of energy transfer in a pipe of 1 m internal diameter with hydrogen at 60 atmospheres pressure flowing at a velocity of 5 m s^{-1}. Assuming the efficiency of the electrolysis to be 60 per cent, *estimate* the overall efficiency of the gas system to the load point (assuming 98 per cent transmission efficiency) and compare it with the use of electricity from a nuclear station. State clearly your assumed values.
(Answer:2.8 GW, 20.6 per cent (gas), 29.7 per cent (electric))

2

Basic Concepts

Before the modelling and analysis of power systems is pursued in detail various basic ideas will be outlined. These include the nature of three-phase systems, the per unit system, and circuit representation. Hopefully, some of this material will be revision for many readers.

2.1 Three-Phase Systems

The rotor flux of alternating current generator induces sinusoidal e.m.f.s, the conductors forming the stator winding, which in a single-phase machine occupies slots over most of the circumference of the stator core. These e.m.f.s are not in phase and the net winding voltage is less than the arithmetic sum of the individual conductor voltages. If this winding is replaced by three separate identical windings as shown in Figure 2.1(a), each occupying one-third of the available slots, then the effective contribution of all the conductors is greatly increased, yielding a greatly enhanced power output for a given machine size.

These three windings give voltages displaced in time or phase by 120° as indicated in Figure 2.1(b). Also because the voltage in the (a) phase reaches its peak 120° before the (b) phase and 240° before the (c) phase, the order of phase voltages reaching their maxima or *phase sequence* is a–b–c. Most countries use a, b & c to denote the phases, however in Britain R (Red), Y (Yellow), B (Blue) is often used. It is seen that the algebraic sum of the winding or phase voltages (and currents if the winding currents are equal) at every instant in time is zero. Hence, if the 'starts' of each winding are connected then the electrical situation is unchanged and the three return lines can be dispensed with, yielding a three-phase, three-wire system as shown in Figure 2.2(a). If the currents from the windings are not equal then it is usual to connect a fourth wire (neutral) to the common connexion or neutral point, as shown in Figure 2.2(b).

This type of winding connexion is called 'wye' or 'star' and two sets of voltages exist:

(a) the winding, phase, or line to neutral voltage, i.e. V_{an}, V_{bn}, V_{cn}; and
(b) the line-to-line voltages, V_{ab}, V_{bc}, V_{ca}. The suffixes here are important, V_{ab} means the voltage of line or terminal (a) with respect to (b), ($V_{ba} = -V_{ab}$). The corresponding phasor diagram is shown in figure 2.3(a) and it can be shown that $V_{ab} = V_{bc} = V_{ca} = \sqrt{3}$ (phase voltage).

60

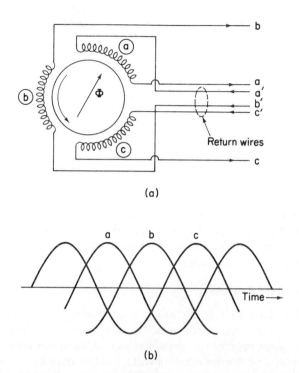

(a)

(b)

Time ⟶

(b)

Figure 2.1(a) Synchronous machine with three separate stator windings a, b, and c displaced physically by 120°. (b) Variation of e.m.f.s
developed in the windings with time.

The phase rotation of a system is very important. Consider the connexion
through a switch of two voltage sources of equal magnitude and both of
rotation a–b–c. When the switch is closed no current flows. If, however, one
source is of reversed rotation (easily obtained by reversing two wires), as shown
in Figure 2.4, i.e. a–c–b, a large voltage ($\sqrt{3} \times$ phase voltage) exists across the
switch contacts c′b and b′c, resulting in very large currents if the switch is closed.
Also, on reversed phase rotation the rotating magnetic field set up by a

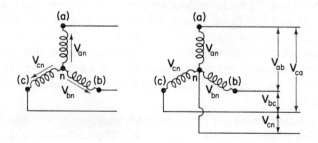

Figure 2.2(a)   Wye or star connexion of windings. (b) Wye connexion
with neutral line.

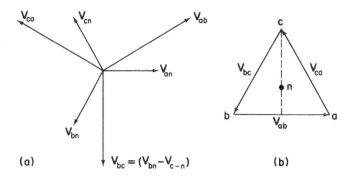

Figure 2.3(a) Phasor diagram for wye connexion. (b) Alternative arrangement of line-to-line voltages. Neutral voltage is at n, geometric centre of equilateral triangle.

three-phase winding is reversed in direction and an induction motor will rotate in the opposite direction, often with disastrous results to its mechanical load, e.g. a pump.

A three-phase load is connected in the same way as the machine windings. The load is *balanced* when each phase takes equal currents, i.e. has equal impedance. With the wye system the phase currents are equal to the current in the lines. The *four-wire* system is of particular use for low-voltage distribution networks in which consumers are supplied with a single-phase supply taken between a line and neutral. This supply is often 240 V and the line-to-line voltage 415 V. In the USA the 240 V supply often comes into a house from a centre-tapped transformer, as shown in Figure 2.5, which in effect gives a choice of 240 V (large domestic appliances) or 120 V (lights, etc.).

The system planner will endeavour to connect the single-phase loads such as to provide balanced (or equal) currents in the supply lines. At any instant in time it is highly unlikely that consumers will take equal loads and at this level of

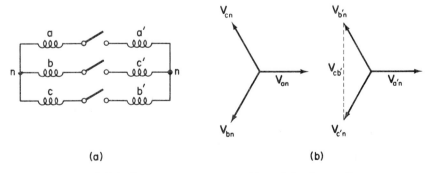

Figure 2.4(a) Two generators connected by switch; phase voltages equal for both sets of windings. (b) phasor diagrams of voltages. $V_{c'b}$ = voltage across switch; $V_{a'a} = 0$.

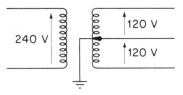

Figure 2.5   Tapped single-phase supply to give 240/120 V. (centre-
tap grounded).

supply considerable unbalance occurs, resulting in currents in the neutral line. If the neutral line has zero impedance this unbalance does not affect the load voltages. Lower currents flow in the neutral and it is usually of smaller cross-sectional area than the main line conductors. The combined or statistical effect of the large number of loads on the low-voltage network is such that when the next higher distribution voltage is considered (say 11 kV (line to line)), which supplies the lower voltage network, the degree of unbalance is small. This and the fact that at this higher voltage large three-phase, balanced motor loads are supplied, allows the three-wire system to be used. It is used exclusively up to the highest transmission voltages, resulting in much reduced line costs and environmental impact.

   In a balanced three-wire system a hypothetical neutral line may be considered and the conditions in one phase only determined. This is illustrated by the phasor diagram of line-to-line voltages shown in Figure 2.3(b). As the system is balanced the magnitudes so derived will apply to the other two phases but the relative phase angles must be adjusted by 120° and 240°. This single-phase approach is very convenient and widely used.

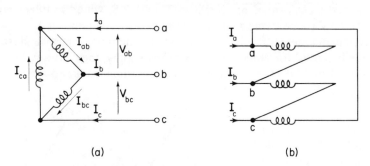

(a)   (b)

Figure 2.6(a)   Mesh or delta-connected load–current relationships.
(b) Practical connexions.

   An alternative method of connexion is shown in Figure 2.6. the individual phases are connected (taking due cognizance of winding polarity in machines and transformers) to form a closed loop. This is known as the mesh or delta connexion. Here the line-to-line voltages are identical to the phase voltages, i.e. $V_{line} = V_{phase}$. The line currents are as follows:

$$\mathbf{I}_a = \mathbf{I}_{ab} - \mathbf{I}_{ca} \qquad \mathbf{I}_b = \mathbf{I}_{bc} - \mathbf{I}_{ab} \qquad \mathbf{I}_c = \mathbf{I}_{ca} - \mathbf{I}_{bc}$$

For balanced currents in each phase it is readily shown from a phasor diagram that, $I_{\text{line}} = \sqrt{3} \times I_{\text{phase}}$. Obviously a fourth or neutral line is not possible with this connexion. It is seldom used for rotating machine stator windings, but is frequently used for the windings of one side of transformers for reasons to be discussed later. A line-to-line voltage transformation ratio of $1 : \sqrt{3}$ is obtained when going from a primary mesh to a secondary wye connexion with the same number of turns per phase. Under balanced conditions the idea of the hypothetical neutral and single-phase solution may still be used (the mesh can be converted to a wye using the Δ–Y transformation).

It should be noted that three-phase systems are usually described by the *line-to-line voltage*.

Analysis of a simple three-phase circuits

(a) *Four-wire systems* If the impedance voltage drops in the lines are negligible then the voltage across each load is the source line or phase voltage. Consider the network in Figure 2.7 with single-phase loads and a balanced

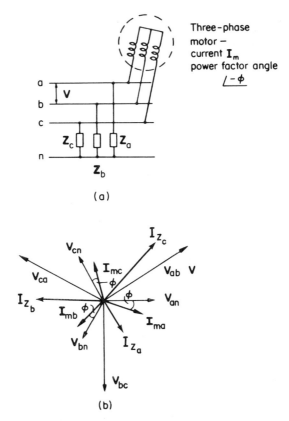

Figure 2.7(a) Four-wire system with single-phase and balanced three-phase loads. Supply line voltage = V. (b) Phasor diagram.

three-phase load (an induction motor). Note that the power factor of the three-phase load is expressed with respect to the phase current and voltage. The line currents are as follows:

$$I_a = \frac{V}{\sqrt{3}} \cdot \frac{1}{Z_a} + I_m(\cos\phi - j\sin\phi)$$

$$I_b = \frac{V}{\sqrt{3}}(-0.5 - j0.866)\frac{1}{Z_b} + I_m[-\cos(60-\phi) - j\sin(60-\phi)]$$

$$I_c = \frac{V}{\sqrt{3}}(-0.5 + j0.866)\frac{1}{Z_c} + I_m[-\cos(60+\phi) + j\sin(60+\phi)]$$

The neutral current, $I_n = I_a + I_b + I_c$.

(b) *Three-wire balanced*   The system may be treated as a single-phase system using phase voltages. It must be remembered, however, that the total three-phase power and reactive power are three times the single-phase values.

(c) *Three-wire unbalanced*   In more complex networks the method of symmetrical components is used, but in simple situations conventional network theory can be applied. Consider the source load arrangement shown in Figure

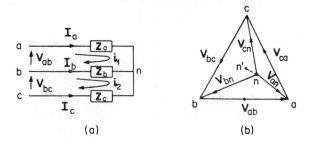

(a)   (b)

Figure 2.8(a)   Three-wire system with unbalanced load. i_1 and i_2 loop currents. (b) Phasor diagram—voltage difference of neutral connexion (n) from ground = n'n; n' = neutral voltage with balance.

2.8, in which $Z_a \neq Z_b \neq Z_c$. From the mesh method of analysis, the following equation is obtained:

$$\begin{bmatrix} Z_a + Z_b & -Z_b \\ -Z_b & Z_b + Z_c \end{bmatrix}\begin{bmatrix} i_1 \\ i_2 \end{bmatrix} = \begin{bmatrix} V \\ V(-0.5 - j0.866) \end{bmatrix}$$

Hence i_1 and i_2 are determined and from these I_a, I_b, and I_c and the voltage of the neutral point (n) from ground (see Figure 2.8(b)). Depending on the severity of the imbalance, the voltage difference from (n) to ground can attain values exceeding the phase voltage, i.e. (n) can lie outside the triangle of line voltages.

The power and reactive power consumed by a *balanced* three-phase load for both wye or mesh connexions are given by

$$P = \sqrt{3} \cdot V \cdot I \cdot \cos \phi \qquad Q = \sqrt{3} \cdot V \cdot I \cdot \sin \phi$$

where,

V = line-to-line voltage;

I = line current;

ϕ = angle between load phase current and load *phase* voltage.

Alternatively,

$$P = 3 V_{ph} \cdot I_{ph} \cdot \cos \phi \quad \text{and} \quad Q = 3 V_{ph} \cdot I_{ph} \cdot \sin \phi$$

As the powers in each phase of a balanced load are equal, a single wattmeter may be used to measure total power. For unbalanced loads two wattmeters are sufficient with 3 wire circuits.

Example 2.1 A three-phase wye-connected load is shown in Figure 2.9. It is supplied from a 200 V, three-phase, four-wire supply of phase sequence a–b–c and the neutral line (n) has a resistance of 5 Ω. Two wattmeters are connected as shown. Calculate the power recorded on each wattmeter.

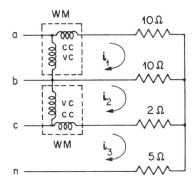

Figure 2.9 Use of two wattmeters to measure three-phase power. Arrangement for Example 2.1; cc = current coil, vc = voltage coil.

Solution The mesh method will be used to determine the currents in the lines and hence in the wattmeter current coils.

The line voltage $\mathbf{V}_{ab}$ is used as a reference phasor. Hence,

$$\mathbf{V}_{bc} = (-0.5 - j0.866)200$$

$$\mathbf{V}_{ca} = (-0.5 + j0.866)200$$

and

$$\mathbf{V}_{cn} = +j\frac{200}{\sqrt{3}}$$

From Figure 2.9,

$$\begin{bmatrix} 20 & -10 & 0 \\ -10 & 12 & -2 \\ 0 & -2 & 7 \end{bmatrix} \cdot \begin{bmatrix} i_1 \\ i_2 \\ i_3 \end{bmatrix} = \begin{bmatrix} 200 \\ -100 - j173.2 \\ j\dfrac{200}{\sqrt{3}} \end{bmatrix}$$

Eliminate i_1,

$$\begin{bmatrix} 7 & -2 \\ -2 & 7 \end{bmatrix}\begin{bmatrix} i_2 \\ i_3 \end{bmatrix} = \begin{bmatrix} -j173.2 \\ j\dfrac{200}{3} \end{bmatrix}$$

From which

$$i_3 = j2.66 \text{ A}$$

$$i_2 = -j24.0 \text{ A}$$

$$i_1 = 10 - j12 \text{ A}$$

Power in wattmeter (1) = real part of $\mathbf{V}_{ab} \cdot \mathbf{I}_a^* = 200(10 + j12)$

$$= 2 \text{ kW}$$

Power in wattmeter (2) = real part of $\mathbf{V}_{cb} \cdot \mathbf{I}_c^* = (-\mathbf{V}_{bc})\mathbf{I}_c^*$

$$= (100 + j173.2)(i_3 - i_2)^*$$

$$= (100 + j173.2)(-j26.66)$$

$$= 4.612 \text{ kW}$$

Actual power consumed $= \mathbf{I}_a^2 \cdot 10 + \mathbf{I}_b^2 \cdot 10 + \mathbf{I}_c^2 \cdot 2 + \mathbf{I}_n^2 \cdot 5$

$$= 6.337 \text{ kW}$$

2.2 Three-Phase Transformers

The usual form of the three-phase transformer, i.e. the core type, is shown in Figure 2.10. If the magnetic reluctances of the three limbs are equal then the sum of the fluxes set up by the three-phase magnetizing currents is zero. In fact, the core is the magnetic equivalent of the wye connected winding. It is apparent from Figure 2.10 that the magnetic reluctances are not exactly equal, but in an introductory treatment may be so assumed. An alternative to the three-limbed core is the use of three separate single-phase transformers. Although more expensive (about 20 per cent extra), this has the advantage of smaller loads for transportation, and this aspect is becoming crucial. Also, with the installation of four single-phase units, a spare is available at reasonable cost.

The wound core, as shown in Figure 2.10, is placed in a steel tank filled with insulating oil. The oil acts both as electrical insulation and as a cooling agent to remove the losses from the windings and core. The low-voltage winding is

Supply

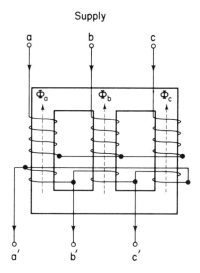

Figure 2.10   Three-phase core-type transformer. Primary connected
in wye, secondary connected in mesh.

situated over the core limb and the high-voltage winding wound over the
low-voltage one. The core comprises steel laminations insulated on one side (to
reduce eddy losses) and clamped together.

   Note that the voltages across the windings are phase voltages and it is these
which are related by the turns ratio.

Auto-transformers (Figure 2.11)   In the auto-transformer only one winding is
used per phase, the secondary voltage being tapped off the primary winding.
There is obviously a saving in size, weight, and cost over a two-windings per
phase transformer. It may be shown that the ratio of the weight of conductor on
an auto-transformer to that on a double-wound one is given by $(1 - N_2/N_1)$.
Hence, maximum advantage is obtained with a relatively small difference
between the voltages on the two sides. There are, however, several requirements in a power system where connexions from one voltage level to another
do not entail large transformer ratios, e.g. 500 kV/345 kV, 500 kV/725 kV,
and the auto-transformer is used. The effective reactance is reduced compared
to the equivalent two-winding transformer and this can give rise to high

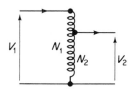

Figure 2.11   One phase of auto-transformer: $V_2/V_1 = N_2/N_1$.

short-circuit currents. The general constructional features of the core and tank are similar to those of double-wound transformers.

2.3 Harmonics in Three-Phase Systems

Harmonics may exist in three-phase systems and are created by non-linear loads, rectifiers, etc. Consider the instantaneous values of phase voltages in a balanced system containing harmonics up to the third.

$$v_a = V_1 \sin \omega t + v_2 \sin 2\omega t + v_3 \sin 3\omega t$$

$$v_b = v_1 \sin \left(\omega t - \frac{2\pi}{3} \right) + v_2 \sin 2 \left(\omega t - \frac{2\pi}{3} \right) + v_3 \sin 3 \left(\omega t - \frac{2\pi}{3} \right)$$

$$v_c = v_1 \sin \left(\omega t - \frac{4\pi}{3} \right) + v_2 \sin 2 \left(\omega t - \frac{4\pi}{3} \right) + v_3 \sin 3 \left(\omega t - \frac{4\pi}{3} \right)$$

The fundamental terms have the normal phase shifts (as do the fourth, seventh and tenth, etc. harmonics). However, the second harmonic terms possess a reversed phase-sequence (also fifth, eighth, eleventh, etc.), and the third harmonic terms are all in phase (also all multiples of three).

When substantial harmonics are present the $\sqrt{3}$ relation between line and phase quantities no longer holds. Harmonic voltages in successive phases are $2\pi n/3$ out of phase with each other and the resulting line voltage becomes

$$v_{ab} = v_n \sin n\omega t - v_n \sin n \left(\omega t - \frac{2\pi}{3} \right)$$

i.e.

$$2v_n \cos n \left(\omega t - \frac{\pi}{3} \right) \sin \left(\frac{n\pi}{3} \right)$$

when $n = 3, 6, 9$, etc. this becomes zero, i.e. no triple harmonics exist in the line voltages. The mesh connexion forms a complete path for the triple harmonic currents which flow in phase around the loop.

When analysing the penetration of harmonics into the power network it is usual to assume that the effective reactance for the nth harmonic is n times the fundamental value.

2.4 Multiphase Systems

Although three-phase is universally used, attention has been given in recent years to the use of more than three phases for transmission purposes. In particular, 6- and 12-phase systems have been studied. Advantages relative to three-phase systems are as follows:

1. Increased thermal loading capacity of lines.
2. The higher the number of phases the smaller the line-to-line voltage becomes relative to the phase voltage, resulting in increased utilization of rights of way because of less phase-to-phase insulation requirement.

3. For a given conductor size and tower configuration the stress on the conductor surface decreases with the number of phases, leading to reduced corona effects.
4. The transmission efficiency is higher. Existing double-circuit lines (two three-phase circuits on each tower) could be converted to single-circuit six-phase lines. It is advantageous to describe multiphase systems in terms of the phase voltage rather than line-to-line, as is the case for three-phase systems.

A six-phase supply is obtainable by suitable arrangement of the secondary windings of a three-phase transformer. Consider Figure 2.12. The windings on

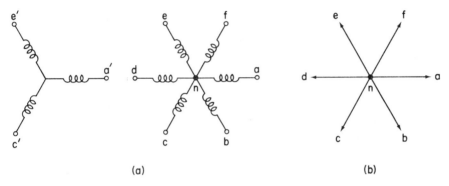

(a)   (b)

Figure 2.12 Six-phase system (a) Transformer connexions. (b) Phasor diagrams.

the three limbs of the transformer are centre-tapped with the taps mutually connected. This gives the phasor diagram shown in Figure 2.12. The line-to-line voltages are equal to the phase voltage in magnitude, i.e. $\mathbf{V}_{af} = \mathbf{V}_{an} - \mathbf{V}_{fn} = \mathbf{V}_{ph}$.

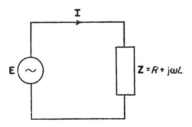

Figure 2.13 Voltage source and load.

2.5 Reactive Power

In the circuit shown in Figure 2.13 let the instantaneous values of voltage and current be

$$e = \sqrt{2}E \sin (\omega t + \phi)$$

and

$$i = \sqrt{2}I \sin(\omega t)$$

The instantaneous power

$$p = ei = EI \cos\phi - EI \cos(2\omega t + \phi)$$

Also,

$$-EI \cos(2\omega t + \phi) = -EI[\cos 2\omega t \cos\phi - \sin 2\omega t \cdot \sin\phi]$$

$$\therefore p = ei = (EI \cos\phi - EI \cos 2\omega t \cos\phi) + (EI \sin 2\omega t \sin\phi)$$

$$= (\text{instantaneous real power}) + (\text{instantaneous reactive power})$$

The mean power $\qquad = EI \cos\phi$.

The mean value of $EI \sin\phi \sin 2\omega t = 0$, but the maximum value $= EI \sin\phi$.

The voltage source is supplying energy to the load in one direction only. At the same time an interchange of energy is taking place between the source and the load of average value zero, but of peak value $EI \sin\phi$. This latter quantity is known as the *reactive power* (Q) and the unit is the VAr (taken from the alternative name, volt-ampere reactive). The interchange of energy between the source and the inductive and capacitive elements (i.e. the magnetic and electric fields) takes place at twice the supply frequency. It is possible to think, therefore, of a power component P (watts) of magnitude $EI \cos\phi$ and a reactive-power component Q (VAr) equal to $EI \sin\phi$, where ϕ is the power factor angle, i.e. the angle between **E** and **I**. It should be stressed, however, that the two quantities P and Q are physically quite different.

The quantity S (volt-amperes), known as the complex power, may be found by multiplying **E** by the conjugate of **I** or vice versa. Consider the case when **I** lags **E**, and assume $S = E^*I$. Referring to Figure 2.14,

$$S = E\,e^{-j\phi_1} \times I\,e^{j\phi_2} = EI\,e^{-(\phi_1-\phi_2)}$$

$$= EI\,e^{-\phi}$$

$$= P - jQ$$

Next assume

$$S = EI^*$$

$$= E\,e^{j\phi_1} = I\,e^{-j\phi_2}$$

$$= P + jQ.$$

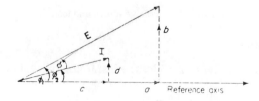

Figure 2.14   Phasor diagram.

Obviously both the above methods give the correct magnitudes for P and Q but the sign of Q is different. The method used is arbitrarily decided and the convention to be adopted in this book is as follows:

The volt-amperes reactive absorbed by an inductive load shall be considered positive, and by a capacitive load negative; hence $\mathbf{S} = \mathbf{EI}^*$. This convention is recommended by the International Electrotechnical Commission.

In a network the net energy is the sum of the various inductive and capacitive stored energies present. The net value of reactive power is the sum of the vars absorbed by the various components present, taking due account of the sign. Lagging vars can be considered as either being produced or absorbed in a circuit; a capacitive load can be thought of as generating lagging vars. Assuming that an inductive load is represented by $R + jX$ and that the $\mathbf{EI}^*$ convention is used then an inductive load *absorbs* positive or lagging vars and a capacitive load *produces* lagging vars. The reactive power flow towards a busbar is positive when the load is lagging; vars exported from a busbar are considered negative for a lagging power factor and positive for leading vars. In this text, reference to the var will imply a lagging power factor and inductive and capacitive vars distinguished by the sign.

The various elements in a network are characterized by their ability to generate or absorb reactive power. Consider a synchronous generator which can be represented by the simple equivalent circuit shown in Figure 2.15. When the generator is over-excited, i.e. its generated e.m.f. is high, it produces lagging vars and a complex power, $P - jQ$. When the machine is under-excited the generated current leads V and the generator produces $P + jQ$. It can also be thought of as absorbing lagging vars. The reactive power characteristics of

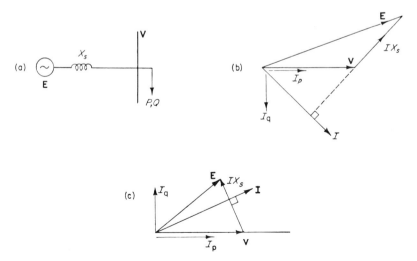

Figure 2.15(a) Line diagram of system. (b) Over-excited generator–phasor diagram. (c) Under-excited generator–phasor diagram.

various power-system components are summarized as follows: reactive power is generated by over-excited synchronous machines, capacitors, cables, lightly loaded overhead lines; and absorbed by under-excited synchronous machines, induction motors, inductors, transformers, heavily loaded overhead lines.

The reactive power absorbed by a reactance of X ohms $= I^2 X$, where I is the current. Note that at a node, $\sum P = 0$ and $\sum Q = 0$.

2.6 The Per Unit System

In the analysis of power networks instead of using actual values of quantities it is usual to express them as fractions of reference quantities, such as rated or full-load values. These fractions are called per unit (denoted by p.u.) and the p.u. value of any quantity is defined as

$$\frac{\text{the actual value (in any unit)}}{\text{the base or reference value in the same unit}}$$

Some authorities express the p.u. value as a percentage. Although the use of p.u. values may at first sight seem a rather indirect method of expression there are in fact great advantages; they are as follows.

(a) The apparatus considered may vary widely in size; losses and volt drops will also vary considerably. For apparatus of the same general type the p.u. volt drops and losses are in the same order regardless of size.

(b) As will be seen later, the use of $\sqrt{3}$'s in three-phase calculations is reduced.

(c) By the choice of appropriate voltage bases the solution of networks containing several transformers is facilitated.

(d) Per unit values lend themselves more readily to automatic computation.

Resistance

$$R_{\text{p.u.}} = \frac{R(\Omega)}{\text{base } R(\Omega)}$$

$$= \frac{R(\Omega)}{\text{base voltage } (V_b)/\text{base current } (I_b)}$$

$$= \frac{R(\Omega) \cdot I_b}{V_b}$$

$$= \frac{\text{voltage drop across } R \text{ at base or rated current}}{\text{base or rated voltage}}$$

Also,

$$R_{\text{p.u.}} = \frac{R(\Omega) I_b^2}{V_b I_b}$$

$$= \frac{\text{power loss at base current}}{\text{base power or volt-amperes}}$$

∴ the power loss (p.u.) at base or rated current $= R_{\text{p.u.}}$
Power loss (p.u.) at $I_{\text{p.u.}}$ current $= R_{\text{p.u.}} \cdot I_{\text{p.u.}}^2$

Similarly,

$$\text{p.u. impedance} = \text{impedance in ohms} \Big/ \dfrac{\text{base voltage}}{\text{base current}}$$

Example 2.2 A d.c. series machine rated at 200 V, 100 A has an armature resistance of 0.1 Ω and field resistance of 0.15 Ω. The friction and windage loss is 1500 W. Calculate the efficiency when operating as a generator.

Solution Total series

$$R_{\text{p.u.}} = \frac{0.25}{200/100} = 0.125 \text{ p.u.}$$

where

$$V_{\text{base}} = 200 \text{ V} \quad \text{and} \quad I_{\text{base}} = 100 \text{ A}$$

Friction and windage loss

$$= \frac{1500}{200 \times 100} = 0.075 \text{ p.u.}$$

At the rated load, the series-resistance loss

$$= 1^2 \times 0.125 \text{ p.u.}$$

and the total loss

$$= 0.125 + 0.075 = 0.2 \text{ p.u.}$$

As the output $= 1$ p.u., the efficiency

$$= \frac{1}{1 + 0.2} = 0.83 \text{ p.u.}$$

Three-phase circuits A p.u. phase voltage has the same numerical value as the corresponding p.u. line voltage. With a line voltage of 100 kV and a rated line voltage of 132 kV, the p.u. value is 0.76. The equivalent phase voltages are $100/\sqrt{3}$ kV and $132/\sqrt{3}$ kV and hence the p.u. value is again 0.76. The actual values of R, X_{L} and X_{c} for lines, cables and other apparatus are phase values. When working with ohmic values it is less confusing to use the phase values of all quantities. In the p.u. system three-phase values of voltage, current and power can be used without undue anxiety about the result being a factor of $\sqrt{3}$ incorrect.

$$I_{\text{base}} = \frac{\text{base } VA}{\sqrt{(3)} V_{\text{base}}}$$

$$\text{base } Z = \frac{V_{\text{base}}/\sqrt{3}}{I_{\text{base}}} \quad \text{assuming a star connected system}$$

$$= \frac{\dfrac{V_{\text{b}}}{\sqrt{3}} \cdot \dfrac{V_{\text{b}}}{\sqrt{3}}}{\dfrac{I_{\text{b}} \cdot V_{\text{b}}}{\sqrt{3}}} = \frac{V_{\text{b}}^2}{\sqrt{(3)} V_{\text{b}} I_{\text{b}}}$$

$$= \frac{(\text{base line voltage})^2}{\text{base } VA} \tag{2.1}$$

It should be noted that the same value for Z_{base} is obtained using purely phase values.

$$\text{Hence, } Z_{\text{p.u.}} = \frac{Z(\Omega) \times \text{base } VA}{(\text{base voltage})^2}$$

i.e. is directly proportional to the volt-ampere base and inversely to (base voltage)^2. Hence, $Z_{\text{p.u.}}$(new bases)

$$= Z_{\text{p.u.}}(\text{original base}) \times \left(\frac{\text{base } V_{\text{old}}}{\text{base } V_{\text{new}}}\right)^2 \times \left(\frac{\text{base } VA_{\text{new}}}{\text{base } VA_{\text{old}}}\right) \tag{2.2}$$

Transformers   Consider a single-phase transformer in which the total series impedance of the two windings referred to the primary is Z_1 (Figure 2.16).

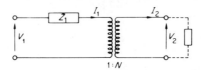

Figure 2.16   Equivalent circuit of single-phase transformer.

Then the p.u. impedance $= I_1 Z_1 / V_1$, where I_1 and V_1 are the rated or base values. The total impedance referred to the secondary

$$= Z_1 N^2 \quad \text{and this in p.u. notation}$$

$$= Z_1 N^2 \left(\frac{I_2}{V_2}\right)$$

$$= Z_1 N^2 \frac{I_1}{N} \frac{1}{V_1 N}$$

$$= \frac{Z_1 I_1}{V_1}$$

Hence the p.u. impedance of a transformer is independent of the winding considered.

In a circuit with several transformers care has to be taken regarding the different voltage levels. Consider the network in Figure 2.17; in this two single-phase transformers supply a 10 kVA resistance load, the load voltage being maintained at 200 V. Hence the load resistance is $(200^2/10,000)$, i.e. 4 Ω. In each of the circuits A, B, and C a different voltage exists, so that each circuit will have its own base voltage, i.e. 100 V in A, 400 V in B, 200 V in C.

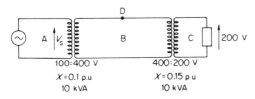

Figure 2.17 Network with two transformers—p.u. approach.

Although it is not essential for rated voltages to be used as bases, it is essential that the *voltage bases used be related by the turns ratios of the transformers.* If this is not so the whole p.u. framework breaks down. The same volt-ampere base is used for all the circuits as $V_1I_1 = V_2I_2$ on each side of a transformer and is taken in this case as 10 kVA.

The base impedance in C

$$= \frac{200^2}{10,000} = 4 \text{ Ω}$$

The load resistance (p.u.) in C

$$= \frac{4}{4} = 1 \text{ p.u.}$$

In B the base impedance

$$= \frac{400^2}{10,000} = 16 \text{ Ω}$$

and the load resistance referred to B

$$= 4 \times 2^2 = 16 \text{ Ω}$$

Hence the p.u. load referred to B

$$= 1 \text{ p.u.}$$

Similarly, the p.u. load resistance referred to A is also 1 p.u. Hence, if the voltage bases are related by the turns ratios the load p.u. value is the same for all circuits. An equivalent circuit may be used as shown in Figure 2.18. Let the

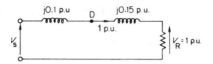

Figure 2.18   Equivalent circuit with p.u. values, of network in Figure 2.17.

volt-ampere base be 10 kVA; the voltage across the load (V_R) is 1 p.u. (as the base voltage in C is 200 V, if the load voltage had been maintained at 100 V, V_R would be 0.5 p.u.). The base current

$$= \frac{(VA)_b}{V_b} = \frac{10,000}{200} = 50 \text{ A in C}$$

The corresponding currents in the other circuits are 25 A in B, and 100 A in A. The actual load current $= 50/50 = 1$ p.u. (in phase with V_R the reference phasor). Hence the supply voltage V_s

$$= 1(j0.1 + j0.15) + 1 \text{ p.u.}$$

$$\therefore \ V_s = 1.03 \text{ p.u.}$$

$$= 1.03 \times 100 = 103 \text{ V.}$$

The voltage at point D in figure 2.17

$$= 1 + j0.15 \times 1 = 1 + j0.15$$

$$= 1.012 \text{ p.u. modulus}$$

$$= 1.012 \times 400 = 404.8 \text{ V.}$$

It is a useful exercise to repeat this example using ohms, volts, and amperes.

Example 2.3   In Figure 2.19 the schematic diagram of a radial transmission system is shown. The ratings and reactances of the various components are shown, along with the nominal transformer line voltages. A load of 50 MW at 0.8 p.f. lagging is taken from the 33 kV substation which is to be maintained at 30 kV. It is required to calculate the terminal voltage of the synchronous machine. The line and transformers may be represented by series reactances. The system is three-phase.

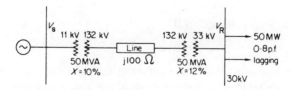

Figure 2.19   Line diagram of system for Example 2.3.

Solution It will be noted that the line reactance is given in ohms; this is usual practice. The voltage bases of the various circuits are decided by the nominal transformer voltages, i.e. 11, 132, and 33 kV. A base of 100 MVA will be used for all circuits. The reactances (resistance is neglected) are expressed on the appropriate voltage and MVA bases. Base impedancê for the line

$$= \frac{132^2 \times 10^6}{100 \times 10^6}$$

$$= 174 \ \Omega$$

Hence the p.u. reactance

$$= \frac{j100}{174} = j0.575$$

Per unit reactance of the sending-end transformer

$$= \frac{100}{50} \times j0.1 = j0.2$$

Per unit reactance of the receiving-end transformer

$$= j0.12 \times \frac{100}{50} = j0.24$$

Load current

$$= \frac{50 \times 10^6}{\sqrt{3} \times 30 \times 10^3 \times 0.8}$$

$$= 1200 \ A$$

Base current for 33 kV, 100 MVA

$$= \frac{100 \times 10^6}{\sqrt{3} \times 33 \times 10^3}$$

$$= 1750 \ A$$

Hence, p.u. load current

$$= \frac{1200}{1750} = 0.685$$

Voltage of load busbar

$$= \frac{30}{33} = 0.91 \ \text{p.u.}$$

The equivalent circuit is shown in Figure 2.20.

78

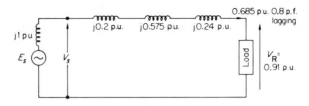

Figure 2.20   Equivalent circuit for Example 2.3.

Also,

$$V_s = 0.685(0.8 - j0.6)(j0.2 + j0.575 + j0.24) + (0.91 + j0)$$

$$= 1.33 + j0.555 \text{ p.u.}$$

$$\therefore \ V_s = 1.44 \text{ p.u. or } 1.44 \times 11 \text{ kV} = 15.84 \text{ kV.}$$

2.7   Power Transfer and Reactive Power

The circuit shown in Figure 2.21 is very important as it represents the simplest electrical model for a synchronous generator feeding into a power system represented by a voltage source (V).

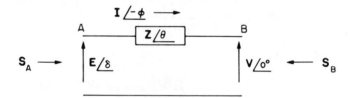

Figure 2.21   Power transfer between sources.

$$S_A = EI^* = E\left(\frac{E - V}{Z}\right)^*$$

$$= E \ e^{j\delta}\left(\frac{E \ e^{-j\delta} - V}{Z \ e^{-j\theta}}\right)$$

$$= \frac{E^2}{Z} e^{j\theta} - \frac{EV}{Z} e^{j(\theta + \delta)}$$

$$P_A = \frac{E^2}{Z} \cos \theta - \frac{EV}{Z} \cos (\theta + \delta)$$

(2.3)

$$Q_A = \frac{E^2}{Z} \sin \theta - \frac{EV}{Z} \sin (\theta + \delta)$$

Similarly,

$$S_B = V\left(\frac{V - E\,e^{-j\delta}}{Z\,e^{-j\theta}}\right)$$

$$P_B = \frac{V^2}{Z}\cos\theta - \frac{EV}{Z}\cos(\theta - \delta)$$

(2.4)

$$Q_B = \frac{V^2}{Z}\sin\theta - \frac{EV}{Z}\sin(\theta - \delta)$$

The power output to V is a maximum when $\cos(\theta - \delta) = 1$, i.e. $\theta = \delta$.

Calculation of sending and received voltages in terms of power and reactive power The determination of the voltages and currents in a network can obviously be achieved by means of complex notation, but in power systems usually power (P) and reactive power (Q) are specified and often the resistance of lines is negligible compared with reactance. For example, if $R = 0.1X$, the error in neglecting R is 0.49 per cent and even if $R = 0.4X$ the error is 7.7 per cent.

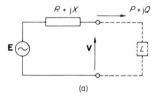

(a)

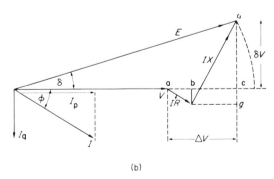

(b)

Figure 2.22   Phasor diagram for transmission of power through a series impedance.

A simple transmission link is shown in Figure 2.22a. It is required to establish the equations for E, V, and δ. From figure 2.22b,

$$E^2 = (V + \Delta V)^2 + \delta V^2$$

$$= (V + RI \cos \phi + XI \sin \phi)^2 + (XI \cos \phi - RI \sin \phi)^2$$

$$\therefore \ E^2 = \left(V + \frac{RP}{V} + \frac{XQ}{V}\right)^2 + \left(\frac{XP}{V} - \frac{RQ}{V}\right)^2 \tag{2.5}$$

Hence,

$$\Delta V = \frac{RP + XQ}{V} \tag{2.6}$$

and

$$\delta V = \frac{XP - RQ}{V} \tag{2.7}$$

If

$$\delta V \ll V + \Delta V$$

$$E^2 = \left(V + \frac{RP + XQ}{V}\right)^2$$

and

$$E - V = \frac{RP + XQ}{V} = \Delta V$$

Hence the arithmetic difference between the voltages is given approximately by

$$\frac{RP + XQ}{V}$$

If

$$R = 0,$$

$$E - V = \frac{XQ}{V}$$

i.e. **voltage depends only on Q,** and from equation (2.5) this is valid if $PX \ll V^2 + QX$. The angle of transmission δ is obtained from $\sin^{-1}(\delta V/E)$, **and depends only on P.**

Equations (2.6) and (2.7) will be used wherever possible owing to their great simplicity.

Consider a 275 kV line (R and X per km = 0.034 and 0.32 Ω respectively), obviously $R \ll X$.

Assume a load of 600 MW, 300 MVAr (600 MVA is the thermal rating for a $2 \times 258 \text{ mm}^2$ line) and taking a base of 100 MVA, the base impedance is

$$\frac{275^2 \times 10^6}{100 \times 10^6} = 755 \ \Omega$$

For line of length 160 km,

$$X = \frac{52}{755} = 0.069 \text{ p.u.}$$

and let the received voltage be

$$275 \text{ kV} = 1 \text{ p.u.}$$

$$PX = 6 \times 0.069 = 0.414$$

and

$$V^2 + QX = 1.207.$$

Hence the use of equation (2.6) will involve some inaccuracy.
    Using equation (2.5) with R neglected,

$$E^2 = \left(1 + \frac{0.069 \times 3}{1}\right)^2 + \left(\frac{0.069 \times 6}{V}\right)^2 = 1.207^2 + 0.414^2$$

$$E = 1.278 \text{ p.u.}$$

by equation (2.6)

$$E - V = \Delta V = \frac{0.069 \times 3}{1} = 0.207$$

$$\therefore \ E = 1.207 \text{ p.u.}$$

i.e. an error of 5.5 per cent.
    With a 80 km length of this line at the same load,

$$PX = 6 \times 0.0345 \quad \text{and} \quad V^2 + QX = (1 + 3 \times 0.0345)$$

$$= 0.207 \qquad\qquad\qquad = 1.103$$

The accurate formula gives

$$E = 1.125 \text{ p.u.}$$

and the approximate one gives

$$E = 1.104 \text{ p.u.}$$

i.e. an error of 1.9 per cent.
    Hence a considerable length of double-circuit line at rated load may be treated with the approximate formula. For a 275 kV, $2 \times 113 \text{ mm}^2$ conductor, line the rated power is 4.3 p.u. and thus a longer length could be considered

than in the above case. With 132 kV lines on a 100 MVA base the reactance of a 160 km 258 mm² line is $160 \times 0.4/174$ or 0.37 p.u. and the rated power 1.5 p.u. ($Q = 0.75$ p.u. for same power factor)

$$PX = 1.5 \times 0.37 \quad \text{and} \quad V^2 + XQ = 1 + 0.277$$
$$= 0.55 \qquad\qquad\qquad = 1.28$$

Hence, an error in the order of 5 per cent could be expected.

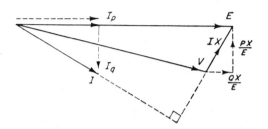

Figure 2.23   Phasor diagram when **E** is specified.

If E is specified and V is required, the phasor diagram in Figure 2.23 is used. From this,

$$V = \sqrt{\left[\left(E - \frac{QX}{E} \right)^2 + \left(\frac{PX}{E} \right)^2 \right]} \tag{2.8}$$

when $R = 0$.
    If

$$\frac{PX}{E} \ll \frac{E^2 - QX}{E}$$

then

$$V = E - \frac{QX}{E} \tag{2.9}$$

2.8   Useful Network Theory

Four terminal networks

A lumped-constant circuit, provided that it is passive, linear, and bilateral, can be represented by the four-terminal network shown in the diagram in Figure 2.24. The complex parameters **A**, **B**, **C**, and **D** describe the network in terms of the sending and receiving end voltages and currents as follows:

$$\mathbf{V_S} = \mathbf{A V_R} + \mathbf{B I_R}$$
$$\mathbf{I_S} = \mathbf{C V_R} + \mathbf{D I_R} \tag{2.10}$$

and it can be readily shown that $\mathbf{AD} - \mathbf{BC} = 1$.

Figure 2.24 Representation of a four-terminal (two-port) network.

$\mathbf{A}$, $\mathbf{B}$, $\mathbf{C}$, and $\mathbf{D}$ may be obtained by measurement and certain physical interpretations made, as follows:

(a) Receiving end short-circuited:

$$\mathbf{V_R} = 0, \quad \mathbf{I_S} = \mathbf{D I_R} \quad \text{and} \quad \mathbf{D} = \frac{\mathbf{I_S}}{\mathbf{I_R}}$$

Also,

$$\mathbf{V_S} = \mathbf{B I_R} \quad \text{and} \quad \mathbf{B} = \frac{\mathbf{V_S}}{\mathbf{I_R}} = \text{short-circuit transfer impedance}$$

(b) Receiving end open-circuited:

Here

$$\mathbf{I_R} = 0, \quad \mathbf{I_S} = \mathbf{C V_R} \quad \text{and} \quad \mathbf{C} = \left(\frac{\mathbf{I_S}}{\mathbf{V_R}}\right)$$

$$\mathbf{V_S} = \mathbf{A V_R}, \quad \mathbf{A} = \left(\frac{\mathbf{V_S}}{\mathbf{V_R}}\right)$$

Expressions for the constants can also be found by measurements carried out solely at the sending end with the receiving end open and short-circuited.

Often it is useful to have a single four-terminal network for two or more items in series or parallel, e.g. a line and two transformers in series.

It is shown in most texts on circuit theory that the generalized constants for the combined network, $\mathbf{A_0}$, $\mathbf{B_0}$, $\mathbf{C_0}$, and $\mathbf{D_0}$ for the two networks (1) and (2) in cascade are as follows:

$$\mathbf{A_0} = \mathbf{A_1 A_2} + \mathbf{B_1 C_2} \quad \mathbf{B_0} = \mathbf{A_1 B_2} + \mathbf{B_1 D_2} \quad \mathbf{C_0} = \mathbf{A_2 C_1} + \mathbf{C_2 D_1}$$

$$\mathbf{D_0} = \mathbf{B_2 C_1} + \mathbf{D_1 D_2}.$$

For two four-terminal networks in parallel it can be shown that the parameters of the equivalent single four-terminal network are:

$$\mathbf{A_0} = \frac{\mathbf{A_1 B_2} + \mathbf{A_2 B_1}}{\mathbf{B_1} + \mathbf{B_2}} \quad \mathbf{B_0} = \frac{\mathbf{B_1 B_2}}{\mathbf{B_1} + \mathbf{B_2}} \quad \mathbf{D_0} = \frac{\mathbf{B_2 D_1} + \mathbf{B_1 D_2}}{\mathbf{B_1} + \mathbf{D_2}}$$

$\mathbf{C_0}$ can be found from $\mathbf{A_0 D_0} - \mathbf{B_0 C_0} = 1$.

(a) Delta–star transformation The delta network connected between the three terminals A, B, and C can be replaced by a star network such that

84

the impedance measured between the terminals is unchanged. The values of the star-connected elements are as follows:

$$\mathbf{Z}_{OA} = \frac{\mathbf{Z}_{AB}\mathbf{Z}_{CA}}{\mathbf{Z}_{AB}+\mathbf{Z}_{CA}+\mathbf{Z}_{BC}}$$

$$\mathbf{Z}_{OB} = \frac{\mathbf{Z}_{AB}\mathbf{Z}_{BC}}{\mathbf{Z}_{AB}+\mathbf{Z}_{CA}+\mathbf{Z}_{BC}} \qquad (2.11)$$

$$\mathbf{Z}_{OC} = \frac{\mathbf{Z}_{BC}\mathbf{Z}_{CA}}{\mathbf{Z}_{AB}+\mathbf{Z}_{CA}+\mathbf{Z}_{BC}}$$

(b) *Star-delta transformation* A star-connected system can be replaced by an equivalent delta connexion if the elements of new network have the following values (Figure 2.25):

$$\mathbf{Z}_{AB} = \frac{\mathbf{Z}_{OA}\mathbf{Z}_{OB}+\mathbf{Z}_{OB}\mathbf{Z}_{OC}+\mathbf{Z}_{OC}\mathbf{Z}_{OA}}{\mathbf{Z}_{OC}}$$

$$\mathbf{Z}_{BC} = \frac{\mathbf{Z}_{OA}\mathbf{Z}_{OB}+\mathbf{Z}_{OB}\mathbf{Z}_{OC}+\mathbf{Z}_{OC}\mathbf{Z}_{OA}}{\mathbf{Z}_{OA}} \qquad (2.12)$$

$$\mathbf{Z}_{CA} = \frac{\mathbf{Z}_{OA}\mathbf{Z}_{OB}+\mathbf{Z}_{OB}\mathbf{Z}_{OC}+\mathbf{Z}_{OC}\mathbf{Z}_{OA}}{\mathbf{Z}_{OB}}$$

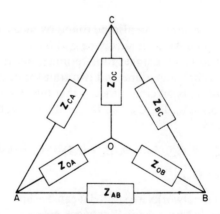

Figure 2.25 Star–delta, delta–star transformation.

Problems

2.1. The star-connected secondary winding of a three-phase transformer supplies 415 V (line to line) at a load point through a four-conductor cable. The neutral conductor is connected to the winding star point which is earthed. The load consists of the following components:

Between a and b conductors	a 1 Ω resistor
Between a and neutral conductors	a 1 Ω resistor
Between b and neutral conductors	a 2 Ω resistor
Between c and neutral conductors	a 2 Ω resistor

Connected to the a, b, and c conductors is an induction motor taking a balanced current of 100 A at 0.866 p.f. lagging. Calculate the currents in the four conductors and the total power supplied.

Take the 'a' to neutral voltage as the reference phasor. The phase sequence is a–b–c. (Answer: 350 kW)

2.2 The wye-connected load shown in Figure 2.26 is supplied from a transformer whose secondary-winding star point is solidly earthed. the line voltage supplied to the load is

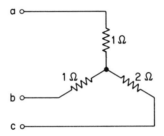

Figure 2.26 Circuit for Problem 2.2.

400 V. Determine (a) the line currents, and (b) the voltage of the load star point with respect to ground. The phase sequence is a–b–c. (Answer: $I_a(200 - j69)$ A; $V_{n-g} - 47$ V)

2.3 Two capacitors, each of 10 μF, and a resistor R, are connected to a 50 Hz three-phase supply, as shown in Figure 2.27. The power drawn from the supply is the same whether the switch S is open or closed. Find the resistance of R. (Answer: 143 Ω)

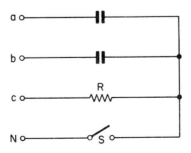

Figure 2.27 Circuit for Problem 2.3.

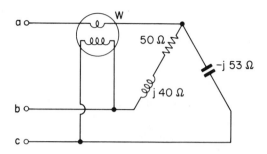

Figure 2.28   Circuit for Problem 2.4.

2.4. The network of Figure 2.28 is connected to a 400 V three-phase supply, with phase sequence a–b–c. Calculate the reading of the wattmeter, W.
(Answer: 2.24 kW)

2.5. A three-phase, three-wire supply feeds a load consisting of three identical resistors which are (a) star-connected, (b) delta-connected. In each case by how much is the power consumption reduced if one of the resistors is open circuited.
(Answer: (a) 50 per cent; (b) 33.3 per cent)

2.6. A 400 V three-phase supply feeds a delta-connected load with the following branch impedances:

$$\mathbf{Z}_{RY} = 100\ \Omega \qquad \mathbf{Z}_{YB} = j100\ \Omega \qquad \mathbf{Z}_{BR} = -j100\ \Omega$$

Calculate the line currents for phase sequences (a) RYB, (b) RBY.
(Answer: (a) 7.73, 7.37, 4 A; (b) 2.07, 2.07, 4 A)

2.7. A synchronous generator represented by a voltage source in series with an inductive reactance X_1, is connected to a load consisting of a fixed inductive reactance X_2 and a variable resistance R in parallel. Show that the generator power output is a maximum when

$$1/R = 1/X_1 + 1/X_2$$

2.8. A single-phase voltage source of 100 kV supplies a load through an impedance j100 Ω. The load may be represented in either of the following ways as far as voltage changes are concerned:

(a) by a constant impedance representing a consumption of 10 MW, 10 MVAr at 100 kV; or
(b) by a constant current representing a consumption of 10 MW, 10 MVAr at 100 kV.

Calculate the voltage across the load using each of these representations.
(Answer: (a) $(90 - j8.2)$ kV; (b) $(90 - j10)$ kV)

2.9. Show that the p.u. impedance (obtained from a short-circuit test) of a star–delta three-phase transformer is the same whether computed from the star-side parameters or from the delta side. Assume a rating of G (volt-amperes), a line-to-line input voltage to the star winding terminals of V volts, a turns ratio of $1:N$ (star to delta) and a short-circuit impedance of $\mathbf{Z}$(ohms) per phase referred to the star side.

2.10. A 11 kV/132 kV, 50 MVA, three-phase transformer has an inductive reactance of j0.005 Ω referred to the primary (11 kV). Calculate the p.u. value of reactance based on the rating. Neglect resistance.
(Answer: 0.002 p.u.)

2.11. Express in p.u. all the quantities shown in The line diagram of a three-phase transmission system in Figure 2.29. Construct the single-phase equivalent circuit. Use a

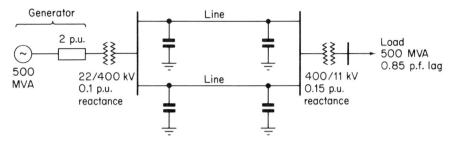

Figure 2.29   Circuit for Problem 2.11.

base of 100 MVA. The line is 80 km in length with resistance and reactance of 0.1 and 0.5 Ω respectively and a capacitive susceptance of $10\,\mu\Omega^{-1}$ per km (split equally between the two ends).

2.12. A wye-connected load is supplied from three-phase 220 V mains. Each branch of the load is a resistor of 20 Ω. Using 220 V and 10 kVA bases calculate the p.u. values of the current and power taken by the load.
(Answer: 0.24 p.u.; 0.24 p.u.)

2.13. A 440 V three-phase supply is connected to three star-connected loads in parallel, through a feeder of impedance $(0.1 + j0.5)\,\Omega$ per phase. The loads are as follows: 5 kW, 4 kVAr; 3 kW, 0 kVAr; 10 kW, 2 kVAr. Determine:

(a) line current;
(b) power and reactive power losses in the feeder per phase;
(c) power and reactive power from the supply and the supply power factor.

(Answer: 24.9 A, 62 W, 310 vars, 18.06 kW, 6.31 kVAr, 0.94)

2.14. Two transmission circuits are defined by the following **ABCD** constants: 1, 50, 0, 1, and $0.9\angle2°$, $150\angle79°$, $9\times10^{-4}\angle91°$, $0.9\angle2°$. Determine the **ABCD** constants of the circuit comprising these two circuits in series.
(Answer: $\mathbf{A} = 45\angle89.86$; $\mathbf{B} = 165.7\angle63.6$)

2.15. A 132 kV, overhead line has a series resistance and inductive reactance per phase per kilometre of 0.156 and 0.4125 Ω respectively. Calculate magnitude of the sending-end voltage when transmitting the full line capability of 125 MVA when the power factor is 0.9 lagging and the received voltage is 132 kV, for 16 km and 80 km lengths of line. Use both accurate and approximate methods.
(Answer: 136.92 kV (accurate), 136.85 kV; 157.95 kV (accurate), 156.27 kV)

2.16. A synchronous generator may be represented by a voltage source of magnitude 1.7 p.u. in series with an impedance of 2 p.u. It is connected to a zero-impedance voltage source of 1 p.u. The ratio of X/R of the impedance is 10. Calculate the power generated and the power delivered to the voltage source if the angle between the voltage sources is 30°.
(Answer: 0.49 p.u.; 0.44 p.u.)

3

Components of a Power System

3.1 Introduction

In this chapter the essential characteristics of the components of a power system will be discussed. It is essential that these be fully understood before the study of large systems of interconnexion of these components is undertaken. In all cases the simplest representation employing an equivalent circuit will be used not only to make the principles clearer but also because these simple models are used in practice. For more sophisticated treatments, especially of synchronous machines, the reader is referred to the more advanced texts given at the end of the chapter.

It is assumed that the reader has completed introductory courses in circuit theory and machines (or electromechanical energy conversion). A knowledge of basic control theory will be needed for the section on the dynamic response of machines, although this may be omitted without essential loss of continuity in the text.

Loads are considered as components even though their exact composition and characteristics are not known with complete certainty. When designing a supply system or extending an existing one the prediction of the loads to be expected is required, statistical methods being used. This is the aspect of power supply known with least precision.

It is stressed that most of the equivalent circuits used are single phase and employ phase-to-neutral values. This assumes that the loads are balanced three phase which is reasonable for normal steady-state operation. When unbalance exists between the phases, full treatment of all phases is required and special techniques for dealing with this are given in Chapter 7.

SYNCHRONOUS MACHINES

3.2 Introduction

In this chapter those characteristics of the synchronous machine pertinent to power systems will be discussed. It is assumed that the reader has a basic knowledge of synchronous-machine theory. The two forms of rotor construction produce characteristics that influence the operation of the system to varying extents. In the round or cyclindrical rotor the field winding is situated in slots cut axially along the rotor length; the diameter is relatively small (in the order of 90 cm) and the machine is suitable for operation at high speeds driven by a steam turbine. Hence it is known as the turbo-generator. In the salient pole rotor the poles project as shown in Figure 3.1, and low-speed operation driven by hydraulic turbines is usual. The frequency of the generated e.m.f. and speed are related by the expression $f = (np/60)$, where n is the speed in revolutions per minute and p the number of pole pairs; a hydro-generator thus needs many poles to generate at normal frequencies.

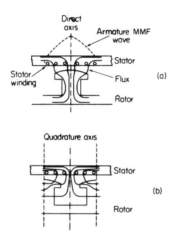

Figure 3.1(a) Rotor pole and associated stator conductors—direct axis. Stator current power-factor such that the MMF wave is in the positions shown. (b) Rotor interpolar gap and associated stator conductors—quadrature axis.

The three-phase currents in the stator winding or armature generate a rotating magnetic field which is stationary relative to the rotor and its field. The effect of this armature MMF wave on the rotor field is referred to as the 'armature reaction'. The direct axis of the rotor pole system is shown in Figure 3.1(a), the reluctance of the flux path is a minimum and the flux linkage of each phase a maximum. The flux linkage per ampere is known as the direct-axis synchronous inductance with a corresponding reactance, X_d. The armature voltage induced by the change in flux linkages is in quadrature with the flux.

The axis through the interpolar gap (Figure 3.1(b)) is called the quadrature axis and here magnetic reluctance is at its maximum. The value of reactance X_q corresponding to this position is therefore less than X_d. In a turbo-alternator rotor X_q is almost equal to X_d and for steady-state operation they may be considered to be equal, which considerably simplifies analysis. With hydro-generators, $X_q < X_d$ and saliency is said to exist; in analysis, fluxes, voltages and currents are resolved into their components along the direct and quadrature axes as will be explained later. The emphasis in this section will be on conditions where $X_q = X_d$, i.e. when only one value of reactance, current and voltage need be considered. This will simplify calculations and make the physical concepts clearer.

In power-system analysis the object is to use an equivalent-circuit model which exhibits the external characteristics of the generator with sufficient accuracy. For a round-rotor machine the circuit in Figure 3.2 is such a model.

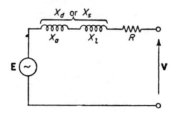

Figure 3.2   Equivalent circuit of a round-rotor synchronous generator (saliency ignored).

The effects of armature reaction and leakage flux are simulated by two reactances in series, the combined reactance being called the synchronous reactance (X_s). The resistance of each phase of the armature winding (R) is often negligible compared with X_s. Figure 3.3 shows the phasor diagram corresponding to the equivalent circuit. Phasors representing the MMFs set up by the field and armature currents are also shown in this diagram. The resultant air-gap MMF, $\mathbf{F} = \mathbf{F}_f + \mathbf{F}_a$; $\mathbf{F}_a$ is in phase with the armature current I producing it.

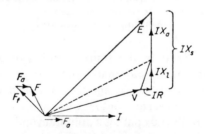

Figure 3.3   Phasor diagram corresponding to the equivalent circuit of a round-rotor machine

Effect of saturation on X_s—the short-circuit ratio

The open-circuit characteristic is the graph of generated voltage against field current with the machine on open circuit and running at synchronous speed. The short-circuit characteristic is the graph of stator current against field current with the terminals short-circuited; curves for a modern machine are shown in Figure 3.4. Here, X_s is equal to the open-circuit voltage produced by the same field current that produces rated current on short circuit, divided by

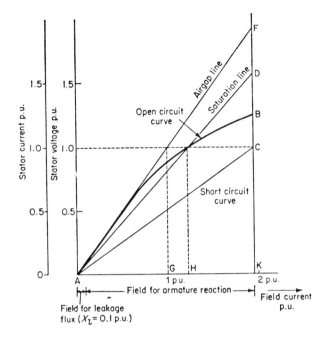

Figure 3.4 Open and short-circuit characteristics of a synchronous machine. Unsaturated value of X_s = FK/CK. With operation near nominal voltage, the saturation line is used to give linear characteristic with saturation.

this rated armature current. This value of X_s is constant only over the linear part of the open-circuit characteristic (the air-gap line) and ignores saturation. The actual value of X_s at full-load current will obviously be less than this value and several methods exist to allow for the effects of saturation[6].

The short-circuit ratio (SCR) of a generator is defined as the ratio between the field current required to give nominal open-circuit voltage and that required to circulate full-load current in the armature when short-circuited. In Figure 3.4 the short-circuit ratio is AH/AK, i.e. 0.63. To allow for saturation it is common practice to assume that the synchronous reactance is 1/SCR which for this machine is 1.58 p.u. Economy demands the design of machines of low SCR and a value of 0.55 is common for modern machines.

Turbo-generators

The ratings of steam-turbine driven round-rotor generators have increased dramatically over the last decade and sets of 1000 MW (1 GW) are now being installed. The advantages of such high-output units lie in increased efficiency (on the steam side) and lower capital costs. The use of machines of 2 GW and above is forecast.

The most striking advances in machine design have been in the area of cooling and heat transfer. By the use of intense cooling methods the physical

92

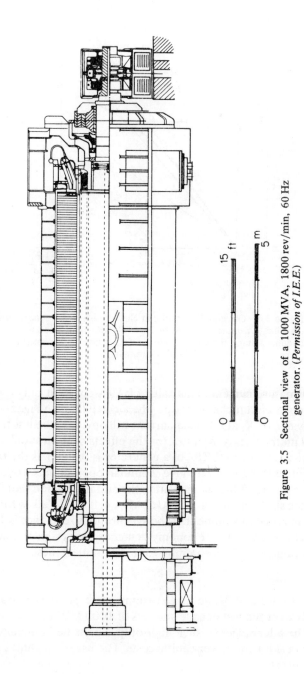

Figure 3.5 Sectional view of a 1000 MVA, 1800 rev/min, 60 Hz generator. (*Permission of I.E.E.*)

dimensions of generators have increased very much less than the corresponding electrical outputs. This is of great importance because of the restrictions in size and weight of plant which must be transported from the manufacturer's works to the generating station by road or rail. Large generators today are cooled by hydrogen or water pumped through the centre of the conductors, the stator winding is usually water-cooled, and the rotor winding hydrogen-cooled. Machines are now appearing with water-cooled rotors as well as stators.

Typical statistics of a large generator are as follows:

500 MW, 588 MVA: winding cooling—rotor, water; stator, water;
rotor diameter 1.12 m; rotor length 6.2 m;
total weight 0.63 kg/kVA;
rotor weight 0.1045 kg/kVA.

A cross-sectional view of a 1000 MVA steam turbo-generator is shown in Figure 3.5.

The most important problems encountered in the development and use of large generators arise from (a) mechanical difficulties due to the rotation of large masses, especially stresses in shafts and rotors, critical speeds, and torsional oscillations; (b) the large forces produced on stator bars which must be withstood by their insulation (as much as 12 t in some designs); (c) the need for more effective cooling. Mica paper and glass bonded with epoxy resin are at present being used as stator insulation, the maximum permissible temperature being 135°C.

Semiconductor rectifiers are employed for excitation. Traditionally, sets had d.c. exciters mounted on the main shaft. This type was followed by an a.c. exciter from which the current is rectified by static mercury arc rectifiers so that a.c. and d.c. slip-rings are required. More recently a.c. exciters with integral fused-diode rectifiers are employed, thus avoiding any brush gear and consequent maintenance problems, the diodes rotating on the main shaft.

Superconducting field windings without ferromagnetic rotors would give magnetic flux densities of about three times the present 2–2.5 T and offer the possibility of reduced machine size and weight and an increase in operating efficiency of about 0.4 per cent; the machine reactance will be smaller (the synchronous reactance may be reduced by as much as four-fifths), but the short-circuit forces will be larger. Superconducting stator windings are impracticable because of the prohibitively large a.c. losses they would produce. Even if such machines are shown to be economic for large sizes their reliability must present major problems. Reliability is all-important, and the loss resulting from an outage of a large conventional generator for one day is roughly equivalent to 1 per cent of the initial cost of the machine.

3.3 Equivalent Circuit Under Balanced Short Circuit Conditions

A typical set of oscillograms of the currents in three armature phases when a synchronous generator is suddenly short-circuited is shown in Figure 3.6(a). In

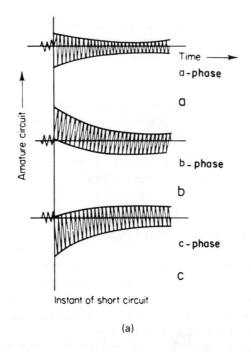

(a)

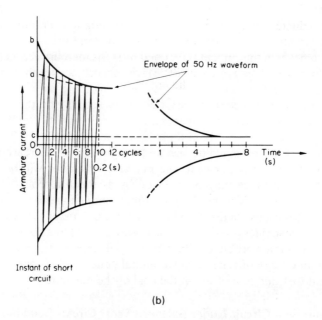

(b)

Figure 3.6(a) Oscillograms of the currents in the three phases of a generator when a sudden short circuit is applied. (b) Trace of a short-circuit current when direct-current component is removed.

all three traces a direct-current component is evident and this is to be expected from a knowledge of transients in R–L circuits. The magnitude of direct current present depends upon the instant at which the short circuit is applied and on the power factor of the circuit. As there are three voltages mutually at $120°$ it is possible for only one to have a zero direct-current component. Often to clarify the physical conditions the direct component is ignored and a trace of short-circuit current as shown in Figure 3.6(b) is considered. Immediately after the application of the short circuit the armature current endeavours to create an armature reaction MMF, but the main air-gap flux cannot change to a new value immediately as it is linked with low-resistance circuits consisting of: (a) the rotor winding which is effectively a closed circuit; and (b) the damper bars, i.e. a winding which consists of short-circuited turns of copper strip set in the poles to dampen oscillatory tendencies. As the flux remains unchanged initially, the stator currents are large and can flow only through the medium of the creation of opposing currents in the rotor and damper windings by what is essentially transformer action. Owing to the higher resistance, the current induced in the damper winding decays rapidly and the armature current commences to fall. After this the currents in the rotor winding and body decay, the armature reaction MMF is gradually established, and the generated e.m.f. and stator current fall until the steady-state condition on short circuit is reached. Here the full armature reaction effect is operational and the machine represented by the synchronous reactance X_s. These effects are shown in Figure 3.6 with the high initial current due to the damper winding and then the gradual reduction until the full armature reaction is established. A detailed analysis is given in reference 6.

To represent the initial short-circuit conditions two new models must be introduced. If in Figure 3.6 the envelopes to the 50 Hz waves are traced, a discontinuity appears. Whereas the natural envelope continues to point 'a' the actual trace finishes at point 'b'; the reasons for this have been mentioned above. To account for both of these conditions, two new reactances are needed to represent the machine, the very initial conditions requiring what is called the *subtransient reactance* (X'') and the subsequent period the *transient reactance* (X'). In the following definitions it is assumed that the generator is on no-load prior to the application of the short-circuit and is of the round-rotor type. Let the no-load phase voltage of the generator be E volts (r.m.s.).

Then from Figure 3.6(b), the subtransient reactance

$$(X'') = \frac{E}{0b/\sqrt{2}}$$

where $0b/\sqrt{2}$ is the r.m.s. value of the subtransient current (I''); the transient reactance

$$(X') = \frac{E}{0a/\sqrt{2}}$$

where $0a/\sqrt{2}$ is the r.m.s. value of the transient current, and finally

$$\frac{E}{0c/\sqrt{2}} = \text{synchronous reactance } X_s.$$

Typical values of X'', X_s, and X_s for various types and sizes of machines are given in Table 3.1.

Table 3.1   Constants of synchronous machines—60 Hz
(*all values expressed as per unit on rating*)

Type of machine	X_s (or X_d)	X_q	X'	X''	X_2	X_0	r_a
Turbo-alternator	1.2–2.0	1–1.5	0.2–0.35	0.17–0.25	0.17–0.25	0.04–0.14	0.003–0.008
Salient pole (hydro electric)	0.16–1.45	0.4–1.0	0.2–0.5	0.13–0.35	0.13–0.35	0.02–0.2	0.003–0.015
Synchronous compensator	1.5–2.2	0.95–1.4	0.3–0.6	0.18–0.38	0.17–0.37	0.03–0.15	0.004–0.01

X_2 = negative sequence reactance
X_0 = zero sequence reactance
X' and X'' are the direct axis quantities
r_a = a.c. resistance of the armature winding per phase.

If the machine is previously on load the voltage applied to the equivalent reactance previously E is now modified due to the initial load volt-drop. Consider Figure 3.7. Initially the load current is I_L and the terminal voltage is V. The voltage behind the transient reactance X' is

$$\mathbf{E}' = I_L(Z_L + jX')$$

$$= V + jI_L X'$$

and hence the transient current on short circuit $= \mathbf{E}'/jX'$.

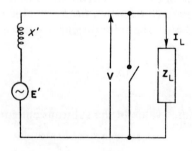

Figure 3.7   Modification of equivalent circuit to allow for initial load current.

3.4   Synchronous Generators in Parallel

Consider two machines A and B the voltages of which have been adjusted to equal values by means of the field regulators and the speeds of which are slightly different. In Figure 3.8(b) the phase voltages are E_{RA}, etc. and the speed of machine A, ω_A radians per second and of B, ω_B radians per second.

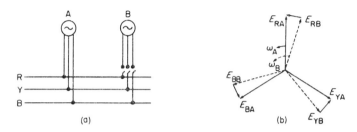

Figure 3.8(a) Generators in parallel. (b) Corresponding phasor diagrams.

If voltage phasors of A are considered stationary those of B rotate at a relative velocity ($\omega_B - \omega_A$) and hence there are resultant voltages across the switch of ($E_{RA} - E_{RB}$), etc. which reduce to zero during each relative revolution. If the switch is closed at an instant of zero voltage the machines are connected (synchronized) without the flow of large currents due to the resultant voltages across the armatures. When the two machines are in synchronism they have a common terminal-voltage, speed, and frequency. Any tendency for one machine to accelerate relative to the other immediately results in a retarding or synchronizing torque being set up due to the current circulated.

Two machines operating in parallel are represented by the equivalent circuits shown in Figure 3.9(a) with $E_A = E_B$ and on no external load. If A tries to gain speed the phasor diagram in Figure 3.9(b) is obtained and $I = E_R/Z_A + Z_B$. The circulating current $\mathbf{I}$ lags $\mathbf{E_R}$ by an angle $\tan^{-1}(X/R)$ and as in most machines $X \gg R$ this angle approaches $90°$. This current is a generating current for A and a motoring current for B, hence A is generating power and tending to slow down and B is receiving power from A and speeding up. A and B therefore remain at the same speed, 'in step', or in synchronism. Figure 3.9(c) shows the state of affairs when B tries to gain speed on A. The quality of a machine to return to its original operating state after a momentary disturbance is measured by the *synchronizing power* and *torque*. It is interesting to note that as the impedance of the machines is largely inductive the restoring powers and torques are large; if the system were largely resistive it would be difficult for synchronism to be maintained.

Normally, more than two generators operate in parallel and the operation of one machine connected in parallel with many others is of great interest. If the remaining machines in parallel are of such capacity that the presence of the generator under study causes no difference to the voltage and frequency

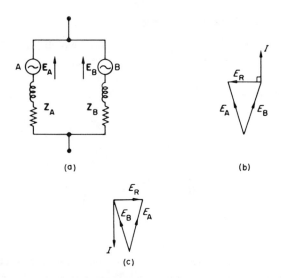

Figure 3.9(a) Two generators in parallel—equivalent circuit. (b) Machine A in phase advance of machine B. (c) Machine B in phase advance of machine A.

of the other they are said to comprise an *infinite busbar system*, i.e. an infinite system of generation. In practice a perfect infinite busbar is never fully realized but if, for example, a 600 MW generator is removed from a 30,000 MW system the difference in voltage and frequency caused will be very slight.

3.5 The Operation of a Generator on Infinite Busbars

In this section in order to simplify the ideas as much as possible the resistance of the generator will be neglected; in practice this assumption is usually reasonable. Figure 3.10(a) shows the schematic diagram of a machine connected to an infinite busbar along with the corresponding phasor diagram. If losses are neglected the power output from the turbine is equal to the power output from the generator. The angle δ between the E and V phasors is known as the load angle and is dependent on the power input from the turbine shaft. With an isolated machine supplying its own load the latter dictates the power required and hence the load angle; when connected to an infinite-busbar system,

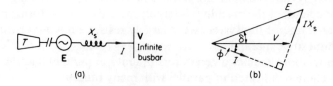

Figure 3.10(a) Synchronous machine connected to an infinite busbar. (b) Corresponding phasor diagram.

however, the load delivered by the machine is no longer directly dependent on the connected load. By changing the turbine output and hence δ the generator can be made to take on any load the operator desires subject to economic and technical limits.

From the phasor diagram in Figure 3.10(b), the power delivered to the infinite busbar $= VI \cos \phi$ per phase but,

$$\frac{E}{\sin (90 + \phi)} = \frac{IX_s}{\sin \delta}$$

hence

$$I \cos \phi = \frac{E}{X_s} \sin \delta$$

$$\therefore \text{ power delivered} = \frac{VE}{X_s} \sin \delta \tag{3.1}$$

This expression is of extreme importance as it governs to a large extent the operation of a power system. It could have been obtained directly from equation 2.3, putting $\alpha = 0$.

Equation (3.1) is shown plotted in Figure 3.11. The maximum power is obtained at $\delta = 90°$. If δ becomes larger than 90° due to an attempt to obtain

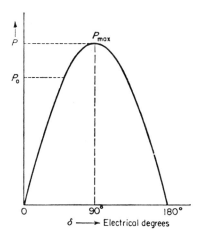

Figure 3.11 Power-angle curve of a synchronous machine. Resistance and saliency neglected.

more than P_{max}, increase in δ results in less power output and the machine becomes unstable and loses synchronism. Loss of synchronism results in the interchange of current surges between the generator and network as the poles of the machine pull into synchronism and then out again.

If the power output of the generator is increased by small increments with the no-load voltage kept constant, the limit of stability occurs at $\delta = 90°$ and is known as the *steady-state stability limit*. There is another limit of stability due to a sudden large change in conditions such as caused by a fault, known as the *transient stability limit*, and it is possible for the rotor to oscillate beyond 90° a number of times. If these oscillations diminish, the machine is stable. The load angle δ has a physical significance; it is the angle between like radial marks on the end of the rotor shaft of the machine and on an imaginary rotor representing the system. The marks are in identical physical positions when the machine is on no-load. The synchronizing power coefficient $= dP/d\delta$ watts per radian and the synchronizing torque coefficient $= (1/\omega_S)/(dP/d\delta)$.

In Figure 3.12(a) the phasor diagram for the limiting steady-state condition is shown. It should be noted that in this condition current is always leading. The following figures, 3.12(b), (c), and (d), show the phasor diagrams for various operational conditions. Another interesting operating condition is variable power and constant excitation. This is shown in Figure 3.13. In this case as V

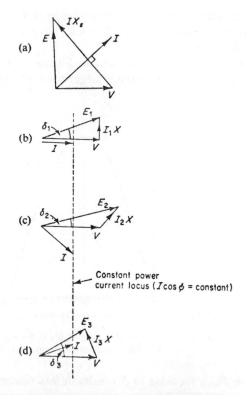

Figure 3.12(a) Phasor diagram for generator at limit of steady-state state stability. (c) and (d) Phasor diagrams for generator delivering constant power to the infinite busbar system but with different excitations. As V is constant the in-phase component of $\mathbf{I}$ must be constant. As $EV/X \sin \delta$ is constant as E changes, δ must change and
$$\delta_3 > \delta_1 > \delta_2.$$

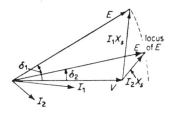

Figure 3.13   Operation at variable power and constant excitation.

and E are constant when the power from the turbine is increased δ must increase and the power factor changes.

It is convenient to summarize the above types of operation in a single diagram or chart which will enable an operator to see immediately whether the machine is operating within the limits of stability and rating.

The performance chart of a synchronous generator

Consider Figure 3.14(a), the phasor diagram for a round-rotor machine ignoring resistance. The locus of constant IX_s, I, and hence MVA, is a circle and the locus of constant E a circle. Hence,

$$0s \text{ is proportional to } VI \text{ or MVA}$$
$$ps \text{ is proportional to } VI \sin \phi \text{ or MVAr}$$
$$sq \text{ is proportional to } VI \cos \phi \text{ or MW}$$

To obtain the scaling factor for MVA, MVAr and MW the fact that at zero excitation, $E = 0$ and $IX_s = V$, is used, from which I is V/X_s at 90° leading to $00'$, corresponding to VAr/phase.

The construction of a chart for a 60 MW machine follows (Figure 3.14(b)).

Machine data   60 MW, 0.8 p.f., 75 MVA
11.8 kV, SCR 0.63, 3000 rev/min
Maximum exciter current 500 A

$$X_s = \frac{1}{0.63} \text{ p.u.} = 2.94 \ \Omega/\text{phase}$$

The chart will refer to complete three-phase values of MW and MVAr. When the excitation and hence E are reduced to zero, the current leads V by 90° and is equal to (V/X_s), i.e. $11,800/\sqrt{3} \times 2.94$. The leading vars correspond to this $= 11,800^2/2.94 = 47$ MVAr.

With centre 0 a number of semicircles are drawn of radii equal to various MVA loadings, the most important being the 75 MVA circle. Arcs with $0'$ as centre are drawn with various multiples of $00'$ (or V) as radii to give the loci for constant excitations. Lines may also be drawn from 0 corresponding to various power factors, but for clarity only 0.8 p.f. lagging is shown. The operational limits are fixed as follows. The rated turbine output gives a 60 MW limit which is drawn as shown, i.e. line efg, which meets the 75 MVA line in g. The MVA

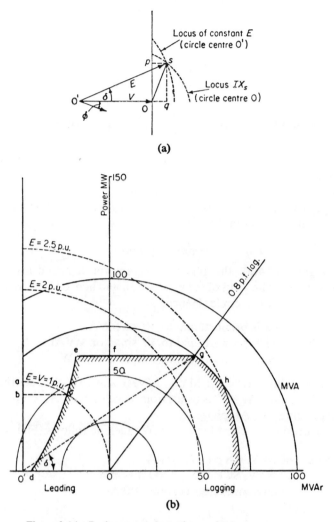

Figure 3.14   Performance chart of a synchronous generator.

arc governs the thermal loading of the machine, i.e. the stator temperature rise, so that over portion gh the output is decided by the MVA rating. At point h the rotor heating becomes more decisive and the arc hj is decided by the maximum excitation current allowable, in this case assumed to be 2.5 p.u. The remaining limit is that governed by loss of synchronism at leading power factors. The theoretical limit is the line perpendicular to 00' at 0' (i.e. $\delta = 90°$), but in practice a safety margin is introduced to allow a further increase in load of either 10 or 20 per cent before instability. In Figure 3.14 a 10 per cent margin is used and is represented by ecd: it is constructed in the following manner. Considering point 'a' on the theoretical limit on the $E = 1$ p.u. arc, the power 0'a is reduced by 10 per cent of the rated power (i.e. by 6 MW) to 0'b; the operating point must, however, still be on the same E arc and b is projected to c which is a point on the new limiting curve. This is repeated for several excitations giving finally the curve ecd.

The complete operating limit is shown shaded and the operator should normally work within the area bounded by this line.

As an example of the use of the chart, the full-load operating point g (60 MW, 0.8 p.f. lagging) will require an excitation E of 2.3 p.u. and the measured load angle δ is 33°. This can be checked by using, power = $VE/X_s \sin \delta$, i.e.

$$60 \times 10^6 = \frac{11,800^2 \times 2.3}{2.94} \sin \delta$$

from which

$$\delta = 33.3°$$

3.6 Salient-pole Generators

So far the direct-axis reactance has been assumed equal to the quadrature-axis value. In the case of turbo-alternators this assumption is reasonably accurate for steady-state conditions, but for salient-pole machines a further axis between the main poles known as the quadrature axis is utilized for accurate representation.

To account for the armature reaction there are two reactances, X_{ad} for the direct axis and X_{aq} for the quadrature axis, and $X_{ad} > X_{aq}$. Let $X_d = X_1 + X_{ad}$ and $X_q = X_{aq}$, where $X_1 =$ leakage reactance; then from the phasor diagram (Figure 3.15)

$$\mathbf{E} = \mathbf{V} + j\mathbf{I_d}X_d + j\mathbf{I_q}X_q$$

$$= \mathbf{V} + j\mathbf{I}X_q + j\mathbf{I_d}(X_d - X_q) \quad \text{or} \quad \mathbf{E_{qd}} + j\mathbf{I_d}(X_d - X_q) \qquad (3.2)$$

As $j\mathbf{I_d}(X_d - X_q)$ as in phase with E knowing V, I, and ϕ, IX_q can be drawn and hence the direction of E determined and consequently $\mathbf{I_d}$ and $\mathbf{I_q}$. Knowing $\mathbf{I_d}$ and $\mathbf{I_q}$ the remaining phasors can be drawn to complete the diagram. Values of X_d and X_q can be found by test.

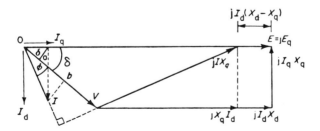

Figure 3.15. Phasor diagram of a synchronous generator—full two-axis representation.

104

Output power

The output power $= VI \cos \phi$. From Figure 3.15,

$$I \cos \phi = \text{ob} = \text{oa} + \text{ab}$$

$$= I_q \cos \delta + I_q \sin \delta.$$

If the armature resistance is negligible,

$$V \sin \delta = I_q X_q$$

$$V \cos \delta = E - I_d X_d$$

$$\therefore \ P = V\left[\frac{V \sin \delta}{X_q}\cos \delta + \frac{E - V \cos \delta}{X_d}\sin \delta\right]$$

$$= \frac{VE}{X_d}\sin \delta + \frac{V^2(X_d - X_q)}{2X_d X_q}\sin 2\delta \tag{3.3}$$

and

$$\frac{dP}{d\delta} = \frac{VE}{X_d}\cos \delta + \frac{V^2(X_d - X_q)}{X_d X_q}\cos 2\delta \tag{3.4}$$

$$= \text{synchronizing power coefficient}$$

The plot of P against δ is shown in Figue 3.16, from which it is evident that the steady-state stability limit occurs at a lower δ than 90°.

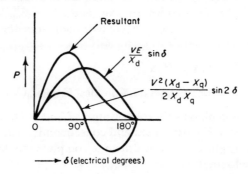

Figure 3.16   Power-angle curve—salient-pole machine.

Equations (3.3) refer to operation in the steady state. Under transient conditions the direct-axis reactance becomes X_d' or X_d'' as previously discussed, in the quadrature axis it is usual to assume $X_q = X_q'$. In the transient state the excitation voltage $E = jE_q$ still lies along the quadrature axis and the voltage 'behind' the transient reactance $E_1 = E_{q1}$ also lies on this axis. The transient power output equation is

$$P = \frac{E_{q1} V}{X_d'}\sin \delta + V^2 \frac{X_q - X_d'}{2X_d' X_q}\sin 2\delta \tag{3.5}$$

In steady-state stability calculations the effects of saliency may usually be ignored, especially compared with the effects of saturation.

Consider a salient-pole machine with the following parameters $X_d = 1.2$ p.u., $X_q = 0.9$ p.u., excitation voltage $E = 1.5$ p.u. The maximum value of power output to an infinite busbar of 1 p.u. voltage is

$$P = \frac{EV}{X_d} \sin \delta + V^2 \frac{X_d - X_q}{2X_d X_q} \sin 2\delta$$

$$= 1.25 \sin \delta + 0.14 \sin 2\delta$$

Obtaining P_{max} graphically or by other means, $P_{max} = 1.25$ p.u. When this exercise is repeated with lower values of E the error introduced by ignoring saliency is seen to increase. Provided that $E > V$, saliency may generally be ignored in steady-state studies.

If the machine is connected to an infinite busbar through a link of reactance X_L then $_TX_d = X_d + X_L$ and $_TX_q = X_q + X_L$, where the prefix T refers to the total effective reactance of the system.

3.7 Automatic Voltage Regulators

In the previous pages the synchronous generator has been considered to operate with a fixed excitation and any changes in excitation have been assumed carried out manually. Any adjustment of the terminal voltage V after a load change takes an appreciable time. In most modern generators the output voltage is controlled by automatic devices so as to remain at a constant prearranged value. In this section the modes of operation of *automatic voltage regulators* (AVR) are discussed and their effect on the operation of synchronous machines.

The original arrangement for the rotor of a synchronous generator is shown in Figure 3.17. The main exciter, a direct current shunt generator, generates a

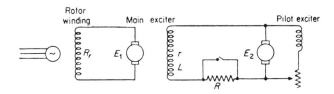

Figure 3.17   Excitation arrangements for a synchronous generator.

voltage E_1 which is supplied direct to the rotor winding through slip-rings. The field of the main exciter is supplied from a further exciter which is self-excited. Both the direct current machine armatures are driven from the main rotor shaft. In series with the main exciter field winding is a resistor R which can be short-circuited as dictated by a voltage-sensing mechanism actuated by the terminal voltage of the synchronous machine. An aspect of major importance is the speed of response of the regulator, i.e. the time elapsing between the voltage deviation and the return to the prescribed value.

Types of automatic voltage regulator

The detailed study of regulators is a specialist field, and it is not intended here to discuss types in any detail but rather to indicate their general effect. There are two broad divisions of automatic regulators, both of which set out to control the output voltage of the synchronous generator by controlling the exciter. In general the deviation of the terminal voltage from a prescribed value is passed to control circuits and thus the field current is varied and it is in the manner and speed in achieving this that the division occurs.

The first and older type can be broadly classed as *electromechanical*. A well-known variety of this is the carbon-pile regulator. In this a voltage proportional to the deviation voltage operates a solenoid assembly to vary the pressure exerted on a carbon-pile resistor in the exciter field, thus varying its resistance. Another type depends upon the conversion of the deviation voltage into a torque by means of a 'torque motor'; according to the angular position of the shaft of the motor, certain resistors in a resistor chain are cut out of circuit and hence the exciter field-current changes. There are various other types including the popular vibrating-reed regulator. All these types suffer from the disadvantages of being relatively slow acting and possessing dead bands, i.e. a certain deviation must occur before the mechanism operates; this is illustrated in Figure 3.18.

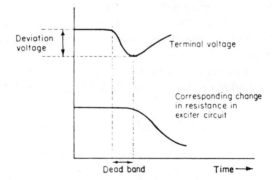

Figure 3.18   The effect of a dead band in a regulator.

The other main group of regulators is known as 'continuously acting' and these are faster than the above and have no dead bands. A general block diagram of a typical control system is given in Figure 3.19.

Automatic voltage regulators and generator characteristics

The equivalent circuit used to represent the synchronous generator can be modified to account for the action of a regulator. Basically there are three conditions to consider.

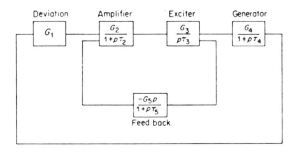

Figure 3.19 Block diagram of a continuously acting closed-loop automatic voltage generator.

1. Operation with fixed excitation and constant no-load voltage (E), i.e. no regulator action. This requires the usual equivalent circuit of E in series with Z_s.
2. Operation with a regulator which is not continuously acting, i.e. the terminal voltage varies with load changes. This can be simulated by E' and a reactance smaller than the synchronous value. It has been suggested by experience in practice that a reasonable value would be the transient reactance, although some authorities suggest taking half of the synchronous reactance. This mode will apply to most modern regulators.
3. Terminal voltage constant. This requires a very fast-acting regulator and the nearest approach to it exists in the forced-excitation regulators used on generators supplying very long lines.

Each of the above representations will give significantly different values of maximum power output. The degree to which this happens depends on the speed of the AVR, and the effect on the operation chart of the synchronous generator is shown in Figure 3.20 which indicates clearly the increase in operating range obtainable. It should be noted, however, that operation in these improved leading power-factor regions may be limited by the heating of

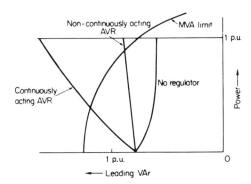

Figure 3.20 Performance chart as modified by the use of automatic voltage regulators

the stator winding. The actual power-angle curve may be obtained by a step-by-step process by using the gradually increasing values of E in $EV/X \sin \delta$.

When a generator has passed through the steady-state limiting angle of $\delta = 90°$ with a fast acting AVR it is possible for synchronism to be retained. The AVR in forcing up the voltage, increases the power output of the machine so that instead of the power falling after $\delta = 90°$ it is maintained and $dP/d\delta$ is still positive. This is illustrated in Figure 3.21 where the P–δ relation for the system in Figure 3.21(a) is shown. Without the AVR the terminal voltage of the machine (V_t) will fall with increased δ, the generated voltage V_g being constant and the power reaching a maximum at 90°. With a perfect AVR, V_t is maintained constant, V_g being increased with increase in δ. As $P = (V_s V_g / X) \sin \delta$, it is evident that power will increase beyond 90° until the excitation limit is reached, as shown in Figure 3.21(b).

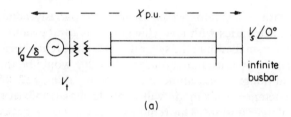

(a)

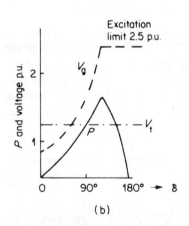

(b)

Figure 3.21(a)   Generator feeding into an infinite busbar. (b) Variations of output power P, generated voltage V_g, and terminal voltage V_t with load (transmission) angle δ. Perfect AVR.

Example 3. Determine the limiting powers for the system shown in Figure 3.22 for the three types of voltage regulation. All values are on 100 MVA, 132 kV bases. It may be assumed that the lines and transformers are each represented by a single series reactance.

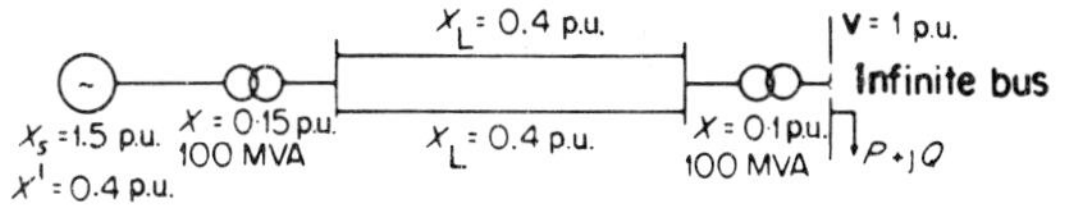

Figure 3.22 Line diagram of system for Example 3.1. Normal operating load $P = 0.8$ p.u., $Q = 0.5$ p.u.

Solution (a) No control, constant excitation voltage.

$$X = 1.5 + 0.15 + 0.1 + \frac{0.4}{2} = 1.95 \text{ p.u.}$$

From equation 2.5

$$E = \sqrt{\left[\left(1 + \frac{0.5 \times 1.95}{1}\right)^2 + \left(\frac{0.8 \times 1.95}{1}\right)^2\right]}$$

$$= 2.52 \text{ p.u.}$$

$$\therefore P_{max} = \frac{EV}{X} = \frac{2.52 \times 1}{1.95} = 1.29 \text{ p.u.}$$

(b) Non-continuously acting AVR.

$$X = 0.4 + 0.15 + 0.1 + 0.2 = 0.85 \text{ p.u.}$$

$$E' = \sqrt{\left[\left(1 + \frac{0.5 \times 0.85}{1}\right)^2 + \left(\frac{0.8 \times 0.85}{1}\right)^2\right]}$$

$$= 1.59 \text{ p.u.}$$

$$\therefore P_{max} = \frac{1.59 \times 1}{0.85} = 1.87 \text{ p.u.}$$

(c)

$$X = 0.15 + 0.1 + 0.2 = 0.45 \text{ p.u.}$$

Terminal voltage of generator V_t is constant.

$$V_t = \sqrt{\left[\left(1 + \frac{0.5 \times 0.45}{1}\right)^2 + \left(\frac{0.8 \times 0.45}{1}\right)^2\right]}$$

$$= 1.28 \text{ p.u.}$$

$$P_{max} = \frac{1.28 \times 1}{0.45} = 2.85 \text{ p.u.}$$

It is interesting to compare the power limit when one of the two lines is disconnected; the transient-reactance representation will be used.

$$E' = \sqrt{\left(1 + \frac{0.5 \times 1.05}{1}\right)^2 + \left(\frac{0.8 \times 1.05}{1}\right)^2}$$

$$= 1.74 \text{ p.u.}$$

$$P_{max} = \frac{1.74 \times 1}{1.05} = 1.65 \text{ p.u.}$$

LINES, CABLES, AND TRANSFORMERS

3.8 Overhead Lines—Types and Parameters

Overhead lines are suspended from insulators which are themselves supported by towers or poles. The span between two towers is dependent upon the allowable sag in the line, and for steel towers with very high-voltage lines the span is normally 370–460 m (1200–1500 ft). Typical supporting structures are shown in Figures 3.23 and 3.24. There are two main types of tower:

- (a) those for straight runs in which the stress due to the weight of the line alone has to be withstood;
- (b) those for changes in route, called deviation towers; these withstand the resultant forces set up when the line changes direction.

When specifying towers and lines, ice and wind loadings are taken into account, as well as extra forces due to a break in the lines on one side of a tower. For lower voltages and distribution circuits wooden or reinforced concrete poles are used with conductors supported in horizontal formations.

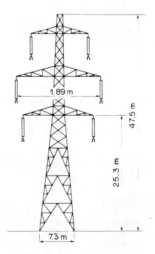

Figure 3.23 400 kV double-circuit overhead line tower. Two conductors per phase (bundle conductors). (*Permission of Institution of Electrical Engineers.*)

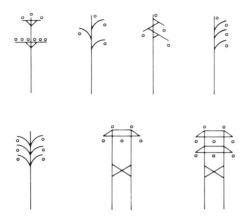

Figure 3.24   Typical pole-type structures.

The live conductors are insulated from the towers by insulators which take two basic forms, the *pin type* and the *suspension type*. The pin-type insulator is shown in Figure 3.25 and it is seen that the conductor is supported on the top of the insulator. This type is used for lines up to 33 kV. The two or three porcelain 'sheds' or 'petticoats' provide an adequate leakage path from the conductor to

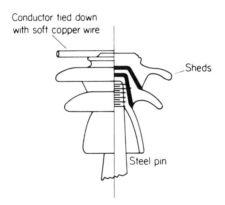

Figure 3.25   Pin-type insulators.

earth and are shaped to follow the equipotentials of the electric field set up by the conductor–tower system. Suspension insulators (Figure 3.26) consist of a string of interlinking separate discs made of glass or porcelain. A string may consist of many discs depending upon the line voltage; for 400 kV lines, 19 discs of overall length 3.84 m (12 ft 7 in) are used. The conductor is held at the bottom of the string which is suspended from the tower. Owing to the capacitances existing between the discs, conductor, and tower, the distribution of voltage along the insulator string is not uniform, the discs nearer the conductor being the more highly stressed. Methods of calculating this voltage

112

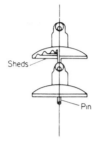

Figure 3.26   Suspension-type insulators.

distribution are available, but of dubious value owing to the shunting effect of the leakage resistance (see Figure 3.27). This resistance depends on the presence of pollution on the insulator surfaces and is considerably modified by rain and fog.

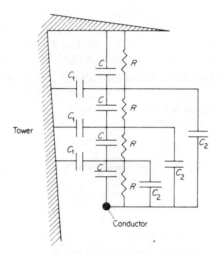

Figure 3.27   Equivalent circuit of a string of four suspension insulators. C = self capacitance of disc; C_1 = capacitance disc to earth; C_2 = capacitance disc to line; R = leakage resistance.

Parameters

The parameters of interest for circuit analysis are inductance, capacitance, resistance, and leakage resistance. The derivation of formulae for the calculation of these quantities is given in reference 12. It is intended here merely to quote these formulae and discuss their application.

The inductance of a single-phase two-wire line

$$= \frac{\mu_0}{4\pi}\left[1 + 4\ln\left(\frac{d-r}{r}\right)\right] \text{H/m} \tag{3.6}$$

where d is the distance between the centres and r is the radius, of the conductors. When performing load flow and balanced-fault analysis on three-phase systems it is usual to consider one phase only with the appropriate angular adjustments made for the other two phases. Therefore, phase voltages are used and the inductances and capacitances are the equivalent phase or line-to-neutral values. For a three-phase line with equilateral spacing (Figure 3.28) the inductance and capacitance with respect to the hypothetical neutral conductor are used and this inductance can be shown to be half the loop inductance of the single-phase line, i.e. the inductance of one conductor.

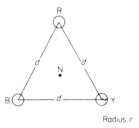

Figure 3.28   Overhead line with equilateral spacing of the conductors.

The line–neutral inductance for equilateral spacing

$$= \frac{\mu_0}{8\pi} \left[1 + 4 \ln \left(\frac{d-r}{r} \right) \right] \text{H/m} \tag{3.7}$$

The capacitance of a single-phase line

$$C = \frac{\pi \varepsilon_0}{\ln \left[(d-r)/r \right]} \text{F/m} \tag{3.8}$$

With three-phase conductors spaced equilaterally the capacitance of each line to the hypothetical neutral is double that for the two-wire circuit, i.e.

$$\frac{2\pi \varepsilon_0}{\ln \left[(d-r)/r \right]} \text{F/m} \tag{3.9}$$

In practice the conductors are rarely spaced in the equilateral formation, and it can be shown that the average value of inductance or capacitance for any formation of conductors can be obtained by the representation of the system by one of equivalent equilateral spacing. The equivalent spacing d_{eq} between conductors is given by

$$d_{eq} = \sqrt[3]{(d_{12} \cdot d_{23} \cdot d_{31})}$$

Often two three-phase circuits are electrically in parallel; if phsysically remote from each other the reactances of the lines are identical. When the two circuits are situated on the same towers, however, the magnetic interaction between them should be taken into account. The use of bundle conductors, i.e. more than one conductor per insulator, reduces the reactance; it also reduces conductor–surface voltage gradients and hence corona loss and radio interference. Unsymmetrical conductor spacing results in different inductances for each phase which causes an unbalanced voltage drop, even when the load currents are balanced. The residual or resultant voltage or current induces unwanted voltages into neighbouring communication lines. This can be overcome by the interchange of conductor positions at regular intervals along the route, a practice known as *transposition*, see Figure 3.29. In practice, lines are rarely transposed at regular intervals and transposition is carried out where physically convenient, for example, at substations. In short lines the degree of unbalance existing without transposition is small and may be neglected in calculations.

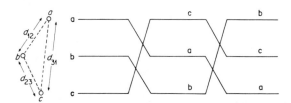

Figure 3.29   Transposition of conductors.

Resistance   Overhead line conductors usually comprise a stranded steel core (for mechanical strength) surrounded by aluminium wires which form the conductor. It should be noted that aluminium and ACSR (aluminium conductor steel reinforced) conductors are sometimes described by area of a copper conductor having the same d.c. resistance, i.e. their *copper equivalent*. In Table 3.2(a) 258 mm^2 (0.4 in^2) ACSR conductor implies the copper equivalent. The

Table 3.2(a)   Overhead line constants at 50 Hz (British)
(*per phase, per km*)

Voltage No. and area of conductors (mm)^2		132 kV 1× 113	1× 258	275 kV 2× 113	2× 258	400 kV 2× 258	4× 258
Resistance (R)	Ohms	0.155	0.068	0.022	0.034	0.034	0.017
Reactance (X_L)	Ohms	0.41	0.404	0.335	0.323	0.323	0.27
Susceptance ($1/X_c$)	Ohms$^{-1} \times 10^{-6}$	7.59	7.59	9.52	9.52	9.52	10.58
Charging current (I_c)	Amps	0.22	0.22	0.58	0.58	0.845	0.945
Surge impedance	Ohms	373	371	302	296	296	258
Natural load	MW	47	47	250	255	540	620
X_L/R ratio	—	2.6	5.9	4.3	9.5	9.5	15.8
Thermal rating:							
Cold weather (below 5°C)	MVA	125	180	525	760	1100	2200
Normal (5–18°C)	MVA	100	150	430	620	900	1800
Hot weather (above 18°C)	MVA	80	115	330	480	790	1580

Table 3.2(b) Typical characteristics of bundled-conductor EHV lines

Number of subconductors in bundle	Country	Line voltage and number of circuits (in parentheses)	Diameter of subconductors	Radius of circle on which subconductors are arranged	Resistance of bundle	Inductive reactance at 50 Hz	Susceptance at 50 Hz
		kV	mm	mm	Ω/km	Ω/km	μmho/km
1	Japan	275 (2)	27.9	—	0.0744	0.511	3.01
	Canada	300 (2)	35.0	—	0.0451	0.492	2.33
	Australia	330 (1)	45.0	—	0.0367	0.422	2.78
	U.S.S.R.	330 (1)	30.2	—	0.065	0.404	2.82
	Italy	380 (1)	50.0	—	0.0294	0.398	2.84
	Average (one conductor)				0.048	0.442	2.75
2	Japan	275 (2)	25.3	200	0.0444	0.374	4.35
	Canada	360 (1)	28.1	299	0.04	0.314	3.57
	Australia	330 (1)	31.8	190.5	0.0451	0.341	3.34
	U.S.A.	345 (1)	30.4	228.6	0.0315	0.325	3.55
	Italy	380 (1)	31.5	200	0.0285	0.315	3.57
	U.S.S.R.	330 (1)	28.0	200	0.04	0.321	3.43
	Average (two conductors)				0.0400	0.323	3.58
3	Sweden	380 (1)	31.68	260	0.018	0.29	3.85
	U.S.S.R.	525 (1)	30.2	230	0.0212	0.294	3.88
4	Germany	380 (2)	21.7	202	0.0316	0.260	4.3
	Germany	500 (1)	22.86	323	0.0285	0.272	4.0
	Canada	735 (1)	35.04	323	0.0121	0.279	3.9

actual area of aluminium is approximately 430 mm^2 and including the steel core the overall cross-section of the line is about 620 mm^2, i.e. a diameter of 28 mm. Electromagnetic losses in the steel increase the effective a.c. resistance, which increases with current magnitude, giving an increase of up to about 10 per cent. With direct current the steel carries 2–3 per cent of the total current. The electrical resistivities of some conductor materials are as follows in

Table 3.2(c)   Overhead line parameters—U.S.A.

Line voltage (kV)	345	345	500	500	735	735
Conductors per phase at 18 in spacing	1	2	2	4	3	4
Conductor code	Expanded	Curlew	Chuker	Parakeet	Expanded	Pheasant
Conductor diameter (inches)	1.750	1.246	1.602	0.914	1.750	1.382
Phase spacing (ft)	28	28	38	38	56	56
GMD (ft)	35.3	35.3	47.9	47.9	70.6	70.6
60 Hz inductive reactance Ω/mile { X_A	0.3336	0.1677	0.1529	0.0584	0.0784	0.0456
X_D	0.4325	0.4325	0.4694	0.4694	0.5166	0.5166
$X_A + X_D$	0.7661	0.6002	0.6223	0.5278	0.5950	0.5622
60 Hz capacitive reactance MΩ-miles { X'_A	0.0777	0.0379	0.0341	0.0126	0.0179	0.0096
X'_D	0.1057	0.1057	0.1147	0.1147	0.1263	0.1263
$X'_A + X'_D$	0.1834	0.1436	0.1488	0.1273	0.1442	0.1359
$Z_0(\Omega)$	374.8	293.6	304.3	259.2	276.4	
Natural loading (MVA)	318	405	822	965	1844	1955
Conductor d.c. resistance at 25°C (Ω/mile)	0.0644	0.0871	0.0510	0.162	0.0644	0.0709
Conductor a.c. resistance (60 Hz) at 50°C (Ω/mile)	0.0728	0.0979	0.0599	0.179	0.0728	0.0805

X_A = component of inductive reactance due to flux within a 1 ft radius.
X_D = component due to other phases.
Total reactance per phase = $X_A + X_D$.
$$X_A = 0.2794 \log_{10} \left(\frac{1}{[N(GMR)\,(A)^{N-1}]^{1/N}} \right) \Omega/\text{mile}.$$
$X_D = 0.2794 \log_{10} (GMD)\ \Omega/\text{mile}.$
X'_A and X'_D are similarly defined for capacitive reactance and
$$X'_A = 0.0683 \log_{10} \left(\frac{1}{[Nr(A)^{N-1}]^{1/N}} \right) M\Omega\text{-mile}; \quad X'_D = 0.0683 \log_{10} (GMD)\ M\Omega\text{-miles}.$$

GMR = geometric mean radius in feet;
GMD = geometric mean diameter in feet;
N = number of conductors per phase;
A = S/2 sin (π/N); N > 1;
A = 0; N = 1;
S = bundle spacing in feet;
r = conductor radius in feet.
Permission of Edison Electric Institute.

ohm-metres at 20°C: copper 1.72×10^{-8}; aluminium 2.83×10^{-8} (3.52×10^{-8} at 80°C); aluminium alloy 3.22×10^{-8}.

Leakage resistance is usually negligible for most calculation purposes and is very difficult to assess because of its dependence on the weather. This resistance represents the combined effect of all the various paths to earth from the line. The main path is that presented by the surfaces of the line insulators, the resistance of which depends on the condition of these surfaces. This varies considerably according to location; in industrial areas there will be layers of dirt and soot whilst in coastal districts deposits of salt occur. The bushings of circuit breakers and transformers form other leakage paths. The leakage losses on 132 kV lines vary between 0.3 and 1 kW per mile.

Table 3.2(a) gives the parameters for various overhead-line circuits for the line voltages operative in Britain, Table 3.2(b) gives values for international lines and Table 3.2(c) relates to the United States.

3.9 Representation of Lines

The manner in which lines and cables are represented depends very much on their length and the accuracy required. There are three broad classifications of length—short, medium and long. The actual line or cable is a distributed-constant circuit, i.e. it has resistance, inductance, capacitance, and leakage resistance distributed evenly along its length as shown in Figure 3.30. Except

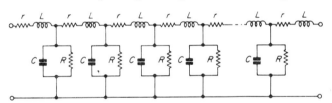

Figure 3.30 Distributed constant representation of a line: $L =$ inductance line to neutral per unit length; $r =$ a.c. resistance per unit length; $C =$ capacitance line to neutral per unit length; $R =$ leakage resistance per unit length.

for long lines the total resistance, inductance, capacitance, and leakage resistance of the line are concentrated to give a lumped-constant circuit. The distances quoted are only a rough guide.

The short line (up to 80 km)

The equivalent circuit is shown in Figure 3.31 and it will be noticed that both shunt capacitance and leakage resistance have been neglected. The four-

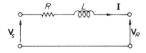

Figure 3.31 Equivalent circuit of a short line—representation under balanced three-phase conditions.

118

terminal network constants are

$$A = 1 \qquad B = Z \qquad C = 0 \quad \text{and} \quad D = 1$$

The drop in voltage along a line is important and the *regulation* is defined as

$$\frac{\text{received voltage on no load} - \text{received voltage on load}}{\text{received voltage on load } (V_R)}$$

It should be noted that if **I** is leading V_r in phase, i.e. a capacitance load, then $V_R > V_S$, as shown in Figure 3.32.

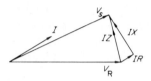

Figure 3.32   Phasor diagram for short line on leading load.

Medium-length lines (up to 240 km)

Owing to the increased length the shunt capacitance is now included to form either a π or a T network. The circuits are shown in Figure 3.33. Of these two

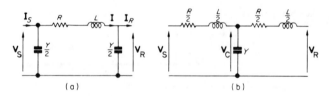

(a)   (b)

Figure 3.33(a) Medium-length line—π equivalent circuit. (b) Medium-length line—T equivalent circuit.

versions the π representation tends to be in more general use but there is little difference in accuracy between the two. For the π network:

$$\mathbf{V_S} = \mathbf{V_R} + \mathbf{IZ} \qquad \mathbf{I} = \mathbf{I_R} + \mathbf{V_R}\frac{\mathbf{Y}}{2} \qquad \mathbf{I_S} = \mathbf{I} + \mathbf{V_S}\frac{\mathbf{Y}}{2}$$

from which $\mathbf{V_S}$ and $\mathbf{I_S}$ are obtained in terms of $\mathbf{V_R}$ and $\mathbf{I_R}$ giving the following constants:

$$\mathbf{A} = \mathbf{D} = 1 + \frac{\mathbf{ZY}}{2} \qquad \mathbf{B} = \mathbf{Z} \quad \text{and} \quad \mathbf{C} = \left(1 + \frac{\mathbf{ZY}}{4}\right)\mathbf{Y}$$

Similarly for the T network:

$$\mathbf{V_S} = \mathbf{V_C} + \frac{\mathbf{ZI_S}}{2} \qquad \mathbf{V_C} = \mathbf{V_R} + \frac{\mathbf{ZI_R}}{2} \qquad \mathbf{I_S} = \mathbf{I_R} + \mathbf{V_C}\mathbf{Y}$$

giving

$$A = D = 1 + \frac{ZY}{2} \qquad B = \left(1 + \frac{ZY}{4}\right)Z \quad \text{and} \quad C = Y$$

The long line (above 240 km)

Here the treatment assumes distributed parameters. The changes in voltage and current over an elemental length Δx of the line, x metres from the sending end are determined and conditions for the whole line obtained by integration.

Let R = resistance/unit length
$\quad L$ = inductance/unit length
$\quad G$ = leakage/unit length
$\quad C$ = capacitance/unit length

$\mathbf{z}$ = impedance/unit length
$\mathbf{y}$ = shunt admittance/unit length
$\mathbf{Z}$ = total series impedance of line
$\mathbf{Y}$ = total shunt admittance of line

The voltage and current x metres from the sending end are given by

$$\left. \begin{aligned} \mathbf{V}_x &= \mathbf{V}_s \cosh \mathbf{P}x - \mathbf{I}_s \mathbf{Z}_0 \sinh \mathbf{P}x \\ \mathbf{I}_x &= \mathbf{I}_s \cosh \mathbf{P}x - \frac{\mathbf{V}_s}{\mathbf{Z}_0} \sinh \mathbf{P}x \end{aligned} \right\} \tag{3.10}$$

where

$$\begin{aligned} \mathbf{P} &= \text{propagation constant} = (\alpha + j\beta) \\ &= \sqrt{(R + j\omega L)(G + j\omega C)} = \sqrt{\mathbf{z} \cdot \mathbf{y}} \end{aligned}$$

and

$$\begin{aligned} \mathbf{Z}_0 &= \text{characteristic impedance} \\ &= \sqrt{\frac{R + j\omega L}{G + j\omega C}} \end{aligned} \tag{3.11}$$

$\mathbf{Z}_0$ is the input impedance of an infinite length of the line; hence if any line is terminated in $\mathbf{Z}_0$ its input impedance is also $\mathbf{Z}_0$.

The propagation constant $\mathbf{P}$ represents the changes occurring in the transmitted wave as it progresses along the line; α measures the attenuation and β the angular phase-shift. With a loss-free line, $\mathbf{P} = j\omega\sqrt{LC}$ and $\beta = \omega\sqrt{(LC)}$. With a velocity of propagation of 3×10^5 km/s the wavelength of the transmitted voltage and current at 50 Hz is 6000 km. Thus lines are much shorter than the wavelength of the transmitted energy.

Usually conditions at the load are required when $x = l$ in equations (3.10).

$$\left. \begin{aligned} \therefore \quad \mathbf{V}_r &= \mathbf{V}_s \cosh \mathbf{P}l - \mathbf{I}_s \mathbf{Z}_0 \sinh \mathbf{P}l \\ \mathbf{I}_r &= \mathbf{I}_s \cosh \mathbf{P}l - \frac{\mathbf{V}_s}{\mathbf{Z}_0} \sinh \mathbf{P}l \end{aligned} \right\} \tag{3.12}$$

and

120

Alternatively,

and

$$V_s = V_r \cosh \mathbf{P}l + \mathbf{I}_R \mathbf{Z}_0 \sinh \mathbf{P}l$$

$$\left. I_s = \frac{V_r}{\mathbf{Z}_0} \sinh \mathbf{P}l + \mathbf{I}_R \cosh \mathbf{P}l \right\}$$

(3.13)

The parameters of the equivalent four-terminal network are thus,

$$\mathbf{A} = \mathbf{D} = \cosh \sqrt{\mathbf{ZY}}$$

$$\mathbf{B} = \sqrt{\frac{\mathbf{Z}}{\mathbf{Y}}} \sinh \sqrt{\mathbf{ZY}}$$

and

$$\mathbf{C} = \sqrt{\frac{\mathbf{Y}}{\mathbf{Z}}} \sinh \sqrt{\mathbf{ZY}}$$

The earliest way to handle the hyperbolic functions is to use the appropriate series.

$$\mathbf{A} = \cosh \sqrt{\mathbf{ZY}} = 1 + \frac{\mathbf{YZ}}{2} + \frac{\mathbf{Y}^2\mathbf{Z}^2}{24} + \frac{\mathbf{Y}^3\mathbf{Z}^3}{720}$$

$$\mathbf{B} = \mathbf{Z}\left(1 + \frac{\mathbf{YZ}}{6} + \frac{\mathbf{Y}^2\mathbf{Z}^2}{120} + \frac{\mathbf{Y}^3\mathbf{Z}^3}{5040}\right)$$

$$\mathbf{C} = \mathbf{Y}\left(1 + \frac{\mathbf{YZ}}{6} + \frac{\mathbf{Y}^2\mathbf{Z}^2}{120} + \frac{\mathbf{Y}^3\mathbf{Z}^3}{5040}\right)$$

Usually not more than three terms are required and for (overhead) lines less than 500 km in length the following expressions for the constants hold approximately:

$$\mathbf{A} = \mathbf{D} = 1 + \frac{\mathbf{ZY}}{2} \qquad \mathbf{B} = \mathbf{Z}\left(1 + \frac{\mathbf{ZY}}{6}\right) \qquad \mathbf{C} = \mathbf{Y}\left(1 + \frac{\mathbf{ZY}}{6}\right)$$

An exact equivalent circuit for the long line can be expressed in the form of the π section shown in Figure 3.34. The application of simple circuit laws will

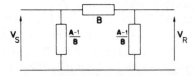

Figure 3.34   Equivalent circuit to accurately represent the terminal conditions of a long line.

show that this circuit yields the correct four-terminal network equations. If only the first term of the expansions is used,

$$\mathbf{B} = \mathbf{Z} \quad \text{and} \quad \frac{\mathbf{A} - 1}{\mathbf{B}} = \frac{\mathbf{Y}}{2},$$

i.e. the medium-length π representation. Figure 3.34 is exact only for conditions at the ends of the line; if intermediate points are to be investigated the full equations must be used.

Example 3.2 The conductors of a 1.6 km (1 mile) long, 3.3 kV, overhead line are in horizontal formation with 762 mm (30 in) between centres. The effective diameter of the conductors is 3.5 mm. The resistance per kilometre of the conductors is 0.41 Ω. Calculate the line to neutral inductance of the line. If the sending-end voltage is 3.3 kV and the load 1 MW at 0.8 p.f. lagging, calculate the voltage at the load busbar, the power loss in the line, and the efficiency of transmission.

Solution The equivalent equilateral spacing is given by

$$d_e = \sqrt[3]{(d_{12}d_{23}d_{31})}$$

In this case

$$d_e = \sqrt[3]{(762 \times 762 \times 1524)} = 762\sqrt[3]{2} = 960 \text{ mm}$$

The inductance (line to neutral)

$$= \frac{\mu_0}{2 \times 4\pi}\left[1 + 4\ln\left(\frac{d-r}{r}\right)\right] \text{H/m}$$

$$= \frac{4\pi}{10^7 \times 2 \times 4\pi}\left[1 + 4\ln\left(\frac{960 - 1.75}{1.75}\right)\right] \text{H/m}$$

$$= \frac{1}{2 \times 10^7}\left(\frac{26.21}{1}\right) \text{H/m}$$

Total inductance

$$= 1.0 \times 5280 \times 0.3048 \times \frac{13.105}{10^7} = 0.00212 \text{ H}$$

Inductive reactance

$$= 2\pi f L = 2\pi \times 50 \times 0.00212 = 0.66 \ \Omega$$

The line is obviously in the category of short and will be treated accordingly.
The load can be expressed as

$$P = 1 \text{ MW} \qquad Q = 0.75 \text{ MVAr}$$

The load can be represented by an equivalent shunt impedance in which case P and $Q \propto V_R^2$. If P and Q are constant regardless of V_R then an iterative procedure must be used. The latter will be used here. As the load voltage is unknown, to obtain the current the nominal voltage of 3.3 kV will be assumed.

The current is then

$$\frac{10^6}{\sqrt{(3)} \times 3300 \times 0.8} = 218\angle -\cos^{-1} 0.8 \text{ A}$$

$$\mathbf{V}_s = \mathbf{V_R} + \mathbf{IZ}$$

$$\frac{3300}{\sqrt{3}} = \mathbf{V_R} + (218 \times 0.8 - \text{j}218 \times 0.6)(0.66 + \text{j}0.66)$$

$$= \mathbf{V_R} + 115.0 + 86.7 + \text{j}(115.0 - 86.7)$$

$$1900 = \mathbf{V_R} + 201.7 + \text{j}28.3$$

$$\mathbf{V_R} = 1698.3 - \text{j}28.3$$

$$\mathbf{V_R} = 1698 \text{ V with negligible phase shift}$$

A quicker method would be to use

$$\Delta V = \frac{RP + XQ}{\mathbf{V_R}}$$

$$= \frac{\frac{1}{3} \times 0.66 \times 10^6 + \frac{1}{3} \times 0.66 \times 0.75 \times 10^6}{1900} = \frac{385}{1.9} = 202 \text{ V}$$

$$\therefore \quad \mathbf{V_R} = 1900 - 202 = 1698 \text{ V}$$

As, however, the load is 1 MW at the receiving end I should be recalculated and the above process repeated. As the approximate method is more convenient, it will be used.

$$\Delta V' = \frac{385}{1.698} = 227 \text{ V}$$

$$V_R' = 1900 - 227 = 1673 \text{ V}$$

Iterating again,

$$\Delta V'' = \frac{385}{1.673} = 230 \text{ V}$$

$$V_R'' = 1670 \text{ V}$$

Again,

$$\Delta V''' = \frac{385}{1670} = 230$$

The final value of V_R is therefore 1670 V.

Here, V_R can be obtained analytically without recourse to iterative procedures, but this is more lengthy in calculation. The type of load in this question in which P and Q remain constant regardless of the voltage is called *stiff*.

$$\text{The line loss} = I^2R = 218^2 \times 0.66 \text{ per phase}$$

$$= 94 \text{ kW for three phases}$$

$$\text{Efficiency of transmission} = \frac{\text{output}}{\text{output} + \text{losses}} = \frac{1000}{1000 + 94} \times 100 \text{ per cent}$$

$$= 91.4 \text{ per cent}$$

Example 3.3   A 275 kV three-phase transmission line of length 482 km is rated at 840 A. The values of resistance, inductance, and susceptance per phase per kilometre are 0.077 Ω, 1.05 mH and 3.68×10^{-6} Ω^{-1} respectively. The receiving-end voltage is 275 kV when full load is transmitted at 0.85 power factor lagging. Calculate the sending-end voltage.

Both the nominal π and long-line methods will be used to solve this problem.

Solution   (a) The equivalent π section is shown in Figure 3.33.

$$\mathbf{Z} = R + jX$$

where

$$R = 482 \times 0.077 = 37.5 \ \Omega$$

$$X = 482 \times 2\pi \times 50 \times \frac{1.05}{1000} = 160 \ \Omega$$

The leakage resistance is infinite.

$$\mathbf{Y} = j\frac{3.68 \times 482}{10^6} \text{ mhos}$$

$$= j1776 \times 10^{-6} \text{ mhos}$$

Take the receiving-end voltage as reference phasor. Load current

$$\mathbf{I_R} = 840 \text{ A, 0.85 p.f. lagging}$$

$$= (720 - j436) \text{ A}$$

$$\mathbf{I_L} = (720 - j436) + \left(\frac{275 \times 10^3}{\sqrt{3}} \times \frac{j1776}{2 \times 10^6}\right)$$

$$= 720 - j436 + j141$$

$$= 720 - j295 \text{ A}$$

$$\mathbf{I_L Z_1} = (720 - j295)(37.5 + j160)$$

$$= 74,300 + j104,100$$

$$\therefore \ \mathbf{V_S} = 159,000 + 74,300 + j104,100$$

$$\mathbf{V_S} = 256 \text{ kV}$$

Line voltage at sending-end

$$= \sqrt{3} \times 256 \text{ kV} = 443 \text{ kV}$$

From the long-line equations,

$$A = D = \cosh \sqrt{ZY}$$

$$B = \sqrt{\frac{Z}{Y}} \sinh \sqrt{ZY}$$

$$C = \sqrt{\frac{Y}{Z}} \sinh \sqrt{ZY}$$

Using the appropriate series,

$$A = D = 1 + \frac{YZ}{2} + \frac{Y^2 Z^2}{24}$$

$$YZ = (j1776 \times 10^{-6})(37.5 + j160) = -0.284 + j0.0666$$

$$(Y^2 Z^2) = 0.0807 - 0.00444 - j0.0375 = 0.08514 - j0.0378$$

$$A = D = 1 + -0.142 + j0.0333 + 0.0035 - j0.001575$$

$$= 1 - 0.1385 + j0.031725 = 0.8615 + j0.0317$$

$$B = Z\left(1 + \frac{YZ}{6} + \frac{Y^2 Z^2}{120}\right)$$

$$= (37.5 + j160)(1 + -0.0463 + j0.0108)$$

$$= 33.97 + j152.9$$

$$C = Y\left(1 + \frac{YZ}{6} + \frac{Y^2 Z^2}{120}\right)$$

$$= j\frac{1776}{10^6}(0.954 + j0.0108)$$

$$= (-19.2 + j1693)10^{-6}$$

$$V_S = AV_R + BI_R$$

$$I_S = CV_R + DI_R$$

Hence

$$V_S = (0.8165 + j0.0317)(159,000) + (33.97 + j152.9)(720 - j436)$$

$$= 221,100 + j100,250$$

$$V_S = 243 \text{ kV (phase)}$$

It is seen that the use of the nominal π model gives a value for V_S which is about 5 per cent high. This error is high and the long-line solution should be used. It is evident, however, that little error would be incurred by the use of only two terms of the series instead of three, i.e. $(1 + \mathbf{YZ}/2)$ and $(1 + \mathbf{YZ}/6)$ thus reducing the work involved.

The drop in voltage along the line is $(243 - 159)$, i.e. 84 kV, giving a regulation of 34.6 per cent. This is a very large voltage drop and could not be tolerated in practice, especially as the range of tap-changing transformers is usually only up to 15 per cent. In practice such a line would normally incorporate series capacitors to reduce the series reactance and the load-current power factor would be increased from 0.9 lag to near unity by the use of shunt capacitors or synchronous compensators at the receiving end.

The natural load

The characteristic impedance Z_0 is also known as the surge impedance. When a line is terminated in its characteristic impedance the power delivered is known as the natural load. For a loss-free line under natural-load conditions the reactive power absorbed by the line is equal to the reactive power generated, i.e.

$$\frac{V^2}{X_c} = I^2 X_{\mathrm{L}}$$

and

$$\frac{V}{I} = Z_0 = \sqrt{X_{\mathrm{L}} X_{\mathrm{C}}} = \sqrt{\frac{L}{C}}$$

At this load V and I are in phase all along the line and optimum transmission conditions obtained; in practice, however, the load impedances are seldom in the order of Z_0. Values of Z_0 for various line voltages are as follows, values of the corresponding natural loads are shown in brackets: 132 kV, 150 Ω (50 MW); 275 kV, 315 Ω (240 MW); 380 kV, 295 Ω (490 MW). The angle of the impedance varies between 0 and $-15°$. For underground cables Z_0 is roughly one-tenth of the overhead-line value. Lines are operated above the natural loading, whereas cables operate below this loading.

3.10 Parameters of Underground Cables

In cables with three conductors contained within the lead or aluminium sheath the electric field set up has components tangential to the layers of impregnated-paper insulation in which direction the dielectric strength is poor. Therefore, at voltages over 11 kV each conductor is separately screened (Höchstadter type) to ensure only radial stress through the paper. The capacitance of single-conductor and individually screened three-conductor cables is readily calculated. For three-conductor unscreened cables resort must be made to empirical design data. A typical high-voltage cable is shown in Figure 3.35. The capacitance (C) of single-core cables may be calculated from design data by

126

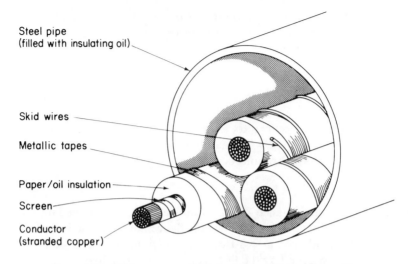

Steel pipe
(filled with insulating oil)

Skid wires

Metallic tapes

Paper/oil insulation

Screen

Conductor
(stranded copper)

Figure 3.35   High-pressure, oil-filled, pipe-type high-voltage cable.

the use of the formula

$$C = \frac{2\pi\varepsilon_0\varepsilon_r}{\ln{(R/r)}}$$ (3.14)

where r and R are the inner and outer radii of the dielectric and ε_r is the relative permittivity of the dielectric, often in the order of 3.5. This also holds for three-core cables with each conductor separately screened.

The various capacitances present in three-core unscreened cables may be represented as shown in Figure 3.36, in which the conductor to conductor capacitance is C_1 and the conductor to sheath capacitance is C_0. The equivalent circuit is shown in Figure 3.36(b) and circuits obtained by star–delta or delta–star transformations in Figure 3.36(c) and (d). The value of

$$C_3 = \frac{C_0}{3}$$

and similarly $C_4 = 3C_1$. The parallel combination of C_4 and C_0 gives the equivalent line-to-neutral value of the cable capacitance. C_0 and C_1 may be measured as follows. The three cores are short-circuited and the capacitance between them and the sheath measured; this measurement gives $3C_0$. Next, the capacitance between two cores is measured, the third being connected to the sheath. It is easily shown that the measured value is now $C_1 + [(C_0 + C_1)/2]$; hènce C_0 and C_1 are obtained.

Owing to the symmetry of the cable the normal phase-sequence values of C and L are the same as the negative phase-sequence values (i.e. for reversed phase rotation). The series resistance and inductance are complicated by the magnetic interaction between the conductor and sheath (see Chapter 11).

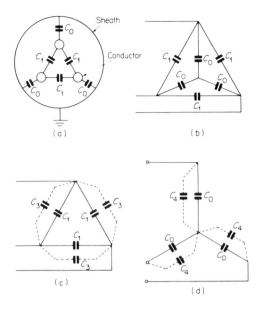

Figure 3.36   Capacitances in a three-core cable.

The effective resistance of the conductor is the direct current resistance modified by the following factors: the skin effect in the conductor; the eddy currents induced by adjacent conductors (the proximity effect); and the equivalent resistance to account for the I^2R losses in the sheath. The determination of these effects is complicated and is discussed further in Chapter 11.

The parameters of the cable having been determined, the same equivalent circuits as for overhead lines are used, paying due regard to the selection of the correct model for the appropriate length of cable. Owing to the high capacitance of cables the charging current, especially at high voltages, is an important factor in deciding the permissible length to be used. Table 3.3 gives the charging currents for self-contained low pressure oil filled cables.

Table 3.3   Underground cable constants at 50 Hz—LPOF
(per km)

		132 kV		275 kV		400 kV
Size	mm^2	355	645	970	1130	1935
Rating (soil $g = 120°C$ cm/W)	Amps	550	870	1100	1100	1600
	MVA	125	200	525	525	1100
Resistance (R) at 85°C	Ohms	0.065	0.035	0.025	0.02	0.013
Reactance (X_L)	Ohms	0.128	0.138	0.22	0.134	0.22
Charging current (I_c)	Amps	7.90	10.69	15.70	17.77	23.86
X_L/R Ratio		2.0	4.0	8.8	6.6	16.8

3.11 Transformers

The equivalent circuit of one phase of a transformer referred to the primary winding is shown in Figure 3.37. The resistances and reactances can be found from the well-known open- and short-circuit tests. In the absence of complete information for each winding, the two arms of the T network can each be assumed to be half the total transformer impedance. Also little accuracy is lost in transferring the shunt branch to the input terminals forming a cantilever circuit.

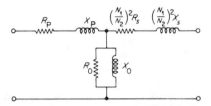

Figure 3.37   Equivalent circuit of a two-winding transformer.

In power transformers the current taken by the shunt branch is usually a small percentage of the load current and can be neglected.

Phase shifts in three-phase transformers

Consider the transformer shown in Figure 3.38(a). The red phases on both sides are taken as reference and the transformation ratio is $1:N$. The corresponding phasor diagrams are shown in Figure 3.38(b). Although no neutral

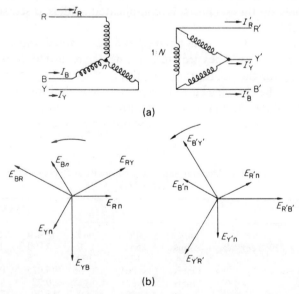

Figure 3.38(a)   Star–delta transformer with turns ratio $1:N$. (b) Corresponding phasor diagrams (N taken as 1 in diagrams).

point is available in the delta side the effective voltages from line to earth are denoted by $E_{R'n}$, $E_{Y'n}$ and $E_{B'n}$. Comparing the two phase diagrams, the following relationships are seen:

$$E_{R'n} = \text{line–earth voltage on the delta side}$$

$$= N E_{Rn} \angle 30°,$$

i.e. the positive sequence or normal balanced voltage of each phase is advanced through 30°. Similarly, it can be shown that the positive sequence currents are advanced through 30°.

By a consideration of the negative phase-sequence phasor diagrams (these are phasors with reversed rotation, i.e. R–B–Y it will readily be seen that the phase currents and voltages are shifted through −30°. When using the per unit system the transformer ratio does not directly appear in calculations. In star–star and delta–delta connected transformers there are no phase shifts; hence transformers having these connexions and those with star–delta should not be connected in parallel. To do so introduces a resultant voltage acting in the local circuit formed by the usually low transformer impedances. In Figure 3.39 is shown the general practice on the British network with regard to

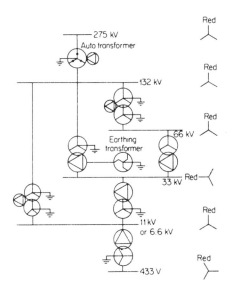

Figure 3.39 Typical phase shifts in a power system—British.

transformers with phase shifts. It is seen that the reference phasor direction is different at different voltage levels. The larger than 30° phase shifts are obtained by suitable rearrangement of the winding connexions. Tertiary or third windings are provided to give sufficient fault current to operate protective gear and to provide a path for third-harmonic currents.

Three-winding transformers Many transformers used in power systems have three windings per phase, the third winding being known as the tertiary. This type can be represented under balanced three-phase conditions by a single-phase equivalent circuit of three impedances star-connected as shown in Figure 3.40. The values of the equivalent impedances Z_p, Z_s and Z_t may be obtained by test. It is assumed that the no-load currents are negligible.

(a)

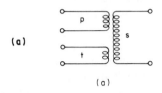

(a)

Figure 3.40(a) Three-winding transformer.

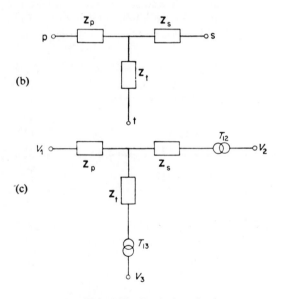

(b)

(c)

Figure 3.40(b) and (c) Equivalent circuits.

Let,

Z_{ps} = impedance of the primary when the secondary is short-circuited and the tertiary open;

Z_{pt} = impedance of the primary when the tertiary is short-circuited and the secondary open;

Z_{st} = impedance of the secondary when the tertiary is short-circuited and the primary open.

The above impedances are in ohms referred to the same voltage base. Hence,

$$\mathbf{Z}_{ps} = \mathbf{Z}_p + \mathbf{Z}_s$$
$$\mathbf{Z}_{pt} = \mathbf{Z}_p + \mathbf{Z}_t$$
$$\mathbf{Z}_{st} = \mathbf{Z}_s + \mathbf{Z}_t$$
$$\left.\begin{aligned}\mathbf{Z}_p &= \tfrac{1}{2}(\mathbf{Z}_{ps} + \mathbf{Z}_{pt} - \mathbf{Z}_{st}) \\ \mathbf{Z}_s &= \tfrac{1}{2}(\mathbf{Z}_{ps} + \mathbf{Z}_{st} - \mathbf{Z}_{pt}) \\ \mathbf{Z}_t &= \tfrac{1}{2}(\mathbf{Z}_{pt} + \mathbf{Z}_{st} - \mathbf{Z}_{ps}) \end{aligned}\right\}$$

$$(3.15)$$

It should be noted that the star point in Figure 3.40b is purely fictitious and that the diagram is a single-phase equivalent circuit. In most large transformers the value of Z_s is very small and can be negative. All impedances must be referred to common volt-ampere and voltage bases. The complete equivalent circuit is shown in Figure 3.40(c).

Auto-transformers The symmetrical auto-transformer may be treated in the same manner as two- and three-winding transformers. This type of transformer shows to best advantage when the transformation ratio is small and it is widely used for the interconnexion of the supply networks working at different voltages, e.g. 275 kV to 132 kV. The neutral point is solidly grounded, i.e. connected directly to earth without intervening resistance.

Earthing (grounding) transformers A means of providing an earthed point or neutral in a supply derived from a delta-connected transformer may be obtained by the use of a zigzag transformer shown schematically in Figure 3.39. By the interconnexion of two windings on each limb a node of zero potential is obtained.

Harmonics Due to the non-linearity of the magnetizing characteristic for transformers the current waveform is distorted and hence contains harmonics; these flow through the system impedances and set up harmonic voltages. In transformers with delta-connected windings the third and ninth harmonics circulate round the delta and are less evident in the line currents. Another source of harmonics is a rectifier load.

On occasions the harmonic content can prove important due mainly to the possibility of resonance occurring in the system, e.g. resonance has occurred with fifth harmonics. Also, the third-harmonic components are in phase in the three conductors of a three-phase line, and if a return path is present these currents add and cause interference in neighbouring communication circuits.

When analysing systems with harmonics it is sufficient to use the normal values for series inductance and shunt capacitance; the effect on resistance is more difficult to assess: however, it is usually only required to assess the presence of harmonics and the possibility of resonance.

Tap-changing transformers A method of controlling the voltages in a network lies in the use of transformers, the turns ratio of which may be changed. In Figure 3.41(a) a schematic diagram of an off-load tap changer is shown; this however requires the disconnexion of the transformer when the tap setting is to be changed. Many transformers now have on-load changers, the basic form of which is shown in Figure 3.41(b). In the position shown the

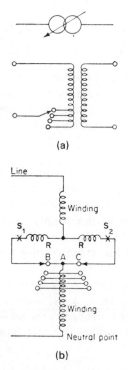

(a)

(b)

Figure 3.41(a)   Tap-changing transformer. (b) On-load tap-changing transformer. S_1, S_2 transfer switches, R centre-tapped reactor.

voltage is a maximum and the current divides equally in the two halves of coil R resulting in zero-resultant flux and minimum impedance. To reduce the voltage, S_1 opens and the total current passes through the other half of the reactor. Selector switch B then moves to the next contact and S_1 closes. A circulating current now flows in R superimposed on the load current. S_2 now opens and C moves to the next tapping; S_2 then closes and the operation is complete. Six switch operations are required for one change in tap position. The voltage change between taps is often 1.25 per cent of the nominal voltage. This small change is necessary to avoid large voltage disturbances at consumer busbars. A schematic block diagram of the on-load tap systems is shown in Figure 3.42. The line drop compensator (LDC) is used to allow for the voltage drop along the feeder to the load point, so that the actual load voltage is seen

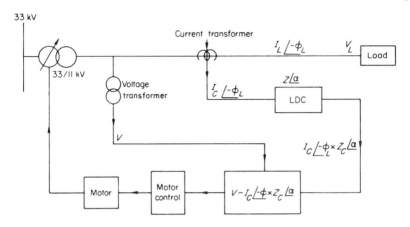

Figure 3.42 Schematic diagram of a control system for an on-load tap-changing transformer incorporating *line drop compensation* (LDC).

and corrected by the transformer. The total range of tapping varies with the transformer usage, a typical figure for generator transformers is +2 to −16 per cent in 18 steps.

Typical parameters for transformers   The leakage reactances of two-winding transformers increase slightly with their rating for a given voltage, i.e. from 3.2 per cent at 20 kVA to 4.3 per cent at 500 kVA at 11 kV. For larger sizes, i.e. 20 MVA upwards, 10 per cent is a typical value at all voltages. For auto-transformers the impedances are usually less than for double wound. Parameters for large transformers are as follows: (a) 400/275 kV auto-transformer, 500 MVA, 12 per cent impedance, tap range +10 to −20 per cent; (b) 400/132 kV double wound, 240 MVA, 20 per cent impedance, tap range +5 to −15 per cent.

Example 3.4   Two three-phase, 275/66/11 kV transformers, are connected in parallel and the windings are employed as follows:

the 11 kV winding supplies a synchronous compensator (i.e. a synchronous motor running light);
the 66 kV winding supplies a network;
the 275 kV winding is connected to the primary transmission system.

The winding data are as shown below. The line diagram is shown in Figure 3.43(a).
   It is required to determine the division of power between the two 275/66 kV windings. Resistance may be neglected.

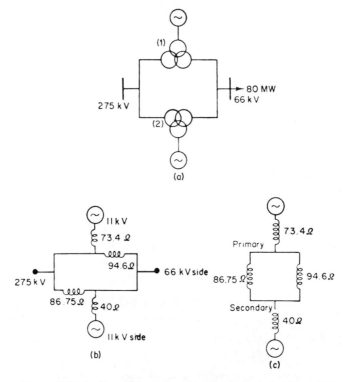

Figure 3.43(a) Line diagram for system in Example 3.6 (b) and (c)
Equivalent circuits: all reactance values refer to bases of 275 kV,
90 MVA.

Solution To obtain the equivalent star circuits the reactances will be expressed on common bases of 90 MVA, 275 kV.

	$X\%$ referred to 90 MVA			MVA of windings		
	275/ 66 kV	66/ 11 kV	11/ 275 kV	275 kV	66 kV	11 kV
Transformer (1)	10.5	20	8	95	90	60
Transformer (2)	10.5	5	15	90	85	45

For transformer (1),

$$X_{275} + X_{66} = \frac{10.5}{100} \times \frac{275^2}{90} = 88.4 \ \Omega$$

$$X_{66} + X_{11} = \frac{20}{100} \times \frac{275^2}{90} = 168 \ \Omega$$

$$X_{11} + X_{275} = \frac{8}{100} \times \frac{275^2}{90} = 67.2 \ \Omega$$

The corresponding quantities for transformer (2) are 88.4 Ω, 42.1 Ω, and 126.3 Ω respectively.

For transformer (1),

$$X_p(275 \text{ kV side}) = \tfrac{1}{2}(X_{ps} + X_{pt} - X_{st})$$
$$= \tfrac{1}{2}(88.4 + 67.2 - 168)$$
$$= -6.2 \ \Omega$$

Similarly,
$$X_s(66 \text{ kV side}) = 94.6 \ \Omega$$
and
$$X_T(11 \text{ kV side}) = 73.4 \ \Omega$$

For transformer (2),

$$X_p = 86.75 \ \Omega \qquad X_s = 1.65 \ \Omega \quad \text{and} \quad X_T = 40 \ \Omega$$

X_p in (1) and X_s in (2) will be neglected.

The equivalent circuit is shown in Figure 3.43(b) which in turn reduces to figure 3.43(c).

The power will divide according to the reactances of the two transformers, see Chapter 6, hence,

$$P_{(1)} = \frac{94.6 \times 80}{94.6 + 86.75} = 41.7 \text{ MW}$$

and

$$P_{(2)} = \frac{86.75 \times 80}{86.75 + 94.6} = 38.3 \text{ MW}$$

3.12 Connexion of Three-phase Transformers

In section 3.11 the phase changes occurring in star–delta transformers are briefly discussed and sufficient treatment given for the following analysis to be understood. From the practical view, however, the connexion of such transformers requires further discussion to be fully appreciated. It is seen that for the connexion shown in Figure 3.38(a) the output line voltages on the delta side are 30° in advance of the input line voltages to the star winding for normal a–b–c phase notation. Also the output line voltage is equal to the input line multiplied by $(N/\sqrt{3})$ and impedances transferred from one side to the other are modified by $(N/\sqrt{3})^2$ and vice versa. The true single-phase equivalent circuit is therefore as shown in Figure 3.44.

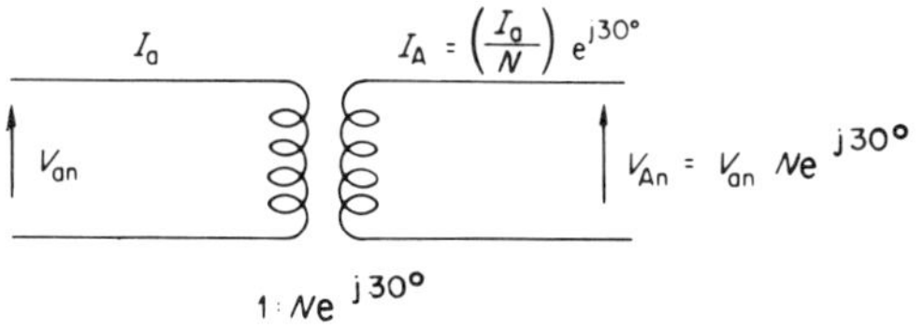

Figure 3.44 Equivalent single-phase circuit of a three-phase transformer with a phase shift from primary to secondary.

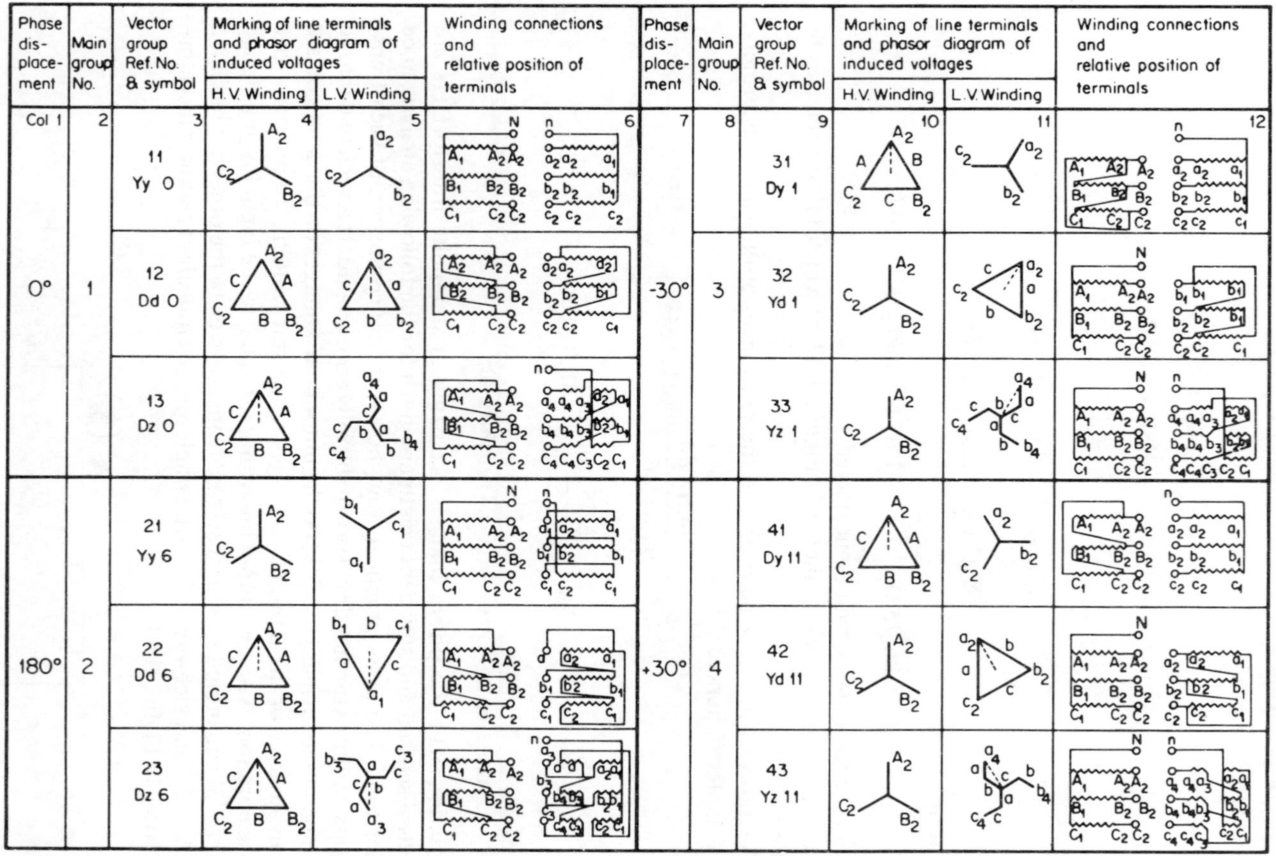

136

Phase displacement	Main group No.	Vector group Ref. No. & symbol	Marking of line terminals and phasor diagram of induced voltages — H.V Winding	L.V Winding	Winding connections and relative position of terminals	Phase displacement	Main group No.	Vector group Ref. No. & symbol	Marking of line terminals and phasor diagram of induced voltages — H.V Winding	L.V Winding	Winding connections and relative position of terminals
Col 1	2	3	4	5	6	7	8	9	10	11	12
0°	1	11 Yy 0				−30°	3	31 Dy 1			
		12 Dd 0						32 Yd 1			
		13 Dz 0						33 Yz 1			
180°	2	21 Yy 6				+30°	4	41 Dy 11			
		22 Dd 6						42 Yd 11			
		23 Dz 6						43 Yz 11			

Figure 3.45 British Standard for connexions of three-phase transformers and resulting phase shifts.

However, by suitable rearrangement of the winding connexions, phase shifts of 30, 90, 150, 210, 270, or 330 electrical degrees may be obtained, and there is a need for standardization of nomenclature and procedure. Should a phase difference exist between the secondary output voltages of transformers connected to a common supply, dangerously large circulating currents will occur and the circuit protection should operate. For uniformity the transformer windings of Figure 3.38(a) shown in Figure 3.45 with capitals referring to the HV winding. It is seen that with the star–delta a shift of −30° can also be produced by a rearrangement of the terminal connexions. British practice* involves the use of a 'vector group reference' number to describe transformer connexions. The first number indicates the phase shift, e.g. 1 indicates 0°, 2—180°, 3—(−30°), 4—30°; the next number indicates the interphase connexion of the secondary, e.g. 1—star, 2—delta, 3—interconnected star. Letters D and y indicate star (y) or delta (D) connexion, the HV winding coming first. Finally comes the phase displacement angle in the form of a clock-reference, the low voltage winding induced voltage being given by the hour hand and the HV voltage by the minute hand, e.g. 0° = 12 o'clock, 180° = 6 o'clock, +30° = 11 o'clock and −30° = 1 o'clock; A_2 in the phasor diagrams is always at 12 o'clock. The application of this principle to the most widely used transformer connexions is shown in Figure 3.45.

From an analytical viewpoint it may be desirable to have a phase shift of 90°, thus facilitating the use of operator 'j'. In this type of work the exact arrangement is not important provided the system adopted is used consistently and the winding arrangement is seldom the cause for concern. However, when dealing with the connexion of actual equipment in the system the correct arrangement of the transformer winding is of vital importance.

3.13 Voltage Characteristics of Loads

The variation of the power and reactive power taken by a load with various voltages is of importance when considering the manner in which such loads are to be represented in load flow and stability studies. Usually in such studies the load on a substation has to be represented and is a composite load consisting of industrial and domestic consumers. A typical composition of a substation load is as follows:

Induction motors	50–70%
Lighting and heating	20–25%
Synchronous motors	10%
(Transmission losses	10–12%)

Lighting This is independent of frequency and consumes no reactive power. The power consumed does not vary as the (voltage)^2, but approximately as (voltage)$^{1.6}$.

* The Americal Standards Association adopt the standard that the high-voltage reference phase is always 30° ahead of the low-voltage reference phase, regardless of whether the three-phase connexions are delta–star or star–delta.

Heating This maintains constant resistance with voltage change and hence the power varies with (voltage)^2.

The above loads may be described as static.

Synchronous motors The power consumed is approximately constant. For a given excitation the vars change in a leading direction with voltage reduction. The P–V, Q–V, generalized characteristics are shown in Figure 3.46.

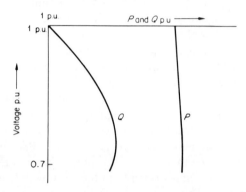

Figure 3.46 P–V and Q–V curves for a synchronous motor.

Induction motors The P–V, Q–V characteristics may be determined by the use of the simplified circuit shown in Figure 3.47. It is assumed that the mechanical load on the shaft is constant.

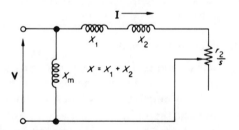

Figure 3.47 Equivalent circuit of an induction motor: X_1 = stator leakage reactance; X_2 = rotor leakage reactance referred to the stator; X_m = magnetizing reactance; r_2 = rotor resistance; s = slip p.u. Magnetizing losses have been ignored and the stator losses lumped in with the line losses.

The electrical power,

$$P_{\text{electrical}} = P_{\text{mechanical}} = P$$

$$= 3I^2 \frac{r_2}{s} = \text{constant}$$

$$\therefore \; s = \frac{3I^2 r_2}{P}$$

The reactive power consumed

$$= \frac{3V^2}{X_m} + 3I^2(X_1 + X_2) \qquad (3.16)$$

Also, from Figure 3.47

$$P = 3I^2 \frac{r_2}{s} = \frac{3V^2}{(r_2/s)^2 + X^2} \cdot \frac{r_2}{s}$$

$$= \frac{3V^2 r_2 s}{r_2^2 + (sX)^2} \qquad (3.17)$$

The graphs of V–P and V–Q are shown in Figure 3.48. The P and Q scales have been so arranged that P and Q are both equal to 1 p.u. at a voltage of

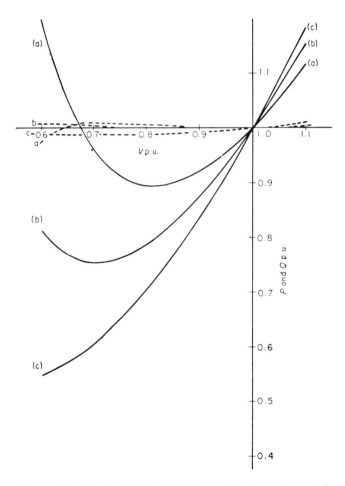

Figure 3.48   Graphs of P–V and Q–V for an induction motor—shaft torque constant. Curve (a) full load, (b) 0.75 full load, (c) 0.5 full load, $---P$, $——Q$.

1 p.u. The effect of shaft load is also shown. Similar analysis can be performed for the induction motor with torque $\propto$ (speed)^2 and torque $\propto$ speed.

The well-known power-slip curves for an induction motor are shown in Figure 3.49. It is seen that for a given mechanical torque there is a critical

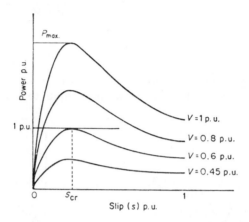

Figure 3.49   Power-slip curves for an induction motor. If voltage falls to 0.6 p.u. at full load $P = 1$, the condition is critical (slip s_c).

voltage and a corresponding critical slip. If the voltage is reduced further the motor becomes unstable and stalls. This critical point occurs when

$$\frac{\mathrm{d}P}{\mathrm{d}s} = 0$$

i.e. when

$$V^2 r_2 \cdot \frac{r_2^2 - (sX)^2}{[r_2^2 + (sX)^2]^2} = 0$$

so that

$$s = \frac{r_2}{X} \quad \text{and} \quad P_{\max} = \frac{V^2}{2X}$$

Alternatively for a given output power P, $V_{\text{critical}} = \sqrt{2PX}$.

Composite loads   Except in isolated cases, such as very large induction motor drives in refineries and steel mills, it is the composite load which is of interest. Tests to obtain P–V and Q–V characteristics of complete substation loads have been few (the consumers object!). The results of one such test on the Polish network[19] are given in Table 3.4. Instead of actual characteristics, values of $\partial P/\partial V$ and $\partial Q/\partial V$ at 1 p.u. voltage are given. As will be seen later these quantities are of considerable interest. Composite load characteristics can be obtained by summing the characteristics of the constituent loads.

Table 3.4 Polish network test results of loads

dP/dV at nominal voltage	Morning	Evening	Night
Large towns	0.9–1.2	1.5–1.7	1.5–1.6
Small towns	0.6–0.7	1.4–1.6	1.4–1.6
Villages	0.5–0.6	1.5	1.5
Street lighting	—	1.55	1.55
Mines	0.6–0.76	0.75–0.89	0.65–0.78
Ironworks	0.6–0.7	0.6–0.75	0.6–0.7
Textile industries	0.5–0.6	0.6–0.7	0.6–0.65
Large chemical industries	0.6–0.7	0.6–0.75	0.6–0.7
Machine industry	0.5–0.55	0.6–0.65	0.6–0.65
Junction points in the HV network	0.6–0.7	0.8–1.1	0.8–1.0

dQ/dV at nominal voltage	Morning	Evening	Night
Large towns	2.5–3.0	2.2–2.5	2.4–3.0
Small towns	2.6–3.5	2.0–2.5	2.5–3.0
Villages	2.6–3.5	2.0–2.5	2.4–3.5
Industry (total)	1.8–2.7	1.8–2.3	1.8–2.7
Junction points in 6 kV network	2.2–2.9	1.8–2.5	2.0–2.6
Junction points in 110 kV network	1.8–2.6	1.5–2.2	1.7–2.3

Source: Boyucki and Wojcik.[19]

In practice, great difficulty is experienced in assessing the degree to which motors are loaded and this materially affects the shape of the characteristics. Loads which require the same power at reduced voltage are called 'stiff'; loads, such as heating, in which the power falls off rapidly at lower voltages are termed 'soft'.

In much analytical work loads are represented by constant impedances and hence: it is assumed that $P \propto V^2$ and $Q \propto V^2$. If the load is $P + jQ$ p.u. the current at voltage V is

$$\frac{P - jQ}{V} \text{ p.u. (remembering } \mathbf{VI}^* = S)$$

Hence the impedance to represent this load

$$= \frac{\mathbf{V}}{\mathbf{I}} = \frac{V^2}{P - jQ} \text{ p.u.}$$

The P–V, Q–V characteristics can be used if they are known, but this process is rather tedious. Some authorities suggest that the representation of the load by a constant-current consuming device gives a good approximation to practical conditions. If analogue or digital methods are used the correct representation of loads presents no great difficulties. When extensive reduction of a network to a simpler form is necessary then the constant impedance representation is usually resorted to, although the load this represents seldom occurs in practice.

Load-frequency characteristics Often voltage changes are accompanied by frequency changes. Again information regarding the characteristics of composite loads with frequency is scarce. With the small frequency changes allowed in practice the effects on the power and reactive power consumed are usually neglected in calculations.

3.14 Switchgear and Protection

Switchgear

The switches by means of which the system is controlled are known as circuit breakers. At high voltages these may have several breaks (contacts) per phase

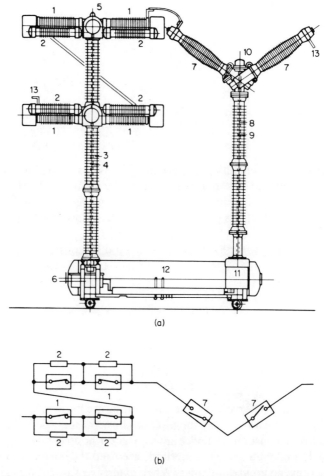

(a)

(b)

Figure 3.50 General arrangement and basic circuit diagram of one pole of a 245 kV airblast circuit breaker. (Permission of Brown Boveri Co.) (a) General circuit arrangement. (b) Circuit (breaker open). 1 Impulse chamber. 2 Parallel resistor to 1. 3 Impulse chamber column. 4 Tie rod. 5 Impulse chamber valve. 6 Impulse chamber control unit. 7 Isolating chamber. 8 Isolating chamber column. 9 Tie rod. 10 Isolating chamber valve. 11 Isolator unit. 12 Air receiver. 13 Terminals.

and are very large and complicated structures (see Chapter 12). The very act of opening a circuit, which at domestic voltages is taken for granted, at very high voltages becomes a major engineering problem. At the higher transmission voltages the fault currents to be interrupted are very high. A diagram of a 245 kV air-blast circuit breaker is shown in Figure 3.50.

Protection

When the system malfunctions in any way, e.g. a short circuit to earth on a line or transformer, rapid automatic means are required to remove the faulty equipment. At the same time the remainder of the system must be left operating intact. System protection embraces the various measuring and relaying schemes used to achieve this aim. In a large highly interconnected supply network efficient protection is of the utmost importance, and its malfunction can bring about a wholesale electrical collapse of the supply system.

Much of the analysis discussed in this text has as its end-product the installation of the correct protection schemes. Protection is intimately connected with switchgear, the former representing the 'nervous system' and the latter the resulting physical action. At the present time in the British 400 kV system it is aimed to clear completely a fault in 140 ms, but with the increasing size and complexity a value of 80 ms is being actively considered in the future. This will allow, say, two cycles of 50 Hz for the protection to detect the fault and two cycles for the circuit breakers to open, a very stringent requirement.

References

BOOKS

1. *Electrical Transmission and Distribution Reference Book*, Westinghouse Electric Corp., East Pittsburgh, Pennsylvania, 1964.
2. Waddicor, H., *Principles of Electric Power Transmission*, Chapman and Hall, London, 5th edn, 1964.
3. McPherson, G., *An Introduction to Electrical Machines and Transformers*, Wiley, New York, 1981.
4. Hindmarsh, J., *Electric Machines*, Pergamon, London, 5th edn, 1977.
5. Say, M. G., *Alternating Current Machines*, Pitman, London, 1977.
6. Kimbark, E. W., *Power System Stability, Vol. 3. Synchronous Machines*, Wiley, New York, 1962.
7. Slemon, G. R., *Magnetoelectric Devices*, Wiley, New York, 1965.
8. Blume, L. F., *et. al.*, *Transformer Engineering*, Wiley, New York, 2nd edn, 1951.
9. Fitzgerald, A. E., C. Kingsley Jr., and A. Kusko, *Electric Machinery*, McGraw-Hill, New York, 3rd end, 1971.
10. Langlois-Berthelot, R., *Transformers and Generators for Power Systems*, (translated by H. M. Clarke), Macdonald, London, 1960.
11. Steven, R. E., *Electromechanics and Machines*, Chapman and Hall, London, 1970.
12. Edison Electric Institute, *EHV Transmission Line Reference Book*, 1968.
13. Adkins, B., *The General Theory of Electrical Machines*, Chapman and Hall, London, 1964.
14. Jones, B., *New Approach to the Design and Economics of EHV Plant* Pergamon, London, 1972.

144

15. Endrenyi, J., *Reliability Modeling in Electric Power Systems*, Wiley, Chichester, 1978.
16. Weedy, B. M., *Underground Transmission of Electric Power*, Wiley, Chichester, 1980.
17. Smeaton, R. *Switchgear and Control Handbook*, McGraw-Hill, New York, 1977.

PAPERS

18. Park, R. H., and P. H. Robinson, 'The reactance of synchronous machines', *Trans. A.I.E.E.*, **47** (1928).
19. Bogucki, A., and M. Wojcik, 'Static characteristics of loads', *Energetyka B.*, No. 3 (Polish), **1959**.
20. Hurley, J. D., *et al.*, 'High response excitation systems on turbine-generators', *I.E.E.E. Trans.*, **PAS-101** (1982), 4211.
21. American Standards Association, *Terminal Markings and Connections for Distribution and Power Transformers*, December 1964.

Problems

(*Note*: All machines, etc. are three-phase unless stated otherwise.)

3.1. When two four pole, 50 Hz synchronous generators are paralleled their phase displacement is 2° mechanical. The synchronous reactance of each machine is 10 Ω/phase and the common busbar voltage is 6.6 kV. Calculate the synchronizing torque.
(Answer: 973.5 Nm)

3.2 A synchronous generator has a synchronous impedance of 2 p.u. and a resistance of 0.01 p.u. It is connected to an infinite busbar of voltage 1 p.u. through a transformer of reactance j0.1 p.u. If the generated (no-load) e.m.f. is 1.1 p.u. calculate the current and power factor for maximum output.
(Answer: 0.708 p.u., 0.74 leading)

3.3. A 6.6 kV synchronous generator has negligible resistance and synchronous reactance of 4 Ω/phase. It is connected to an infinite busbar and gives 2000 A at unity power factor. If the excitation is increased by 25 per cent find the maximum power output and the corresponding power factor. State any assumptions made.
(Answer: 31.6 MW; 0.96 leading.)

3.4. A synchronous generator whose characteristic curves are given in Figure 3.4 delivers full load at the following power factors; 0.8 lagging, 1.0, and 0.9 leading. Determine the percentage regulation at these loads.

3.5. A salient-pole, 75 MVA, 11 kV synchronous generator is connected to an infinite busbar through a link of reactance 0.3 p.u. and has $X_d = 1.5$ p.u. and $X_q = 1$ p.u. and negligible resistance. Dtermine the power output for a load angle 30° if the excitation e.m.f. is 1.4 times the rated terminal voltage. Calculate the synchronizing coefficient in this operating condition. All p.u. values are on a 75 MVA base.
(Answer: $P = 0.48$ p.u., $\partial P/\partial \delta = 0.78$ p.u.)

3.6. A synchronous generator of open circuit terminal voltage 1 p.u. is on no-load and then suddenly short-circuited; the trace of current against time is shown in Figure 3.6(b). In Figure 3.6(b) the current oc = 1.8 p.u., oa = 5.7 p.u., and ob = 8 p.u. Calculate the values of X_s, X' and X''. Resistance may be neglected. If the machine is delivering 1 p.u. current at 0.8 power factor lagging at the rated terminal voltage before the short circuit occurs, sketch the new envelope of the 50 Hz current waveform.
(Answer: X_s 0.8 p.u.; $X' = 0.25$ p.u.; $X'' = 0.18$ p.u.)

3.7. Construct a performance chart for a 22 kV, 500 MVA, 0.9 p.f. generator having a short-circuit ratio of 0.55.

3.8. A 275 kV three-phase transmission line of length 96 km is rated at 800 A. The values of resistance, inductance, and capacitance per phase per kilometre are 0.078 Ω,

1.056 mH and 0.029 μF, respectively. The receiving end voltage is 275 kV when full load is transmitted at 0.9 power factor lagging. Calculate the sending end voltage and current, and the transmission efficiency and compare with the answer obtained by the short line representation. Use the nominal π and T methods of solution. The frequency is 60 Hz.

(Answer: $V_s = 178$ kV per phase.)

3.9. A 220 kV, 60 Hz three-phase transmission line is 320 km long and has the following constants per phase per km:

Inductance	1.3 mH
Capacitance	0.22 μF
Resistance	0.55 Ω

Ignore leakage conductance.

If the line delivers a load of 300 A, 0.8 power-factor lagging, at a voltage of 220 kV, calculate the sending-end voltage. Determine the π circuit which will represent the line. Calculate the error resulting from the use of the medium-length π section.

(Answer: Series $(11 + j116)$ Ω; shunt $(0.44 + j0.72) \times 10^{-3}$ Ω^{-1}.)

3.10. In a three-core cable, the capacitance between the three cores short-circuited together and the sheath is 87 μF/km, and that between two cores connected together and the third core is 0.84 μF/km..

Determine the kVA required to keep 16 km of this cable charged when the supply is 33 kV, three phase, 50 Hz.

(Answer: 270 kVA).

3.11. Calculate the **A B C D** constants for a 275 kV overhead line of length 83 km. The parameters per kilometre are as follows:

Resistance	0.078 Ω
Reactance	0.33 Ω
Admittance (shunt capacitative)	9.53 $\times 10^{-6}$ mhos

The shunt conductance is zero.

(Answer:

$$[A = 0.8615 + j0.0317, \ B = 34.0 + j153, \ C = (-19.2 + j1693) \ 10^{-6})]$$

3.12. A 132 kV, 60 Hz transmission line has the following generalized constants:

$$A = 0.9696\angle 0.49°$$

$$B = 52.88\angle 74.79° \ \Omega$$

$$C = 0.001177\angle 90.15° \ \text{mhos}$$

If the receiving-end voltage is to be 132 kV when supplying a load of 125 MVA 0.9 p.f. lagging, calculate the sending-end voltage and current.

(Answer: 95.9 kV phase, 554 A)

3.13. Two identical transformers each have a nominal or no-load ratio of 33/11 kV and a reactance of 2 Ω referred to the 11 kV side; resistance may be neglected. The transformers operate in parallel and supply a load of 9 MVA, 0.8 p.f. lagging. Calculate the current taken by each transformer when they operate five tap steps apart (each step is 1.25 per cent of the nominal voltage).

(Answer: 307 A, 194 A for three-phase transformers.)

3.14. An induction motor, the equivalent circuit of which is shown in Figure 3.51 is connected to supply busbars which may be considered as possessing a voltage and frequency which is independent of the load. Determine the reactive power consumed for various busbar voltages and construct the Q–V characteristic. Calculate the critical

146

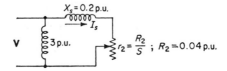

Figure 3.51 Equivalent circuit of 500 kW, 6.6 kV induction motor in Problem 3.14. All p.u. values refer to rated voltage and power ($P = 1$ p.u. and $V = 1$ p.u.).

voltage at which the motor stalls and the critical slip, assuming that the mechanical load is constant.

(Answer: $S_{cr} = 0.2$, $V_c = 0.63$.)

3.15. In the system shown in Figure 2.19, Example 2.3, the generator has a synchronous reactance of 1.0 p.u. The internal voltage (no-load) of the generator is held at 17 kV (line). Determine the voltage at the load busbar for the same load as before, i.e. 50 MW, 0.8 p.f. lagging. The load may be represented by a constant impedance equal to the initial value in Example 2.3.

(Answer: 0.76 p.u.)

3.16. A 100 MVA round-rotor generator of synchronous reactance 1.5 p.u. supplies a substation (L) through a step-up transformer of reactance 0.1 p.u., two lines each of reactance 0.3 p.u. in parallel and a step-down transformer of reactance 0.1 p.u. The load taken at L is 100 MW at 0.85 lagging. L is connected to a local generating station which is represented by an equivalent generator of 75 MVA and synchronous reactance of 2 p.u. All reactances are expressed on a base of 100 MVA.

Draw the equivalent single-phase network. If the voltage at L is to be 1 p.u. and the 75 MVA machine is to deliver 50 MW, 20 MVAr, calculate the internal voltages of the machines.

(Answer: $E_1 = 2$ p.u.; $E_2 = 1.72$ p.u.; $\delta_{2L} = 35.45°$.)

3.17. The following data applies to the power system shown in Figure 3.52:

Generating station A. Four identical turbo-alternators each rated at 16 kV, 125 MVA, and of synchronous reactance 1.5 p.u. Each machine supplies a 125 MVA, 0.1 p.u. transformer connected to a busbar sectioned as shown.

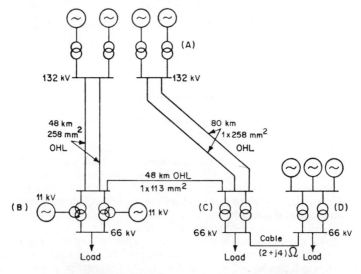

Figure 3.52 Line diagram of system in Problem 3.17.

Substation B. Two identical 150 MVA, three-winding transformers each having the following reactances between windings: 132/66 kV windings 10 per cent; 66/11 kV windings 20 per cent; 132/11 kV windings 20 per cent; all on a 150 MVA base.

The secondaries supply a common load of 200 MW at 0.9 p.f. lagging. To each tertiary winding is connected a 30 MVA synchronous compensator of synchronous reactance 1.5 p.u.

Substation C. Two identical 150 MVA transformers each of 0.15 p.u. reactance supply a common load of 300 MW at 0.85 p.f. lagging.

Generating station D. Three identical 11 kV, 75 MVA generators each of 1 p.u. synchronous reactance supply a common busbar which is connected to an outgoing 66 kV cable through two identical 100 MVA transformers. Load 50 MW, 0.8 p.f lagging.

Determine the equivalent circuit for balanced operation giving component values on a base of 100 MVA. Treat the loads as impedances.

4

Control of Power and Frequency

4.1 Introduction

Although to a certain extent the control of power and frequency is interrelated to the control of reactive power and voltage, it is hoped that by dealing with power and frequency separately from voltage control a better appreciation of the operation of power systems will be obtained. In a large interconnected system many generation stations large and small are synchronously connected and hence all have the same frequency. The following remarks refer mainly to networks in which the control of power is carried out by the decisions and actions of engineers, as opposed to systems in which the control and allocation of load to machines is effected completely automatically. The latter are sometimes based on a continuous load-flow calculation by analogue or digital computers. The allocation of the required power among the generators has to be decided before the advent of the load which must therefore be predicted. An analysis is made of the loads experienced over the same period in previous years, account is also taken of the value of the load immediately previous to the period under study and of the weather forecast. The probable load to be expected, having been decided, is allocated to the various turbine-generators.

A daily load cycle is shown in Figure 1.4. It is seen that the rate of growth of the load is very high and varies between 2 MW/min per 1000 MW of peak demand in the early morning and about 8 MW/min per 1000 MW of peak demand between 07.00 and 08.00 hours. Hence the ability of machines to increase their output quickly from zero to full load is important. It is extremely unlikely that the output of the machines at any instant will exactly equal the load on the system. If the output is higher than the demand the machines will tend to increase in speed and the frequency will rise, and vice versa. Hence the frequency is not a constant quantity but continuously varies; these variations are normally small and have no noticeable effect on most consumers. The frequency is continuously monitored against standard time-sources and when long-term tendencies to rise or fall are noticed the control engineers take appropriate action by regulating the generator outputs.

Should the total generation available be insufficient to meet the demand the frequency will fall. If the frequency falls by more than 1 Hz the reduced speed of power station pumps, fans, etc. may reduce the station output and a serious situation arises. In this type of situation, although the reduction of frequency will cause a reduction in power demand, voltage must be reduced, and if this is not sufficient loads will have to be disconnected and continue to be disconnected until the frequency rises to a reasonable level.

When a permanent increase in load occurs on the system the speed and frequency of all the interconnected generators fall, since the increased energy requirement is met from the kinetic energy of the machines. This causes an increase in steam admitted to the turbines due to the operation of the governors and hence a new load balance is obtained. Initially the boilers have a thermal reserve by means of which sudden changes can be supplied until the new firing rate has been established.

4.2 The Turbine Governor

In the following discussion of governor mechanisms both steam and water turbines will be considered together and points of difference indicated where required. A simplified schematic diagram of a traditional governor system is shown in Figure 4.1. The sensing device, which is sensitive to change of speed,

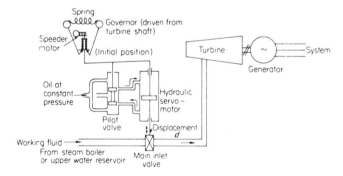

Figure 4.1   Governor control system employing the Watt governor as sensing device and a hydraulic servo-system to operate main supply valve. Speeder-motor gear determines the initial setting of the governor position.

is the time-honoured Watt centrifugal governor. In this two weights move radially outwards as their speed of rotation increases and thus move a sleeve on the central spindle. This sleeve movement is transmitted by a lever mechanism to the pilot-valve piston and hence the servo-motor is operated. A dead band is present in this mechanism, i.e. the speed must change by a certain amount before the valve commences to operate, due to friction and mechanical backlash. The time taken for the main steam valve to move due to delays in the hydraulic pilot-valve and servo-motor systems is appreciable, 0.2–0.3 s. The governor characteristic for a large steam turbo-alternator is shown in Figure 4.2 and it is seen that there is a 4 per cent drop in speed between no load and full load. Because of the requirement for high response speed, low dead band, and

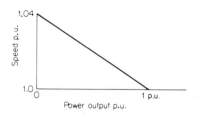

Figure 4.2   Idealized governor characteristic of a turbo-alternator.

accuracy in speed and load control the mechanical governor has been replaced in large modern turbo-generators by electro-hydraulic governing. The method usually used to measure the speed is based on a toothed wheel (on shaft) and magnetic-probe pickup. The use of electronic controls requires an electro-hydraulic conversion stage, using conventional servo valves.

An important feature of the governor system is the mechanism by means of which the governor sleeve and hence the main-valve positions can be changed and adjusted quite apart from when actuated by the speed changes. This is accomplished by the speed changer or 'speeder motor', as it is sometimes termed. The effect of this adjustment is the production of a family of parallel characteristics, as shown in Figure 4.3. Hence the power output of the generator at a given speed may be adjusted at will and this is of extreme importance when operating at optimum economy.

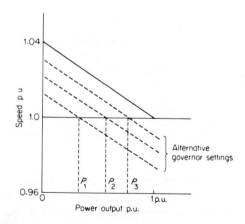

Figure 4.3   Effect of speeder gear on governor characteristics P_1, P_2, and P_3 = outputs at various settings but at same speed.

The torque of the turbine may be considered to be approximately proportional to the displacement d of the main inlet valve. Also, the expression for the change in torque with speed may be expressed approximately by the equation:

$$T = T_0(1 - kN) \qquad (4.1)$$

where T_0 is the torque at speed N_0 and T the torque at speed N; k is a constant for the governor system. As the torque depends on both the main-valve position and the speed, $T = f(d, N)$.

There is a time delay between the occurrence of a load change and the new operating conditions. This is due not only to the governor mechanism but also the fact that the new flow rate of steam or water must accelerate or decelerate the rotor in order to attain the new speed. In Figure 4.4 typical curves are shown for a turbo-generator which has a sudden decrease in the electrical

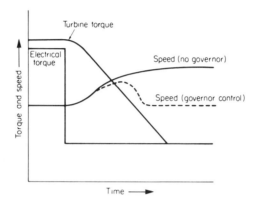

Figure 4.4   Graphs of turbine torque, electrical torque, and speed against time when the load on a generator suddenly falls.

power required, and hence the retarding torque on the turbine shaft is suddenly much smaller. In the ungoverned case the considerable time-lag between the load change and the attainment of the new steady speed is obvious. With the regulated or governed machine, due to the dead band in the governor mechanism, the speed–time curve starts to rise, the valve then operates and the fluid supply is adjusted. It is possible for damped oscillations to be set up after the load change; this will be discussed in Chapter 8.

An important factor regarding turbines is the possibility of overspeeding, when the load on the shaft is lost completely, with possible drastic mechanical breakdown. To avoid this special valves are incorporated to automatically cut off the energy supply to the turbine. In a turbo-generator normally running at 3000 rev/min this overspeed protection operates at about 3300 rev/min.

Example 4.1   An isolated 75 MVA synchronous generator feeds its own load and operates initially at no-load at 3000 rev/min, 50 Hz. A 20 MW load is suddenly applied and the steam valves to the turbine commence to open after 0.5 s due to the time-lag in the governor system. Calculate the frequency to which the generated voltage drops before the steam flow meets the new load. The stored energy for the machine is 4 kW-s per kVA of generator capacity.

Solution   For this machine the stored energy at 3000 rev/min

$$= 4 \times 75,000 = 300,000 \text{ kW-s}$$

Before the steam valves start to open the machine loses $20,000 \times 0.5 = 10,000$ kW-s of the stored energy in order to supply the load.

The stored energy $\propto$ (speed)^2. Therefore the new frequency

$$= \sqrt{\frac{300,000 - 10,000}{300,000}} \times 50 \text{ Hz}$$

$$= 49.2 \text{ Hz.}$$

4.3   Control Loops

The machine and its associated governor and voltage-regulator control systems may be represented by the block diagram shown in Figure 4.5. The nature of the voltage control loop has been discussed already in Chapter 3; accurate machine representation is required involving two-axis expressions. Here again the scope of the problem will be outlined and the reader referred to the references for further information.

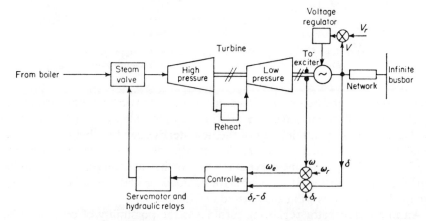

Figure 4.5   Block diagram of complete turbo-alternator control systems. The governor system is more complicated that that shown in Figure 4.1 owing to the inclusion of the load-angle δ in the control loop. Suffixes r refer to reference quantities and suffixes e to error quantities. The controller modifies the error signal from the governor by taking into account the load angle.

Two factors have a large influence on the dynamic response of the prime mover: (a) entrained steam between the inlet valves and the first stage of the turbine (in large machines this can be sufficient to cause loss of synchronism after the valves have closed); (b) the storage action in the reheater which causes the output of the low-pressure turbine to lag behind that of the high-pressure side. The transfer function $\dfrac{\text{prime mover torque}}{\text{valve opening}}$ accounting for both these effects is

$$\frac{G_1 G_2}{(1 + \tau_t p)(1 + \tau_r p)} \qquad (4.2)$$

where

$\quad G_1 =$ entrained steam constant,
$\quad G_2 =$ reheater gain constant,
$\quad \tau_t =$ entrained steam time constant,
$\quad \tau_r =$ reheater time constant.

The transfer function relating steam-valve opening to changes in speed due to the governor feedback loop is

$$\frac{\Delta d}{\Delta \omega}(p) = \frac{G_3 G_4 G_5}{(1 + \tau_g p)(1 + \tau_1 p)(1 + \tau_2 p)} \qquad (4.3)$$

where

$\quad \tau_g =$ governor-relay time constant;
$\quad \tau_1 =$ primary-relay time constant;
$\quad \tau_2 =$ secondary-valve-relay time constant;
$\quad (G_3 G_4 G_5) =$ constant relating system-valve lift to speed change.

By a consideration of the transfer function of the synchronous generator with the above expressions the dynamic response of the overall system may be obtained.

4.4 Division of Load between Generators

The use of the speed changer enables the steam input and electrical power output at a given frequency to be changed as required. The effect of this on two machines can be seen in Figure 4.6. The output of each machine is not therefore determined by the governor characteristics but can be varied to meet economical and other considerations by the operating personnel. The governor characteristics only completely decide the outputs of the machines when a sudden change in load occurs or when machines are allowed to vary their outputs according to speed within a prescribed range in order to keep the frequency constant. This latter mode of operation is known as *free-governor action*.

It has been shown in Chapter 2 that the voltage difference between the two ends of an interconnector of total impedance $R + jX$ is given by

$$\Delta V = E - V = \frac{RP + XQ}{V}$$

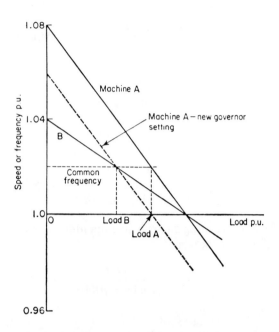

Figure 4.6 Two machines connected to an infinite busbar. The speeder gear of machine A is adjusted so that the machines load equally.

Also the angle between the voltage phasors (i.e. the transmission angle) δ is given by

$$\sin^{-1} \frac{\delta V}{E}$$

where

$$\delta V = \frac{XP - RQ}{V}$$

When $X \gg R$, i.e. for most transmission networks,

$$\delta V \propto P \quad \text{and} \quad \Delta V \propto Q.$$

Hence, (a) *the flow of power between two nodes is determined largely by the transmission angle,* (b) *the flow of reactive power is determined by the scalar voltage difference between the two nodes.*

These two facts are of fundamental importance to the understanding of the operation of power systems.

The angular advance of G_A (Figure 4.7) is due to a greater relative energy input to turbine A than to B. The provision of this extra steam (or water) to A is possible because of the action of the speeder gear without which the power

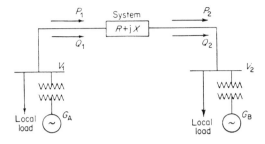

Figure 4.7 Two generating stations linked by an interconnector of impedance $(R + jX)\Omega$. The rotor of A is in phase advance of B and $V_1 > V_2$.

outputs of A and B would be determined solely by the nominal governor characteristics. The following simple example illustrates these principles.

Example 4.2 Two synchronous generators operate in parallel and supply a total load of 200 MW. The capacities of the machines are 100 MW and 200 MW and both have governor droop characteristics of 4 per cent from no load to full load. Calculate the load taken by each machine, assuming free governor action.

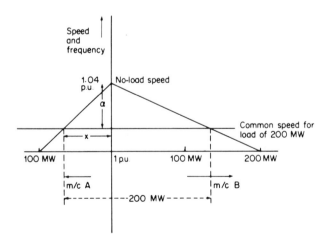

Figure 4.8 Speed–load diagram for Example 4.3.

Solution Let x megawatts be the power supplied from a 100 MW generator. Referring to Figure 4.8,

$$\frac{4}{100} = \frac{\alpha}{x}$$

For the 200 MW machine,

$$\frac{4}{200} = \frac{\alpha}{200 - x}$$

$$\therefore \quad \frac{4x}{100} = \frac{800 - 4x}{200}$$

and $x = 66.6$ MW = load on the 100 MW machine. The load on the 200 MW machine = 133.3 MW.

It will be noticed that when the governor droops are the same the machines share the total load in proportion to their capacities or ratings. Hence it is advantageous for the droops of all turbines to be equal.

Example 4.3 Two units of generation maintain 66 kV and 60 kV (line) at the ends of an interconnector of inductive reactance per phase of 40 Ω and with negligible resistance and shunt capacitance. A load of 10 MW is to be transferred from the 66 kV unit to the other end. Calculate the necessary conditions between the two ends, including the power factor of the current transmitted.

Solution

$$\delta V = \frac{XP - RQ}{V_R} = \frac{40 \times 3.33 \times 10^6}{60,000/\sqrt{3}} = 3840 \text{ V (phase)}$$

Also,

$$\frac{\delta V}{66,000/\sqrt{3}} = \sin \delta = 0.101$$

$$\therefore \quad \delta = 5° \, 44'$$

Hence the 66 kV busbars are 5° 44′ in advance of the 60 kV busbars.

$$\Delta V = \frac{6000}{\sqrt{3}} = \frac{RP + XQ}{V_R} = \frac{40Q}{60,000/\sqrt{3}}$$

Hence

$$Q = 3 \text{ MVAr per phase}$$

The p.f. angle $\phi = \tan^{-1} Q/P = 42°$ and hence the p.f. = 0.74.

4.5 The Power/Frequency Characteristic of an Interconnected System

The change in power for a given change in the frequency in an interconnected system is known as the *stiffness* of the system. The smaller the change in frequency for a given load change the stiffer the system. The power-frequency characteristic may be approximated by a straight line and $dP/df = K$, where K is a constant (MW per Hz) depending on the governor and load characteristics.

Let dP_G be the change in generation with the governors operating 'free acting' resulting from a sudden increase in load dP_L. The resultant out-of-balance power in the system

$$dP = dP_L - dP_G \tag{4.4}$$

and therefore

$$K = \frac{\mathrm{d}P_\mathrm{L}}{\mathrm{d}f} - \frac{\mathrm{d}P_\mathrm{G}}{\mathrm{d}f} \tag{4.5}$$

$\mathrm{d}P_\mathrm{L}/\mathrm{d}f$ measures the effect of the frequency characteristics of the load and $\mathrm{d}P_\mathrm{G} \propto (P_\mathrm{T} - P_\mathrm{G})$, where P_T is the turbine capacity connected to the network and P_G the output of the associated generators. When steady conditions are again reached the load P_L is equal to the generated power P_G (neglecting losses), hence, $K = K_1 P_\mathrm{T} - K_2 P_\mathrm{L}$, where K_1 and K_2 are the coefficients relevant to the turbines and load respectively.

Here, K can be determined experimentally by connecting two large separate systems by a single link, breaking the connexion and measuring the frequency change. For the British system, tests show that $K = 0.8P_\mathrm{T} - 0.6P_\mathrm{L}$ and lies between 2000 and 5500 MW per Hz i.e. a change in frequency of 0.1 Hz requires a change in the range 200–550 MW, depending on the amount of plant connected. In smaller systems the change in frequency for a reasonable load change is relatively large and electrical governors have been introduced to improve the power–frequency characteristic.

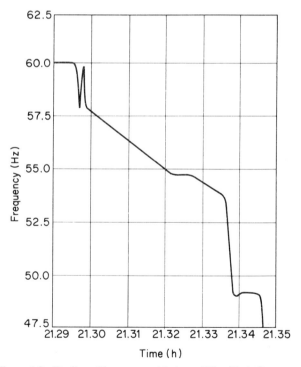

Figure 4.9   Decline of frequency with time of New York City system when isolated from external supplies. (*Copyright © 1977 by the Institute of Electrical and Electronics Engineers, Inc. Reprinted by permission from I.E.E.E. Spectrum, Vol. 15, No. 2 (Feb. 1978) pp. 38–46.*)

In 1977 owing to a series of events triggered off by lightning, New York City was cut off from external supplies and the internal generation available was much less than the city load (see Chapter 12). The resulting fall in frequency with time is shown in Figure 4.9 illustrating the power–frequency characteristics of an isolated power system.

4.6 Systems Connected by Lines of Relatively Small Capacity

Let K_A and K_B be the respective power–frequency constants of two separate power systems A and B shown in Figure 4.10. Let the change in the power

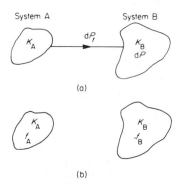

(a)

(b)

Figure 4.10(a) Two interconnected power systems connected by a tie-line. (b) The two systems with the tie-line open.

transferred from A to B when a change resulting in an out-of-balance power dP occurs in system B, be dP_t, where dP_t is positive when power is transferred from A to B. The change in frequency in system B due to an extra load dP and an extra input of dP_t from A, is $-(dP-dP_t)/K_B$ (the negative sign indicates a fall in frequency). The drop in frequency in A due to the extra load dP_t is $-dP_t/K_A$, but the changes in frequency in each system must be equal as they are electrically connected. Hence,

$$\frac{-(dP-dP_t)}{K_B} = \frac{-dP_t}{K_A}$$

$$\therefore \ dP_t = +\left(\frac{K_A}{K_A+K_B}\right) dP \qquad (4.6)$$

Next, consider the two systems operating at a common frequency f with A exporting dP_t to B. The connecting link is now opened and A is relieved of dP_t and assumes a frequency f_A, and B has dP_t more load and assumes f_B. Hence,

$$f_A = f + \frac{dP_t}{K_A} \quad \text{and} \quad f_B = f - \frac{dP_t}{K_B}$$

from which

$$\frac{dP_t}{f_A - f_B} = \frac{K_A K_B}{K_A + K_B} \qquad (4.7)$$

Hence, by opening the link and measuring the resultant change in f_A and f_B the values of K_A and K_B can be obtained.

When large interconnected systems are linked electrically to others by means of tie-lines the power transfers between them are usually decided by mutual agreement and the power controlled by regulators in accordance. As the capacity of the tie-lines is small compared with the systems, care must be taken to avoid excessive transfers of power.

Effect of governor characteristics

A fuller treatment of the performance of two interconnected systems in the steady state requires a further consideration of the control aspects of the generation process.

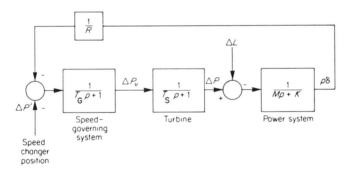

Figure 4.11 Block diagram for turbine-generator connected to a power system. (*From Economic Control of Interconnected Systems, by L. K. Kirchmayer. Copyright © 1959 John Wiley & Sons Ltd. By permission of John Wiley & Sons Inc.*)

A more complete block diagram for the steam turbine-generator connected to a power system is shown in Figure 4.11. For this system the following equation holds:

$$Mp^2\delta + Kp\delta = \Delta P - \Delta P_L$$

where M is a constant depending on inertia (see Chapter 8) in radians; K is the stiffness or damping coefficient (i.e. change of power with speed), in MW/Hz or MW per rad/s

$$= \frac{\partial \text{ (load power)}}{\partial(p\delta)} - \frac{\partial \text{ (turbine power)}}{\partial(p\delta)}$$

$p \equiv d/dt$;

ΔP and ΔP_L = change in prime mover and load powers;

δ = change from initial angular position;

R = preset speed regulation (or governor droop), i.e. drop in speed or frequency when combined machines of an area change from no load to full load, expressed as p.u. or Hz or rad/s per MW;

$\Delta P'$ = change in speed-changer setting.

Therefore, change from normal speed or frequency,

$$p\delta = \frac{1}{Mp + K}(\Delta P - \Delta P_\mathrm{L})$$

This analysis holds for steam turbine generation; for hydro turbines, the large inertia of the water must be accounted for and the analysis is more complicated.

The representation of two systems connected by a tie-line is shown in Figure 4.12. The general analysis is as before except for the additional power terms due to the tie-line. The machines in the individual power systems are considered to be closely coupled and to possess one equivalent rotor.

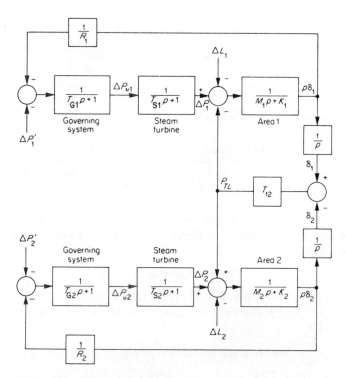

Figure 4.12   Block control diagram of two power systems connected by a tie-line. (*From Economic Control of Interconnected Systems by L. K. Kirchmayer. Copyright © 1959 John Wiley & Sons Ltd. By permission of John Wiley & Sons Inc.*)

For system (1),

$$M_1 p^2 \delta_1 + K_1 p \delta_1 + T_{12}(\delta_1 - \delta_2) = \Delta P_1 - \Delta P_{L1} \qquad (4.8)$$

where T_{12} is the synchronizing torque coefficient of the tie-line.

For system (2),

$$M_2 p^2 \delta_2 + K_2 p \delta_2 + T_{12}(\delta_2 - \delta_1) = \Delta P_2 - \Delta P_{L2} \qquad (4.9)$$

The steady-state analysis of two interconnected systems may be obtained from the transfer functions given in the block diagram.

The speed governor general response is given by

$$\Delta P_V = \frac{-1}{(T_G p + 1)} \left(\frac{1}{R} p \delta + \Delta P' \right) \qquad (4.10)$$

In the steady state from (4.10),

$$\Delta P_1 = -\frac{1}{R_1} p \delta_1 \qquad (4.11)$$

and

$$\Delta P_2 = -\frac{1}{R_2} p \delta_2 \qquad (4.12)$$

Similarly, from equations (4.8) and (4.9) in the steady state,

$$\left(K_1 + \frac{1}{R_1} \right) p \delta_1 + T_{12}(\delta_1 - \delta_2) = -\Delta P_{L1} \qquad (4.13)$$

and

$$\left(K_2 + \frac{1}{R_2} \right) p \delta_2 + T_{12}(\delta_2 - \delta_1) = -\Delta P_{L2} \qquad (4.14)$$

Adding (4.13) and (4.14),

$$\left(K_1 + \frac{1}{R_1} \right) p \delta_1 + \left(K_2 + \frac{1}{R_2} \right) p \delta_2 = -\Delta P_{L1} - \Delta P_{L2} \qquad (4.15)$$

In a synchronous system $p\delta_1 = p\delta_2 = p\delta =$ angular frequency and (4.15) becomes

$$\left[K_1 + K_2 + \left(\frac{1}{R_1} + \frac{1}{R_2} \right) \right] p\delta = -\Delta P_{L1} - \Delta P_{L2}$$

and

$$p\delta = \frac{-\Delta P_{L1} - \Delta P_{L2}}{(K_1 + K_2) + (1/R_1 + 1/R_2)} \qquad (4.16)$$

From (4.13) and (4.14)

$$T_{12}(\delta_1 - \delta_2) = \frac{-\Delta P_{L1}[K_2 + (1/R_2)] + \Delta P_{L2}[K_1 + (1/R_1)]}{[K_2 + (1/R_2) + K_1 + (1/R_1)]} \quad (4.17)$$

It is normally required to keep the system frequency constant and to maintain the interchange through the tie-line at its scheduled value. To achieve this, additional controls are necessary to operate the speed-changer settings as follows.

For area (1),

$$p\Delta P_1^1 \propto \gamma_{1t} T_{12}(\delta_1 - \delta_2) + \gamma_{1f} p\delta_1$$

$$\propto \gamma_{1t}\left[T_{12}(\delta_1 - \delta_2) + \frac{\gamma_{1f}}{\gamma_{1t}} p\delta_1 \right] \quad (4.18)$$

Similarly for area (2),

$$p\Delta P_2' \propto \gamma_{2t}\left[T_{12}(\delta_2 - \delta_1) + \frac{\gamma_{2f}}{\gamma_{2t}} p\delta_2 \right] \quad (4.19)$$

where γ_t and γ_f refer to the control constants for power transfer and frequency respectively. When a load change occurs in a given area the changes in tie-line power and frequency have opposite signs, i.e. the frequency falls for a load increase and the power transfer increases and vice versa. In the interconnected area, however, the changes have the same sign. Typical system parameters have the following orders of magnitude:

K, 0.75 p.u. on system capacity base.
T_{12}, 0.1 p.u. (10 per cent of system capacity results in 1 rad displacement between areas (1) and (2)).
R, 0.04 p.u. on a base of system capacity.
γ_f, 0.005
γ_t, 0.0009.

Example 4.4   Two power systems, A and B, each have a regulation (R) of 0.1 p.u. (on respective capacity bases) and a stiffness K of 1 p.u. The capacity of system A is 1500 MW and of B 1000 MW. The two systems are interconnected through a tie-line and are initially at 60 Hz. If there is a 100 MW load change in system A, calculate the change in the steady-state values of frequency and power transfer.

Solution

$$K_A = 1 \times 1500 \text{ MW per Hz}$$
$$K_B = 1 \times 1000 \text{ MW per Hz}$$
$$R_A = \frac{\Delta f(\text{no load to full load})}{\text{full load capacity}} = \frac{0.1 \times 60}{1500}$$

$$= 6/1500 \frac{\text{Hz}}{\text{MW}}$$

$$R_B = (6/1000) \frac{\text{Hz}}{\text{MW}}$$

From (4.16),

$$\Delta f = \frac{-\Delta P_1}{(K_1 + 1/R_1) + (K_2 + 1/R_2)}$$

$$= \frac{-100}{1500 + \dfrac{1500}{6} + 1000 + \dfrac{1000}{6}}$$

$$= \frac{-600}{17,500} = -0.034 \text{ Hz}$$

$$P_{12} = T_{12}(\delta_1 - \delta_2) = \frac{-\Delta P_1(K_2 + 1/R_2)}{(K_1 + 1/R_1) + (K_2 + 1/R_2)}$$

$$= \frac{-100\left(\dfrac{7000}{6}\right)}{10,500/6} = \frac{-7000}{105} = -6 \text{ MW}$$

Note, without the participation of governor control,

$$P_{12} = \left(\frac{-K_B}{K_A + K_B}\right) \qquad dP = -\frac{1000}{2500} \times 100 = -40 \text{ MW}$$

Frequency-bias–tie-line control

Consider three interconnected power systems A, B and C, as shown in Figure 4.13, the systems being of similar size. Assume that initially A and B export to

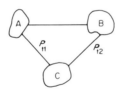

Figure 4.13   Three power systems connected by tie-lines.

C their previously agreed power transfers. If C has an increase in load the overall frequency tends to decrease and hence the generation in A, B, and C increases. This results in increased power transfers from A and B to C. These transfers, however, are limited by the tie-line power controller to the previously agreed values and therefore instructions are given for A and B to reduce generation and hence C is not helped. This is a severe drawback of what is known as straight tie-line control, which can be overcome if the systems are controlled by using a consideration of both load transfer and frequency, such that the following equation holds:

$$\sum \Delta P + K \Delta f = 0 \qquad (4.20)$$

where $\sum\Delta P$ is the net transfer error and depends on the size of the system and the governor characteristic, and Δf is the frequency error and is positive for high frequency. In the case above, after the load change in C the frequency error is negative (i.e. low frequency) for A and B and the sum of ΔP for the line AC and BC is positive. For correct control,

$$\sum\Delta P_A + K_A\Delta f = \sum\Delta P_B + K_B\Delta f = 0$$

Systems A and B take no regulating action despite their fall in frequency. In C, $\sum\Delta P_C$ is negative as it is importing from A and B and therefore the governor speeder motors in C operate to increase output and restore frequency. This system is known as frequency-bias–tie-line control.

4.7 Phase-shift Transformers

Although the generators constitute the only source of power in the system the distribution of this power around the network can be influenced by the provision of phase-changing equipment. Often in practice lines or cables in parallel would have currents which are unrelated to their thermal ratings, giving restricted capacity.

The distribution of power and reactive power in the network is decided by the impedances of the various paths. As has been already seen, voltage magnitudes decide the reactive power flow but not, to any large extent, the flow of power. To influence the power in a line a phase shift is needed and this can be obtained by the connexions shown in Figure 4.14(a). The booster arrangement shows only the injection of voltage into one phase; it is repeated for the other two phases. In Figure 4.14(b) the corresponding phasor diagram is shown and the nature of the angular shift of the voltage boost V'_{YB} indicated. By the use of tappings on the energizing transformer several values of phase shift may be obtained.

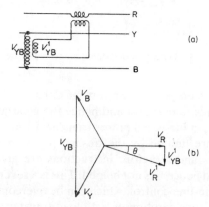

Figure 4.14(a) Connexions for one phase of a phase-shifting transformer. Similar connexions for the other two phases. (b) Corresponding phasor diagram.

4.8 Optimization of Power System Operation

Introduction

The extensive interconnexion of power sources has made the operation of a system in the most economical manner a complex subject. Economy must be balanced against considerations such as security of supply. The use of the *merit order* ensures that as far as possible the most economical generating sets are used. A knowledge of the flows of real and reactive power and other parameters in the network and effective means of dealing with the analysis of large systems is required for the operation to attain an economic optimum, although experienced operators certainly approach this aim.

The rapid expansion in the use of digital computers for load flows and fault calculations and the development of optimization techniques in control theory has resulted in much attention being given to this topic.

Apart from financial considerations, it is becoming difficult for operators to cope with the information produced by large complex systems in times of emergency such as with major faults. Computers with on-line facilities can more readily digest this information and take correct action by instructing control gear or by the suitable display of relevant information to enable human operators to take correct action. In the attempt to obtain an economic optimization the limitations of the system such as plant ratings and stability limits must be obeyed.

Optimization may be considered in a number of ways according to the time scale involved, namely: daily, yearly (especially with hydro stations) and over much longer periods when planning for future developments, although this latter is not strictly operational optimization. In an existing system the various factors involved are the fixed and variable costs. The former includes labour, administration, interest and depreciation, etc. and the latter mainly fuel. A major problem is the effective prediction of the future load whether it occurs in ten minutes, a few hours or in several years' time. It is not intended here to describe the mathematical theory of the solutions so far achieved as this would be beyond the scope and level intended in this text, but the problem will be outlined and several references given and further discussed in Appendix 1.

The following information is required.

For each generator:

(a) maximum and economic output capacities;
(b) fixed and incremental heat rates;
(c) minimum shut-down time;
(d) minimum stable outputs, maximum run-up and run-down rates.

For each station:

(a) cost and calorific value of fuel (thermal stations);
(b) factor reflecting recent operational performance of the station;

(c) minimum time between loading and unloading successive generators;

(d) constraints in station output.

For the system:

(a) load demand at given intervals for the specified period,

(b) specified constraints imposed by transmission capability,

(c) running-spare requirements,

(d) effect of transmission losses.

The input–output characteristic of a turbine is of great importance when economical operation is considered. A typical characteristic is shown in Figure 4.15. The *incremental heat rate* is defined as the slope of the input–output curve

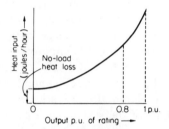

Figure 4.15 Input–output characteristic of a turbo-generator set. Often the curve above an output of 0.8 p.u. is steeper than the remainder and the machine has a maximum rating at 1 p.u. and a maximum economic rating at, say, 0.8 p.u.

at any given output. The graph of the incremental heat rate against output is known as the Willans line. For large turbines with a single valve the incremental heat-rate is approximately constant over the operating range (most steam turbines in Britain are of this type); with multi-valve turbines (as used in the U.S.A.) the Willans line is not horizontal but curves upwards and is often represented by the closest linear law. The value taken for the incremental heat rate of a generating set is sometimes complicated because if only one or two shifts are being operated (there are normally three shifts per day) heat has to be expended in banking boilers when they are not required.

Instead of plotting incremental heat rate or fuel consumption against power output for the turbine-generator the incremental fuel cost may be used. This is advantageous when allocating load to generators for optimum economy as it incorporates differences in the fuel costs of the various generating stations. Usually the graph of incremental fuel cost against power output can be approximated by a straight line (Figure 4.16). Consider two turbine-generator sets having the following different incremental fuel costs, dC_1/dP_1 and dC_2/dP_2, where C_1 is the cost of the fuel input to unit number 1 for a power output of P_1, and similarly C_2 and P_2 relate to unit number 2. It is required to load the generators to meet a given requirement in the most economical

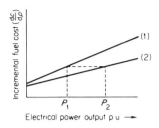

Figure 4.16 Idealized graphs of incremental fuel cost against output
for two machines sharing a load equal to P_1 and P_2.

manner. Obviously the load on the machine with the higher dC/dP will be
reduced by increasing the load taken by the machine with the lower dC/dP.
This transfer will be beneficial until the values of dC/dP for both sets are equal,
after which the machine with the previously higher dC/dP now becomes the
one with the lower value and vice versa. There is no economic advantage in
further transfer of load, and the condition when $dC_1/dP_1 = dC_2/dP_2$ therefore
gives optimum economy; this can be seen by considering Figure 4.16. The
above argument can be extended to several machines supplying a load.
Generally for optimum economy the *incremental fuel cost should be identical for
all contributing turbine-generator sets*.

The above reasoning must be modified when the distances of generating
stations from the common loads are different; here the cost of different
transmission losses will affect the argument.

As important as the transmission losses, is the optimum method of trans-
porting fuel from the production centres to the generating stations. The
transport of both electrical energy and fuel in the optimum manner forms
transportation problems which may be dealt with by special techniques (e.g. the
north west rule) or by the general method of linear programming. A good
introduction to these methods is given in reference 6.

Basic formulation of the short-term optimization problem for thermal stations

Kirchmayer[4] uses Lagrange multipliers in formulating equations including
transmission losses.

Let

P_i = power output of unit i (MW);
P_R = total load on system (MW);
P_L = transmission losses (MW);
F_T = total cost of units (money/hour);
λ = Lagrange multiplier ((money per MW-h);
n = number of generating units.

The total input to the system from all generators $= P_T = \sum_{i=1}^{n} P_i$, and

$$\left(\sum_{1}^{n} P_i\right) - P_L - P_R = 0$$

Using Lagrangian multipliers the expression,

$$\gamma = F_T - \lambda\left(\sum_1^n P_i - P_L - P_R\right) \text{ is obtained.}$$

For minimum cost (F_T), $\partial\gamma/\partial P_i = 0$ for all values of i.

$$\therefore \quad \frac{\partial\gamma}{\partial P_i} = \frac{dF_i}{dP_i} - \lambda + \lambda\frac{\partial P_L}{\partial P_i} = 0$$

i.e.
$$\frac{dF_i}{dP_i} + \lambda\frac{\partial P_L}{\partial P_i} = \lambda \tag{4.21}$$

In equation (4.21), dP_L/dP_i is the incremental transmission loss. One way of solving the equations described by (4.21) is known as the penalty factor method in which (4.21) is rewritten as

$$\frac{dF_i}{dP_i}L_i = \lambda \tag{4.22}$$

where

$$L_i = \left(\frac{1}{1 - \partial P_L/\partial P_i}\right) = \text{penalty factor of plant } i$$

An approximate penalty factor, $L_i = 1 + \partial P_L/\partial P_i$. Using this and (4.22)

$$\frac{dF_i}{dP_i} + \left(\frac{dF_i}{dP_i} \cdot \frac{\partial P_L}{\partial P_i}\right) = \lambda$$

where $i = 1 \ldots n$ (number of plants). In practice the determination of dP_L/dP_i is difficult and the use of the so-called loss or 'B' coefficients is made, i.e.

$$\frac{\partial P_L}{\partial P_i} = \sum_i 2B_{mi} P_m + B_{i0}$$

where the B coefficients are determined from the network.[4]

There are many drawbacks to the above treatment, e.g. limitations on power flows by equipment ratings, transformer settings, and maximum phase angles allowable across transmission lines on stability grounds. Also it is concerned only with active power, reactive power being neglected. A brief introduction to a more comprehensive approach using linear programming is given in Appendix 1.

4.9 Computer Control of Load and Frequency

Control of tie-lines

Automatic control of area power systems connected by tie-lines has been already discussed. The methods used will now be extended to include optimum economy as well as power transfer and frequency control. The basic systems described are typical of United States practice and have been comprehensively discussed by Kirchmayer.[4] In the previous section, methods for economic analysis and optimization as developed by Kirchmeyer[2] have been summarized. The choice of generating units to be operated is largely decided by spinning reserve, voltage control, stability, and protection requirements. The methods discussed decide the allocation of load to particular machines.

If transmission losses be neglected, it has been shown that optimum economy results when $dF_n/dP_n = \lambda$. Control equipment to adjust the governor speed-change settings such that all units comply with the appropriate value of dF_n/dP_n is incorporated in the control loops for frequency and power-transfer adjustment, as shown in Figure 4.17. The frequency and load-transfer control acts

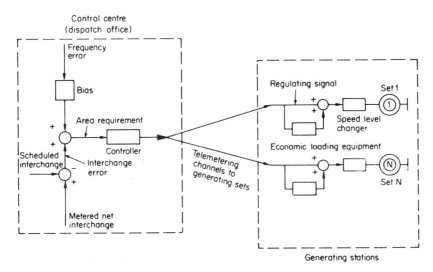

Figure 4.17 Schematic diagram of automatic control arrangements covering frequency, tie-line power flows, and economic loading of generating sets. (*From Economic Control of Interconnected Systems by L. K. Kirchmayer. Copyright © John Wiley & Sons Ltd. By permission of John Wiley & Sons Inc.*)

quickly, and once these quantities have been decided the slower acting economic controls act. For example, if an increase in load occurs in the controlled area, a signal requiring increased generation transmits through the control system. These changes alter the value of λ and cause the economic control apparatus to call for generation to be operated at the same incremental cost. Eventually the system is again in the steady state, the load change having been absorbed and all units operate at an identical value of incremental loss.

If transmission losses are included, the basic economic requirement calls for $dF_n/dP_n = \lambda/L_n$, where L_n is the penalty factor; $(1/L_n)$ signals are generated by a computer from a knowledge of system parameters in the form of the so-called B coefficients. A complete and detailed account of United States practice in the economic control of interconnected systems is given in references 2 and 4.

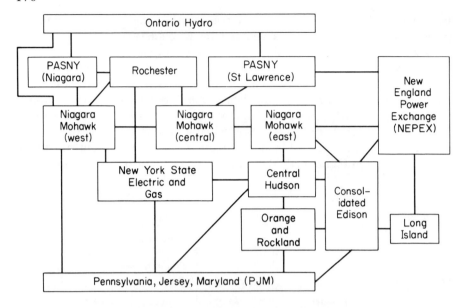

Figure 4.18   Block diagram showing the eight power companies
within the New York Power Pool. (*Copyright © 1973 by the Institute of
Electrical and Electronics Engineers, Inc. Reprinted by permission from
I.E.E.E. Spectrum, Vol. 10, No. 3, March 1973, pp. 54–61.*)

In the U.S.A. the many separate power companies are connected into power
pools. The aim is optimal control of the participating machines in each pool
within security contraints. In Figure 4.18 is shown a block diagram of the New
York Power Pool. Among the advantages of such pools are; the more economic
use of large generators, emergency assistance to neighbouring utilities, reduced
spinning reserves, and lower overall generation costs. Each pool is connected
to others by tie-lines, e.g. the New York Pool to Ontario Hydro, New England
Power Exchange and the Pennsylvania–New Jersey–Maryland interconnexion.
Control of the generators in the pool is achieved either centrally or from each of
the constituent areas (i.e. area dispatch centres) as indicated in Figure 4.19.

The central control mechanisms for pools are basically the same as for areas
and the systems of Figure 4.17 may be used with 'pool' substituted for 'area'.
Allocation between generators in an area or between areas in a pool may be
accomplished by the use of base points or loadings and participation factors.
The former gives the economic allocation for a specified total generation and
are normally established every few minutes or when loadings change. When the
generation allocations are established they are compared with the actual values
being generated and a control error formed. The unit participation factor (K_n)
for any unit (n) in the pool is given by:

$$K_n = \left(\frac{1}{F_n L_n}\right)\bigg/ \sum \frac{1}{F_n L_n} \qquad n = 1, 2, 3, \ldots$$

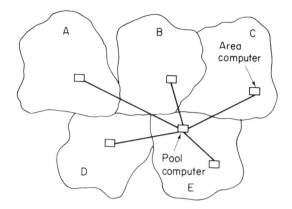

Figure 4.19 Multicomputer configuration formed by a five-area pool. (*Copyright © 1973 by the Institute of Electrical and Electronics Engineeers, Inc. Reprinted by permission from I.E.E.E. Spectrum, Vol. 10, No. 3, March 1973, pp. 54–61.*)

where F_n = slope of the incremental cost curve for unit n and L_n = penalty factor for unit n. The above method may be extended with the control of areas from the pool centre replacing that of the individual units; the areas in turn controlling the units. Area base points are the sum of the unit base points in that area and represent the economic area allocation for a specified total pool generation. This is achieved by a multi-area dispatch program which is run every few minutes and accounts for limits on interchanges and tie-lines, as well as the usual parameters. In this case the additional pool generation (ΔG) required between updating base points is allocated among areas according to the area participation factors defined as follows:

$$K_a = \frac{\Delta P_a}{\Delta G}$$

where ΔP_a is the power allocated to area a. At the present time the electric capacity of some pools approaches the capacities of state-run centrally controlled European supply organizations. It is anticipated that with further increase in loads that in the U.S. super-pools may be set up, forming another layer in the hierarchy.

In Britain, using the present non-automatic methods, there is a frequency error of less than 0.1 Hz for 90 per cent of the day. For the frequency to be automatically controlled a range of controlled generation of $\pm(300\text{–}500)$ MW would be required to correct the frequency error and this must be spread over several generating stations to avoid uneconomic power flows. A possible scheme would be for the selected machines to be fitted with turbine regulators which control the speeder motor gear and hence the power output. These regulators would be controlled by means of telemetering pulses transmitted from a central system controller.

172

References

BOOKS

1. Sullivan R. L., *Power System Planning*, McGraw-Hill, New York, 1977.
2. Kirchmayer, L. K., *Economic Control of Interconnected Systems*. Available from L. K. Kirchmayer, 27 St. Stephens Lane, Scotia, New York, 12302, U.S.A.
3. Chestnut, H., and R. W. Mayer, *Servomechanisms and Regulating System Design*, Vol. 1, Wiley, New York, 1951.
4. Kirchmayer, L. K., *Economic Operation of Power Systems*, Wiley, New York, 1958.
5. Kingsley, C., A. E. Fitzgerald, and A. Kusko, *Electric Machinery*, McGraw-Hill, New York, 3rd edn., 1971.
6. Metzer, K. W., *Elementary Mathematical Programming*, Wiley, New York, Science edn., 1963.
7. Scravulescu S. C. (Ed.) *Computerized Operation of Power Systems*, Elsevier, Amsterdam, 1976.

ARTICLES AND PAPERS

8. Kidd, W. L., 'Load and frequency control', *Elec. J.*, London, **1961**.
9. Moran, F., 'Power systems automatic frequency control techniques', *Proc. I.E.E.*, **106A** (1959).
10. Walker, P. A. W., and A. S. Aldred, 'Frequency—response analysis of displacement governing in synchronous power systems', *Proc. I.E.E.*, **108C** (1961).
11. Broadbent, D., and K. N. Stanton, 'An analytical review of power-system frequency, time and tie-line control', *Proc. I.E.E.*, **108C** (1961).
12. Coles, H. E., 'Effects of prime-mover governing and voltage regulation on turbo-alternator performance', *Proc. I.E.E.*, **112** (1965).
13. 'Automation in the electricity supply industry', *I.E.E. Conf. Rept.*, **1962**.
14. Carpentier, J., 'Contribution à l'étude du dispatching economique', *Bulletin de la Société Française des Electriciens*, Ser. 8, **3** (1962).
15. Carpenteir, J., and J. Sirioux, 'L'optimization de la production à l'électricité de France', *Bulletin de la Société Française des Electriciens*, **1963**.
16. Dopazo, J. F., O. A. Klitin, G. W. Stagg, and M. Watson, 'An optimization technique for real and reactive power allocation', *Trans. I.E.E.E.*, *P.A.S.*, **55**, No. 11 (1967).
17. Kuhn, H. W., and A. W. Tucker, 'Non-linear programming', *Proc. of Second Berkeley Symposium on Math. Stat. & Prob.*, **1951**, University of California Press, Berkeley, U.S.A.
18. Moskalev, A. G., 'Principles of the most economical distribution of the active and reactive loads in automatically controlled power systems', *Elektrichestvo*, **1963**, No. 12, 24–33.
19. Fouad, A. A. *et al.*, 'Effect of coordinated correction of tie line bias control in interconnected power system operation', *I.E.E.E. Trans.*, **PAS-101** (1982), 1134.
20. Lukie, V. P., 'Optimal operating policy for energy storage', *I.E.E.E. Trans.*, **PAS-101** (1982), 3295.
21. Wells, D. W., 'Method for economic secure loading of a power system', *Proc. I.E.E.*, **115**, No. 8 (1968) 1190.
22. Kennedy, T., and A. G. Hoffman, 'On-line digital computer application techniques for complex electric systems dispatch', *I.E.E.E. Trans.*, *P.A.S.*, **87** (1968), 67.
23. Sasson, A. M., 'Non-linear programming solutions for load-flow, minimum-loss, and economic dispatching problems', *I.E.E.E. Trans.*, *P.A.S.*, **88** (1969), 399.
24. El-Abiad, A. H., and F. J. Jaimes, 'A method for optimum scheduling of power and voltage magnitude', *I.E.E.E. Trans.*, *P.A.S.*, **88** (1969), 413.

25. Rindt, L. J., 'Economic scheduling of generation for planning studies', *I.E.E.E. Trans., P.A.S.*, **88** (1969) 801.
26. Cohn, N., S. B. Biddle, Jr., R. G. Lex, Jr., E. H. Preston, C. W. Ross, and D. R. Whitten, 'On-line computer applications in the electric power industry', *Proc. I.E.E.E.*, **58**, No 1 (1970), 78.
27. Happ, H. H., 'Optimal power dispatch—a comprehensive survey', *I.E.E.E. Trans.*, **PAS-96** (1977), 841–853.
28. Happ, H. H. 'Power pools and superpools', *I.E.E.E. Spectrum* (March 1973) 54–61.
29. Isoda, H., 'On-line despatching method considering load variation characteristics', *I.E.E.E. Trans.*, **PAS-101** (1982), 2925.
30. Billington, R., and S. S. Sachdeva, 'Real and reactive power optimization by suboptimum techniques', *I.E.E.E. Trans.*, **PAS-92** (1973) 950–956.
31. Happ, H. H. *et al.*, 'Large-scale hydro-thermal unit commitment', *I.E.E.E. Trans.*, **PAS-90** (1971), 1373–1384.
32. Delson, J. K., 'Controlled emission dispatch, *I.E.E.E. Trans.*, **PAS-93** (1974), 1359–1366.
33. Ham, P. A. L., 'Electronics in the control of turbine generators', *I.E.E. Electronics & Power*, **May 1978,** 365–369.

Problems

4.1 Two identical 60 MW synchronous generators operate in parallel. The governor settings on the machines are such that they have 4 per cent and 3 per cent droops (no-load to full-load percentage speed drop). Determine (a) the load taken by each machine for a total of 100 MW; (b) the percentage adjustment in the no-load speed to be made by the speeder motor if the machines are to share the load equally.
(Answer: (a) 42.8 and 57.2 MW; (b) 0.83 per cent increase in no-load speed on the 4 per cent droop machine)

4.2. Two generating stations A and B are linked by a line and two transformers of total reactance 20 Ω referred to 132 kV and negligible resistance. A load of 100 MW, 0.9 p.f. lagging is taken from the busbars of A and 200 MW, 0.85 p.f. lagging from B. Determine the phase angle between the busbars of A and B and the voltage adjustment required to equalize the load on each station. Initially both stations have busbar voltages of 11 kV which are in phase.
(Answer: 3° 9'; 5.8 kV increase on 132 kV side of A)

4.3 The incremental fuel costs of two units in a generating station are as follows:

$$\frac{dF_1}{dP_1} = 0.003P_1 + 0.7$$

$$\frac{dF_2}{dP_2} = 0.004P_2 + 0.5$$

where F is in £/h and P is in MW.

Assuming continuous running with a total load of 150 MW calculate the saving per hour obtained by using the most economical division of load between the units as compared with loading each equally. The maximum and minimum operational loadings are the same for each unit and are 125 MW and 20 MW.
(Answer: $P_1 = 57$ MW, $P_2 = 93$ MW; saving £1.14 per hour)

4.4 For the two generating units in Problem 4.3, plot incremental fuel cost for maximum economy against total load, over the range 40–250 MW. (Note the maximum and minimum limits to machine outputs.)

4.5 Two power systems A and B are interconnected by a tie-line and have power-frequency constants K_A and K_B MW per Hz. An increase in load of 500 MW on system A causes a power transfer of 300 MW from B to A. When the tie-line is open the frequency of system A is 49 Hz and of system B 50 Hz. Determine the values of K_A and K_B, deriving any formulae used.

(Answer: K_A 500 MW/Hz; K_B 750 MW/Hz)

4.6 Two power systems, A and B, having capacities of 3000 and 2000 MW respectively, are interconnected through a tie-line and both operate with frequency-bias–tie-line control. The frequency bias for each area is 1 per cent of the system capacity per 0.1 Hz frequency deviation. If the tie-line interchange for A is set at 100 MW and for B set (incorrectly) at 200 MW, calculate the steady state change in frequency.

(Answer: 0.6 Hz. Use $\Delta P_A + \sigma_A \Delta f = \Delta P_B + \sigma_B \Delta f$)

Control of Voltage and Reactive Power

5.1 Introduction

The approximate relationship between the scalar voltage difference between two nodes in a network and the flow of reactive power was shown in Chapter 2 to be

$$\Delta V = \frac{RP + XQ}{V} \tag{2.6}$$

Also it was shown that the transmission angle is proportional to

$$\frac{XP - RQ}{V} \tag{2.7}$$

For networks where $X \gg R$, i.e. most power circuits, ΔV, the voltage difference, determines Q. Consider the simple interconnector linking two generating stations A and B as shown in Figure 4.7. The machine at A is in phase advance of that at B and V_1 is greater than V_2, hence there is a flow of power and reactive power from A to B. This can be seen from the phasor diagram shown in Figure 5.1. It is seen that I_d and hence P is determined by $\angle\delta$

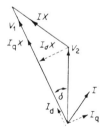

Figure 5.1 Phasor diagram for system shown in Figure 4.7; $V_1 > V_2$. Resistance of line zero, inductive reactance X ohms. I_d and I_q in-phase and quadrature components of the current **I**.

and I_q and hence Q mainly by $V_1 - V_2$. In this case $V_1 > V_2$ and reactive power is transferred from A to B; if by varying the generator excitations such that $V_2 > V_1$, the direction of the reactive power is reversed as shown in Figure 5.2.

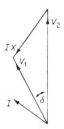

Figure 5.2　Phasor diagram for system in Figure 4.7. $V_2 > V_1$.

Hence power can be sent from A to B or B to A by suitably adjusting the amount of steam (or water) admitted to the turbine and reactive power can be sent in either direction by adjusting the voltage magnitudes. These two operations are approximately independent of each other if $X \gg R$ and the flow of reactive power can be studied almost independently of the power flow. The phasor diagrams show that if a scalar voltage difference exists across a largely reactive link the reactive power flows towards the node of lower voltage. From another point of view, if in a network there is a deficiency of reactive power at a point, this has to be supplied from the connecting lines and hence the voltage at that point falls. Conversely if there is a surplus of reactive power generated (e.g. lightly loaded cables absorb leading or negative vars and hence generate positive vars) then the voltage will rise. This is a convenient way of expressing the effect of the power factor of the transferred current, and although it may seem unfamiliar initially, the ability to think in terms of var flows instead of exclusively with power factors and phasor diagrams, will make the study of power networks much easier.

If it can be arranged that Q_2 in the system in Figure 4.7 is zero, then there will be no voltage drop between A and B, a very satisfactory state of affairs. Assuming that V_1 is constant, consider the effect of keeping V_2 and hence the voltage drop ΔV constant. From equation (2.6)

$$Q_2 = \frac{V_2 \cdot \Delta V - R \cdot P_2}{X} = K - \frac{R}{X} P_2 \qquad (5.1)$$

where K is a constant.

If this value of Q_2 does not exist naturally in the circuit then it will have to be obtained by artificial means such as the connexion at B of capacitors or inductors. If the value of the power changes from P_2 to P_2^1 and if V_2 remains constant then the reactive power at B must change to Q_2^1 such that

$$Q_2^1 - Q_2 = \frac{R}{X}(P_2^1 - P_2)$$

i.e. an increase in power causes an increase in reactive power. The change, however, is proportional to (R/X) which is normally small. It is seen that voltage can be controlled by the injection into the network of reactive power of the correct sign. Other methods of a more obvious kind for controlling voltage are the use of tap-changing transformers and voltage boosters.

5.2 The Generation and Absorption of Reactive Power

In view of the findings in the previous section a review of the characteristics of a power system from the viewpoint of reactive power is now appropriate.

Synchronous generators

These can be used to generate or absorb reactive power. The limits on the capability for this can be seen in Figure 3.14. The ability to supply reactive power is determined by the short-circuit ratio (1/synchronous reactance) as the distance between the power axis and the theoretical stability-limit line in Figure 3.14 is proportional to the short-circuit ratio. In modern machines the value of this ratio is made low for economic reasons and hence the inherent ability to operate at leading power factors is not large. For example a 200 MW, 0.85 p.f. machine with a 10 per cent stability allowance has a capability of 45 MVAr at full power outputs. The var capacity can, however, be increased by the use of continuously acting voltage regulators, as explained in Chapter 3. An over-excited machine, i.e. one with greater than normal excitation, generates reactive power whilst an under-excited machine absorbs it (or generates negative or leading vars). The generator is the main source of supply to the system of both positive and negative vars.

Overhead lines and transformers

When fully loaded, lines absorb reactive power. With a current I amperes for a line of reactance per phase X ohms the vars absorbed are $I^2 X$ per phase. On light loads the shunt capacitances of longer lines may become predominant and the lines become var generators.

Transformers always absorb reactive power. A useful expression for the quantity may be obtained for a transformer of reactance X_T p.u. and a full-load rating of $3V \cdot I_{rated}$.

The ohmic reactance

$$= \frac{V \cdot X_T}{I_{rated}}$$

Therefore the vars absorbed

$$= 3 \cdot I^2 \cdot \frac{V \cdot X_T}{I_{rated}}$$

$$= 3 \cdot \frac{I^2 V^2}{(IV)_{rated}} \cdot X_T = \frac{(VA \text{ of load})^2}{\text{Rated } VA} \cdot X_T$$

Cables

Cables are generators of reactive power owing to their high capacitance. A 275 kV, 240 MVA cable produces 6.25 to 7.5 MVAr per km; a 132 kV cable roughly 1.9 MVAr per km and a 33 kV cable, 0.125 MVAr per km.

Loads

A load at 0.95 power factor implies a reactive power demand of 0.33 kVAr per kW of power, which is more appreciable than the mere quoting of the power factor would suggest. In planning a network it is desirable to assess the reactive power requirements to ascertain whether the generators are able to operate at the required power factors for the extremes of load to be expected. An example of this is shown in Figure 5.3 where the reactive losses are added for each item until the generator power factor is obtained.

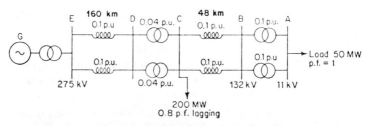

Figure 5.3   Radial transmission system with intermediate loads. Calculation of reactive-power requirement.

Example 5.1   In the radial transmission system shown in Figure 5.3 all p.u. values are referred to the voltage bases shown and 100 MVA. Determine the power factor at which the generator must operate.

Solution   Voltage drops in the circuits will be neglected and the nominal voltages assumed.

Busbar A,

$$P = 0.5 \text{ p.u.} \qquad Q = 0$$

I^2X loss in 132 kV lines and transformers

$$= \frac{P^2 + Q^2}{V^2} X_{CA} = \frac{0.5^2}{1^2} \cdot 0.1$$

$$= 0.025 \text{ p.u.}$$

Busbar C,

$$P = 2 + 0.5 = 2.5 \text{ p.u.}$$

$$Q = 1.5 + 0.025 \text{ p.u.}$$

$$= 1.525 \text{ p.u.}$$

I^2X loss in 275 kV lines and transformers

$$= \frac{2.5^2 + 1.525^2}{1^2} 0.07$$

$$= 0.6 \text{ p.u.}$$

The I^2X loss in the large generator-transformer will be negligible so that the generator must deliver $P = 2.5$ and $Q = 2.125$ p.u. and operate at a power factor of 0.76 lagging. It is seen in this example that starting with the consumer load the vars for each circuit in turn are added to obtain the total.

5.3 Relation Between Voltage, Power, and Reactive Power at a Node

The phase voltage V at node is a function of P and Q at that node, i.e.

$$V = \phi(P, Q)$$

The voltage is also dependent on adjacent nodes and the present treatment assumes that these are infinite buses.

The total differential of V,

$$dV = \frac{\partial V}{\partial P} \cdot dP + \frac{\partial V}{\partial Q} \cdot dQ$$

and using

$$\frac{\partial P}{\partial V} \cdot \frac{\partial V}{\partial P} = 1 \quad \text{and} \quad \frac{\partial Q}{\partial V} \cdot \frac{\partial V}{\partial Q} = 1$$

$$dV = \frac{dP}{(\partial P / \partial V)} + \frac{dQ}{(\partial Q / \partial V)} \tag{5.2}$$

It can be seen from equation (5.2) that the change in voltage at a node is defined by the two quantities,

$$\left(\frac{\partial P}{\partial V}\right) \quad \text{and} \quad \left(\frac{\partial Q}{\partial V}\right).$$

As an example consider a line with series impedance $(R + jX)\Omega$ and zero shunt admittance. From equation (2.6),

$$(V_1 - V)V - PR - XQ = 0 \tag{5.3}$$

where V_1, the sending-end voltage, is constant and V the receiving-end voltage depends on P and Q (Figure 5.4).

From equation (5.3)

$$\frac{\partial P}{\partial V} = \frac{V_1 - 2V}{R} \tag{5.4}$$

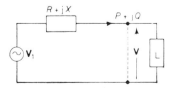

Figure 5.4   Single-phase equivalent circuit of a line supplying a load of $P + jQ$ from an infinite busbar of voltage V_1.

Also,

$$\frac{\partial Q}{\partial V} = \frac{V_1 - 2V}{X} \qquad (5.5)$$

Hence,

$$dV = \frac{dP}{\partial P/\partial V} + \frac{dQ}{\partial Q/\partial V}$$

$$= \frac{dP \cdot R + dQ \cdot X}{V_2 - 2V} \qquad (5.6)$$

For constant V and ΔV, $R\,dP + X\,dQ = 0$ and $dQ = -(R/X)dP$ which is obtainable directly from (5.1).

Normally $\partial Q/\partial V$ is the quantity of greater interest. It can be found experimentally using a network analyser (see Chapter 6) by the injection of a known quantity of vars at the node in question and measuring the difference in voltage produced. From the results obtained,

$$\frac{\Delta Q}{\Delta V} = \frac{Q_{\text{after}} - Q_{\text{before}}}{V_{\text{after}} - V_{\text{before}}}$$

ΔV should be small for this test, a few per cent of the normal voltage.

From the expression,

$$\frac{\partial Q}{\partial V} = \frac{V_1 - 2V}{X}$$

proved for the line, it is evident that the smaller the reactance associated with a node the larger the value of $\partial Q/\partial V$ for a given voltage drop, i.e. the voltage drop is inherently small. The greater the number of lines meeting at a node the smaller the resultant reactance and the larger the value of $\partial Q/\partial V \cdot \partial Q/\partial V$ obviously depends on the network configuration, but a high value would lie in the range 10–15 MVAr/kV. If the natural voltage drop at a point without the artificial injection of vars is, say, 5 kV and the value of $\partial Q/\partial V$ at this point is 10 MVAr/kV, then to maintain the voltage at its no-load level would require 5×10 or 50 MVAr. Obviously the greater the value of $\partial Q/\partial V$ the more expensive it becomes to maintain voltage levels by injection of reactive power.

$\partial Q/\partial V$ *and the short-circuit current at a node*   It has been shown that for a connector of reactance X ohms with a sending-end voltage V_1 and a received voltage V per phase,

$$\frac{\partial Q}{\partial V} = \frac{V_1 - 2V}{X} \qquad (5.5)$$

If the three phases of the connector are now short-circuited at the receiving end (i.e. three-phase symmetrical short circuit applied) the current flowing in the lines

$$= \frac{V_1}{X} \text{ amperes, assuming } R \ll X$$

With the system on no load

$$V = V_1 \quad \text{and} \quad \frac{\partial Q}{\partial V} = -\frac{V_1}{X}$$

Hence the magnitude of $(\partial Q / \partial V)$ is equal to the short-circuit current; the sign decides the nature of the reactive power. With normal operation V is within a few per cent of V_1 and hence the value of $\partial V / \partial Q$ at $V = V_1$ gives useful information regarding reactive power/voltage characteristics for small excursions from the nominal voltage. This relationship is especially useful as the short-circuit current will normally be known at all substations.

Example 5.2   Three supply points A, B, and C are connected to a common busbar M. Supply point A is maintained at a nominal 275 kV and is connected to M through a 275/132 kV transformer (0.1 p.u. reactance) and a 132 kV line of reactance 50 Ω. Supply point B is nominally at 132 kV and is connected to M through a 132 kV line of 50 Ω reactance. Supply point C is nominally at 275 kV and is connected to M by a 275/132 kV transformer (0.1 p.u. reactance) and a 132 kV line of 50 Ω reactance.

If at a particular system load the line voltage at M falls below its nominal value by 5 kV, calculate the magnitude of the reactive volt ampere injection required at M to re-establish the original voltage.

The p.u. values are expressed on a 500 MVA base and resistance may be neglected throughout.

Solution   The line diagram and equivalent single-phase circuit are shown in Figures 5.5 and 5.6.

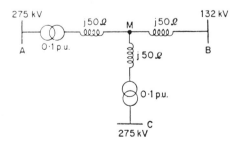

Figure 5.5   Schematic diagram of the system for Example 5.2.

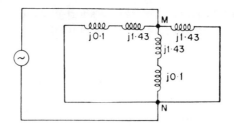

Figure 5.6 Equivalent single-phase network with the node M short-circuited to neutral (refer to Chapter 7 for full explanation of the derivation of this circuit).

It is necessary to determine the value of $\partial Q/\partial V$ at the node or busbar M; hence the current flowing into a three-phase short-circuit at M is required.

The base value of reactance in the 132 kV circuit

$$=\frac{132^2 \times 1000}{500,000} = 35\,\Omega$$

Therefore the line reactances

$$=\frac{j50}{35} = j1.43 \text{ p.u.}$$

The equivalent reactance from M to N $= j0.5$ p.u.

Hence the fault MVA at M

$$=\frac{500}{0.5} = 1000 \text{ MVA}$$

and the fault current

$$=\frac{1000 \times 10^6}{\sqrt{(3)} \times 132,000}$$

$$= 4380 \text{ A at zero power factor lagging.}$$

It has been shown that $\partial Q_M/\sqrt{(3)}\partial V_M =$ three-phase short-circuit current, when Q_M and V_M are three-phase and line values

$$\therefore \frac{\partial Q_M}{\partial V_M} = 7.6 \text{ MVAr/kV}$$

The natural voltage drop at M $= 5$ kV. Therefore the value of the injected vars required, ΔQ_M, to offset this drop

$$= 7.6 \times 5 = 38 \text{ MVAr}$$

5.4 Methods of Voltage Control (i)—Injection of Reactive Power

The background to this method has been given in the previous sections. This is the most fundamental method, but in transmission systems it lacks the flexibility and economy of transformer tap changing. Hence it is only used in schemes when transformers alone will not suffice. The provision of static capacitors to improve the power factors of factory loads has been long established. The capacitance required for the power-factor improvement of loads for optimum economy is determined as follows.

Let the tariff of a consumer be

$$\text{\pounds}A \times \text{kVA} + B \times \text{kWh}$$

A load of P_1 kilowatts at power factor ϕ_1 lagging has a kVA of $P_1/\cos\phi_1$. If this power factor is improved to $\cos\phi_2$, the new kVA is $P_1/\cos\phi_2$. The saving is therefore

$$\text{\pounds}P_1 A\left(\frac{1}{\cos\phi_1} - \frac{1}{\cos\phi_2}\right).$$

The reactive power required from the correcting capacitors

$$= P_1 \tan\phi_1 - P_1 \tan\phi_2 \text{ kVAr}$$

Let the cost per annum in interest and depreciation on the capacitor installation be $\text{\pounds}C$ per kVAr or

$$\text{\pounds}C(P_1 \tan\phi_1 - P_1 \tan\phi_2)$$

The net saving

$$= \text{\pounds}\left[AP_1\left(\frac{1}{\cos\phi_1} - \frac{1}{\cos\phi_2}\right) - CP_1(\tan\phi_1 - \tan\phi_2)\right]$$

This saving is a maximum when

$$\frac{d\,(\text{saving})}{d\phi_2} = 0$$

i.e. when $\sin\phi_2 = C/A$.

It is interesting to note that the optimum power factor is independent of the original one. The improvement of load power factors in such a manner will obviously alleviate the whole problem of var flow in the transmission system.

The main effect of transmitting power at non-unity power factors is as follows. It is evident from equation (2.6) that the voltage drop is largely determined by the reactive power (Q). The line currents are larger giving increased I^2R losses and hence reduced thermal capability. One of the obvious places for the artificial injection of reactive power is at the loads themselves. In general three methods of injection are available, involving the use of:

(a) static shunt capacitors;

(b) static series capacitors;

(c) synchronous compensators.

184

Shunt capacitors and reactors

Shunt capacitors are used for lagging power-factor circuits, whereas reactors are used on those with leading power factors such as created by lightly loaded cables. In both cases the effect is to supply the requisite reactive power to maintain the values of the voltage. Capacitors are connected either directly to a busbar or to the tertiary winding of a main transformer and are disposed along the route to minimize the losses and voltage drops. Unfortunately, as the voltage falls the vars produced by a shunt capacitor or reactor fall, thus when needed most their effectiveness falls. Also on light loads when the voltage is high the capacitor output is large and the voltage tends to rise to excessive levels.

Series capacitors

These are connected in series with the line conductors and are used to reduce the inductive reactance between the supply point and the load. One major drawback is the high overvoltage produced when a short-circuit current flows through the capacitor and special protective devices are incorporated (e.g. spark gaps). The phasor diagram for a line with a series capacitor is shown in Figure 5.7(b).

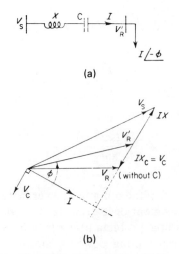

Figure 5.7(a)   Line with series capacitor. (b) Phasor diagram.

The relative merits between shunt and series capacitors may be summarized as follows:

(a) If the load var requirement is small, series capacitors are of little use.

(b) With series capacitors the reduction in line current is small, hence if thermal considerations limit the current little advantage is obtained and shunt compensation should be used.

(c) If voltage drop is the limiting factor, series capacitors are effective; also voltage fluctuations due to arc furnaces, etc. are evened out.

(d) If the total line reactance is high, series capacitors are very effective and stability is improved.

Synchronous compensators

A synchronous compensator is a synchronous motor running without a mechanical load, and depending on the value of excitation it can absorb or generate reactive power. As the losses are considerable compared with static capacitors the power factor is not zero. When used with a voltage regulator the compensator can automatically run overexcited at times of high load and underexcited at light load. A typical connexion of a synchronous compensator is shown in Figure 5.8 and the associated voltage–var output characteristic in

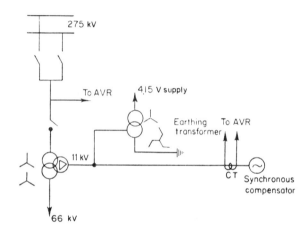

Figure 5.8 Typical installation with synchronous compensator connected to tertiary (delta) winding of main transformer. A neutral point is provided by the earthing transformer shown. The automatic voltage regulator on the compensator is controlled by a combination of the voltage on the 275 kV system and the current output; this gives a droop to the voltage–var output curve which may be varied as required.

Figure 5.9. The compensator is run up as an induction motor in 2.5 min and then synchronized.

A great advantage is the flexibility of operation for all load conditions. Although the cost of such installations is high in some circumstances it is justified, for example, at the receiving-end busbar of a long high-voltage line where transmission at power factors less than unity cannot be tolerated. Static shunt capacitors would then be used on the lower voltage network as required.

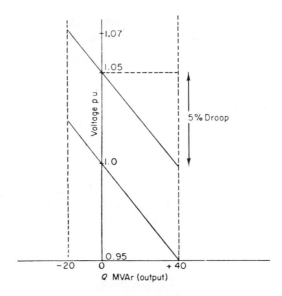

Figure 5.9   Voltage-reactive power output of a typical synchronous
compensator.

5.5   Methods of Voltage Control (ii)—Tap-changing Transformers

The basic operation of the tap-changing transformer has been discussed in
Chapter 3; by changing the transformation ratio the voltage in the secondary
circuit is varied and voltage control obtained. This constitutes the most popular
and widespread form of voltage control at all voltage levels.

Consider the operation of a radial transmission system with two tap-
changing transformers, as shown in the equivalent single-phase circuit of
Figure 5.10. Here, t_s and t_r are fractions of the nominal transformation ratios,
i.e. the tap ratio/nominal ratio. For example, a transformer of nominal ratio
6.6 to 33 kV when tapped to give 6.6 to 36 kV has a $t_s = 36/33 = 1.09$. V_1
and V_2 are the nominal voltages, at the ends of the line the actual voltages are
$t_s V_1$ and $t_r V_2$. It is required to determine the tap-changing ratios required to
completely compensate for the voltage drop in the line. The product $t_s t_r$ will be

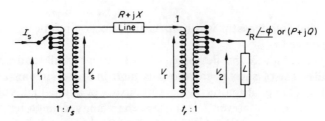

Figure 5.10 (a)   Coordination of two tap-changing transformers in a
radial transmission link.

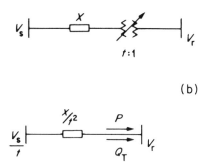

(b)

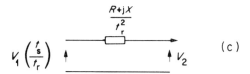

$$\frac{R+jX}{r_r^2}$$

$V_1\left(\frac{t_s}{t_r}\right)$ ↑   ↑ V_2 (c)

Figure 5.10 (b) and (c) Equivalent circuits for dealing with off-nominal tap ratio. (b) single transformer (c) two transformers.

made unity; this ensures that the overall voltage level remains in the same order and that the minimum range of taps on both transformers is used.

Note all values are in per unit. t is the off-nominal tap ratio.

Transfer all quantities to the load circuit. The line impedance becomes $(R+jX)/t_r^2$; $V_s = V_1 t_s$ and as the impedance has been transferred, $V_r = V_1 t_s$. The input voltage to the load circuit becomes $V_1 t_s/t_r$ and the equivalent circuit is as shown in Figure 5.10(c). The arithmetic voltage drop

$$= (V_1 t_s/t_r) - V_2 \doteq \frac{RP + XQ}{t_r^2 V_2}$$

When, $t_r = 1/t_s$,

$$t_s^2 V_1 V_2 - V_2^2 = (RP + XQ)t_s^2$$

and

$$V_2 = \tfrac{1}{2}[t_s^2 V_1 \pm t_s(t_s^2 V_1^2 - 4(RP + XQ))^{\frac{1}{2}}] \qquad (5.10)$$

Hence if t_s is specified there are two values of V_2 for a given V_1.

Example 5.3 A 132 kV line is fed through an 11/132 kV transformer from a constant 11 kV supply. At the load end of the line the voltage is reduced by another transformer of nominal ratio 132/11 kV. The total impedance of the line and transformers at 132 kV is $(25+j66)\ \Omega$. Both transformers are equipped with tap-changing facilities which are so arranged that the product of the two off-nominal settings is unity. If the load on the system is 100 MW at 0.9 p.f. lagging calculate the settings of the tap-changers required to maintain the voltage of the load busbar at 11 kV. Use a base of 100 MVA.

188

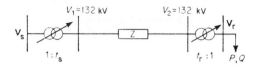

Figure 5.11   Schematic diagram of system for Example 5.3.

Solution   The line diagram is shown in Figure 5.11. As the line voltage drop is to be completely compensated $V_1 = V_2 = 132$ kV = 1 p.u. Also $t_s \times t_r = 1$. The load is 100 MW, 48.3 MVAr., i.e. $1 + j0.483$ p.u.

Using equation (5.10),

$$1 = \tfrac{1}{2}[(t_s^2 \cdot 1) \pm t_s(t_s^2 - 4(0.14 \times 1 + 0.38 \times 0.48))^{\frac{1}{2}}]$$

$$\therefore \ 2 = t_s^2 \pm t_s(t_s^2 - 1.28)^{\frac{1}{2}}$$

$$\therefore \ (2 - t_s^2)^2 = t_s^2(t_s^2 - 1.28)$$

Hence, $t_s = 1.21$ and $t_r = 1/1.21 = 0.83$

These settings are large for the normal range of tap-changing transformers (usually not more than ± 20 per cent tap range). It would be necessary in this system to inject vars at the load end of the line to maintain the voltage at the required value.

It is important to note that the transformer does not improve the var-flow position and also that the current in the supplying line is increased if the ratio is increased. From Figure 5.12 it can be seen that initially the line current $(I) = 100$ A (ratio 1:1). After the tap change the current $(I) = 100 \times 1.1$, and the volt drop in $Z_L = 1.1 \times$ the original value. Hence the secondary voltage increase is somewhat offset by the increased drop in the line. Obviously if the line impedance is high it is possible for the voltage drop to be too large for the transformer to maintain voltages with the tap range available.

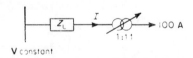

V constant

Figure 5.12   Effect of tap changing on line current and voltage drop.

5.6   Combined Use of Tap-changing Transformers and Reactive Power Injection

The usual practical arrangement is shown in Figure 5.13 where the tertiary winding of a three-winding transformer is connected to a synchronous compensator. For given load conditions it is proposed to determine the necessary transformation ratios with certain outputs of the compensator.

The transformer is represented by the equivalent star connexion and any line impedances from V_1 or V_2 to the transformer can be lumped together with the

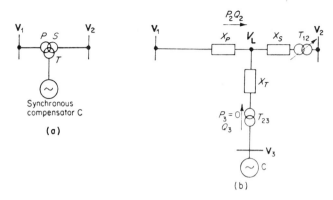

Figure 5.13 (a) Schematic diagram with combined tap changing and synchronous compensation. (b) Equivalent network.

transformer branch impedances. V_n is the phase voltage at the star-point of the equivalent circuit in which the secondary impedance (Z_s) is usually approaching zero and hence neglected. Resistance and losses are neglected. The allowable ranges of voltage for V_1 and V_2 are specified and the values of P_2, Q_2, P_3, and Q_3 given, P_3 is usually taken as zero.

The volt drop V_1 to V_n

$$= \Delta V \doteq X_\text{p} \frac{Q_2/3}{V_n}$$

or

$$X_\text{p} \frac{Q_2}{V_\text{L}\sqrt{3}},$$

where V_L is the line voltage $= \sqrt{(3)} V_n$ and Q_2 is the total vars. Also,

$$\partial V = X_\text{p} \frac{P_2}{V_\text{L}\sqrt{3}}$$

$$\therefore (V_n + \Delta V)^2 + (\delta V)^2 = V_1^2$$

(see phasor diagram of 2.22; phase values used) and

$$\left(V_n + X_\text{p} \frac{Q_2}{V_\text{L}\sqrt{3}} \right)^2 + X_\text{p}^2 \left(\frac{P_2^2}{3 V_\text{L}^2} \right) = V_1^2$$

$$\therefore (V_\text{L}^2 + X_\text{p} Q_2)^2 + X_\text{p}^2 P_2^2 = V_{1\text{L}}^2 V_\text{L}^2$$

where $V_{1\text{L}}$ is the line voltage $= \sqrt{(3)} V_1$

$$\therefore V_\text{L}^2 = \frac{V_{1\text{L}}^2 - 2 X_\text{p} Q_2}{2} \pm \tfrac{1}{2} \sqrt{[V_{1\text{L}}^2 (V_{1\text{L}}^2 - 4 X_\text{p} Q_2) - 4 \cdot X_\text{p}^2 P_2^2]}$$

Once V_L is obtained, the transformation ratio is easily found. The procedure is best illustrated by an example.

Example 5.4 A three-winding grid transformer has windings rated as follows: 132 kV (line), 75 MVA, star connected; 33 kV (line), 60 MVA, star connected; 11 kV (line), 45 MVA, delta connected. A synchronous compensator is available for connexion to the 11-kV winding.

The equivalent circuit of the transformer may be expressed in the form of three windings, star connected, with an equivalent 132 kV primary reactance of 0.12 p.u., negligible secondary reactance, and an 11 kV tertiary reactance of 0.08 p.u., both values expressed on a 75 MVA base.

In operation, the transformer must deal with the following extremes of loading:

(a) Load of 60 MW, 30 MVAr with primary and secondary voltages governed by the limits 120 kV and 34 kV; synchronous compensator disconnected.

(b) No load. Primary and secondary voltage limits 143 kV and 30 kV; synchronous compensator in operation and absorbing 20 MVAr.

Calculate the range of tap changing required. Ignore all losses.

Solution The value of X_p the primary reactance in ohms

$$= 0.12 \times 132^2 \times 1000/75 \times 1000 = 27.8\,\Omega$$

Similarly the effective reactance of the tertiary winding is 18.5 Ω. The equivalent star circuit is shown in Figure 5.14.

The first operating conditions are as follows:

$$P_1 = 60\ \text{MW} \qquad Q_1 = 30\ \text{MVAr}, \qquad V_{1L} = 120\ \text{kV}.$$

Hence,

$$V_L^2 = \tfrac{1}{2}(120{,}000^2 - 2 \times 27.8 \times 30 \times 10^6)$$
$$\pm \tfrac{1}{2}\sqrt{[120{,}000^2(120{,}000^2 - 4 \times 27.8 \times 30 \times 10^6) - 4 \times 27.8^2 \times 60^2}$$
$$\times 10^{12}]$$

$$= \left(63.61 \pm \frac{122}{2}\right)10^8 = 124.4 \times 10^8$$

$$\therefore \ V_L = 111\ \text{kV}$$

The second set of conditions are:

$$V_{1L} = 143\ \text{kV} \qquad P_2 = 0 \quad \text{and} \quad Q_2 = 20\ \text{MVAr}.$$

Again using the formula for V_L,

$$V_L = 138.5\ \text{kV}$$

The transformation ratio under the first condition

$$= 111/34 = 3.27$$

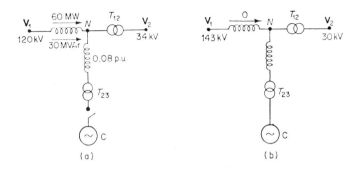

Figure 5.14 (a) Example 5.4. System with loading condition (a).
(b) System with loading condition (b).

and for the second condition,

$$\frac{138.5}{30} = 4.61.$$

The actual ratio will be taken as the mean of these extremes, i.e. 3.94, varying by ±0.67 or 3.94 ± 17 per cent. Hence the range of tap changing required is ±17 per cent.

A further method of var production is the use of adjustment of tap settings on transformers connecting large interconnected systems. Consider the situation in Figure 5.15(a) in which V_s and V_r are constant voltages representing the two connected systems. The circuit may be rearranged as in Figure 5.15(b) where t

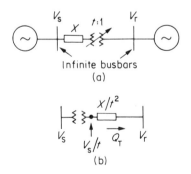

Figure 5.15 (a) Two power systems connected via a tap-change transformer. (b) Equivalent circuit with impedance transferred to receiver side.

is the off-nominal (per unit) tap setting; resistance is zero. The voltage drop between busbars

$$= \left(\frac{V_s}{t}\right) - V_r = \frac{X}{t^2} \cdot \frac{Q_r}{V_r}$$

hence,

$$(V_s V_r t - V_r^2 t^2)1/X = Q_T$$

and

$$t(1-t)V^2/X = Q_T \quad \text{when} \quad V_s = V_r = V$$

Also $Q_T = t(1-t)S$, where $S =$ short-circuit level, i.e. V^2/X. When

$t < 1$, Q_T is positive, i.e. a flow of lagging vars to V_r

$t > 1$, Q_T is negative, a flow of leading vars to V_r

Thus, by suitable adjustment of the tap setting an appropriate injection of reactive power is obtained.

The idea can be extended to two transformers in parallel between networks. If one transformer is set to an off-nominal ratio of say $1:1.1$ and the other to $1:0.8$ (i.e. in opposite directions) then a circulation of reactive power occurs round the loop resulting in a net absorption of vars. This is known as 'tap-stagger' and is a comparatively inexpensive method of var absorption.

5.7 Booster Transformers

It may be desirable on technical or economic grounds to increase the voltage at an intermediate point in a line rather than at the ends as with tap-changing transformers, or the system may not warrant the expense of tap changing. Here booster transformers are used as shown in Figure 5.16. The booster can be

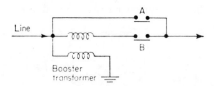

Figure 5.16   Connexion of in-phase booster transformer. One phase only shown.

brought into the circuit by the closure of relay B and the opening of A and vice versa. The mechanism by which the relays are operated can be controlled either from a change in the voltage or in the current. The latter method is the more sensitive, as from no load to full load represents a 100 per cent change in current, but only in the order of a 10 per cent change in voltage. This booster gives an in-phase boost as does a tap-changing transformer. An economic advantage is that the rating of the booster is the product of the current and the injected voltage, and is hence only about 10 per cent of that of a main transformer. Boosters are often used in distribution feeders where the cost of tap-changing transformers is not warranted.

Example 5.5   In the system shown by the line diagram in Figure 5.17 each transformer T_A and T_B have tap ranges of ± 10 per cent in 14 steps of 1.43 per

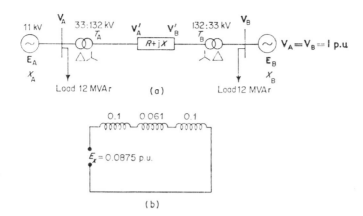

Figure 5.17 (a) Line diagram of system for Example 5.5. (b) Equivalent network with voltage boost E_x acting.

cent. Initially $\mathbf{V}'_A = \mathbf{V}'_B$ and hence no power or var transfer takes place through the line. It is required to calculate the magnitude of the circulating current resulting from an in-phase boost of 8.75 per cent on T_A, with the busbar voltages maintained constant by automatic voltage regulators on generators A and B. The plant data are as follows expressed on a 20 MVA base, resistance ignored. Generators A and B: 20 MVA, $X = 0.2$ p.u. Transformers T_A and T_B: 20 MVA, 132/33, $X = 0.1$ p.u.; 132 kV line, 8 km, $X = 3.85\ \Omega = 0.061$ p.u.

Solution   The equivalent circuits of the network are shown in Figure 5.17. The only voltage source is the boost $E_x = 0.0875$ p.u. which produces a current of 0.336 p.u. As the busbar voltages are maintained constant at 33 kV by voltage regulators no circulating current flows in the loads. The reactive power in generator G_A is now $12 + (0.336 \times 1 \times 20)$, i.e. 18.7 MVAr and in G_B is $12 + (-0.336 \times 1 \times 20)$, i.e. 5.3 MVAr.

It is seen that G_A is heavily loaded with vars whilst G_B is loaded lightly; this could lead to a loss of stability. In-phase boosts redistribute currents in a network and this has been used to de-ice lines in the winter, the extra current produces sufficient I^2R loss heating.

Circulating current is also used to automatically make tap-changing transformers in parallel, return to the same tap point if they tend to operate several taps apart.

Example 5.6   In the system shown in Figure 5.18 it is required to keep the nominally 11 kV busbar at constant voltage. The range of taps is not sufficient and it is proposed to use shunt capacitors connected to the tertiary winding. The data are as follows, per unit quantities being referred to a 15 MVA base: line 16 km, 115 mm², OHL, 33 kV, $Z = (0.0304 + j0.0702)$; Z referred to 33 kV side $= (2.2 + j5.22)\Omega$.

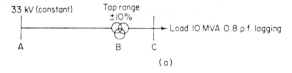

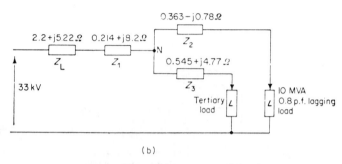

Figure 5.18 (a) Line diagram for Example 5.6. (b) Equivalent network—referred to 33 kV.

Three-winding transformer

Winding	MVA	Voltage kV	p.u. Z referred to nameplate MVA	p.u. Z on 15 MVA base	$(Z(\Omega))$ referred to 33 kV	Equivalent $(Z(\Omega))$ referred to 33 kV
P–S	15	33/11	$0.008 + j0.1$	$0.008 + j0.1$	$0.57 + j7.3$	$Z_1 =$ $0.214 + j8.2$
P–T	5	33/1.5	$0.0035 + j0.0595$	$0.0105 + j.179$	$0.76 + j4.32$	$Z_2 =$ $0.363 - j0.78$
S–T	5	11/1.5	$0.0042 + j0.0175$	$0.0126 + j0.0525$	$0.915 + j1.27$	$Z_3 =$ $0.545 + j4.77$

For the three-winding transformer the measured impedances between the windings and the resulting equivalent star impedances Z_1, Z_2 and Z_3 are given. The equivalent circuit referred to 33 kV is shown in Figure 5.18(b).

The voltage across the receiving-end load

$$= \frac{33,000}{\sqrt{3}} - \Delta V$$

where

$$\Delta V \doteqdot \frac{RP + XQ}{V_C}$$

$$\therefore \ \Delta V \doteqdot \frac{2.77 \times 8/3 \times 10^6 + 12.64 \times 6/3 \times 10^6}{33,000/\sqrt{3}}$$

As V_C referred to the 33 kV base is not known because of the system volt drop, 33 kV is assumed initially. The revised value is then used and the process repeated.

$$\Delta V = \frac{7.4 + 25.28}{19} \text{kV} = 1.715 \text{ kV} \quad \text{and} \quad V_C = 17.285 \text{ kV}$$

Repeating the calculation for ΔV with the new V_C,

$$\Delta' V = \frac{2.77 \times 8/3 \times 10^6 + 12.64 \times 6/3 \times 10^6}{17.285} = 1.89 \text{ kV}$$

Hence

$$V'_C = 19 - 1.89 = 17.11 \text{ kV}.$$

$$\Delta'' V = 1.9 \quad \text{and} \quad V''_C = 17.1 \text{ kV}.$$

This will be the final value of V_C.

V_C referred to 11 kV $= 17.1/3 = 5.7$ kV (phase) or 9.9 kV (line). In order to maintain 11 kV at C the voltage is raised by tapping down on the transformer. Using the full range of 10 per cent, i.e. $t_r = 0.9$ the voltage at C is

$$\frac{29.7}{(33 \times 0.9)/11} = 11 \text{ kV}$$

The true voltage will be less than this as the primary current will have increased by $(1/0.9)$ because of the change in transformer ratio. It is evident that the tap-changing transformer is not able to maintain 11 kV at C and the use of a static capacitor connected to the tertiary will be investigated.

Consider a shunt capacitor of capacity 5 MVAr (the capacity of the tertiary).

Assume the transformer to be at its nominal ratio 33/11 kV. The voltage drop

$$= \frac{2.414 \times 8/3 \times 10^6 + 13.42 \times 1/3 \times 10^6}{V_N (\doteqdot 19 \text{ kV})}$$

$$= 0.587 \text{ kV}$$

$$V'_N = 19 - 0.587 = 18.413 \text{ kV (phase)}$$

$$\therefore \Delta' V_N = 0.606 \quad \text{and} \quad V'_N = 18.394 \text{ kV}.$$

Therefore the volt drop N to C

$$\Delta V_C = \frac{0.363 \times 8/3 - 0.78 \times 6/3}{18.394} \text{kV}$$

$$= -0.032 \text{ kV}$$

$$\therefore V_C = 18.394 + 0.032$$

$$= 18.426 \text{ kV (phase)}.$$

As ΔV_C is so small there is no need to iterate further.

Referred to 11 kV, $V_C = 10.55$ kV (line). Hence to have 11 kV the transformer will tap such that $t_r = (1 - 0.35/11) = 0.97$, i.e. a 3 per cent tap change, which is well within the range and leaves room for load increases. On *no load*

$$\Delta V = \frac{2.959 \times 0 + 18.19 \times (-5/3)}{19} \text{kV}$$

$$= -\frac{30.3}{19} = -1.595 \text{ kV (phase)}$$

The shunt capacitor is a constant impedance load and hence as V_N rises the current taken increases causing further volt increase.

Ignoring this effect initially,

$$V_C = 10 + 1.6 = 20.6 \text{ kV (phase); at 11 kV}$$

$$V_C = 11.85 \text{ kV (line)}$$

therefore the tap change will have to be at least 7.15 per cent which is well within the range. Further refinement in the value of V_C will be unnecessary. If an accurate value of V_C is required then the reactance of the capacitor must be found and the current evaluated.

5.8 Voltage Stability

Voltage stability is essentially an aspect of load stability to be discussed in Chapter 8. As the voltages to be maintained in a system are influenced by voltage stability it is appropriate to discuss this subject here.

Consider the circuit shown in Figure 5.19(a). If V_S is fixed (i.e. an infinite busbar supply) the graph of V_R against P for given power factors is as shown in Figure 5.19(b). In Figure 5.19(b), Z represents the series impedance of a 160 km long, double circuit, 400 kV, 260 mm^2 conductor overhead line. The fact that two values of voltage exist for each value of power is easily demonstrated by considering the analytical solution of this circuit. At the lower voltage a very high current is taken to produce the power. The seasonal thermal ratings of the line are also shown and it is apparent that for loads of power factor less than unity the possibility exists that before the thermal rating is reached the operating power may be on that part of the characteristic where small changes in load cause large voltage changes and *voltage instability* will have occurred. In this condition the action of tap-changing transformers is interesting. If the receiving end transformers 'tap up' to maintain the load voltage the line current increases, thereby causing further increase in the voltage drop. It would in fact be more profitable to 'tap down', thereby reducing the current and voltage drop. It is feasible therefore for a 'tapping-down' operation to result in increased secondary voltage, and vice versa.

The possibility of an actual voltage collapse depends upon the nature of the load. If this is stiff (constant power), e.g. induction motors, the collapse is aggravated. If the load is soft, e.g. heating, the power falls off rapidly with voltage and the situation is alleviated. Referring to Figure 5.19 it is evident that

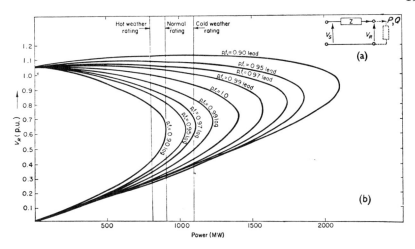

Figure 5.19 (a) Equivalent circuit of a line supplying a load $P + jQ$.
(b) Relation between load voltages and received power at constant power factor for a 400 kV, 2×260 mm^2 conductor line, 160 km in length. Thermal ratings of the line are indicated.

a critical quantity is the power factor; at full load a change in lagging power factor from 0.99 to 0.90 will precipitate voltage collapse. On long lines, therefore, for reasonable power transfers it is necessary to keep the power factor of transmission approaching unity, certainly above 0.97 lagging and it is economically justifiable to employ var injection by static capacitors or synchronous compensators at the load.

A problem arises with the operation of two or more lines in parallel, for example the system shown in Figure 5.20 in which the shunt capacitance has

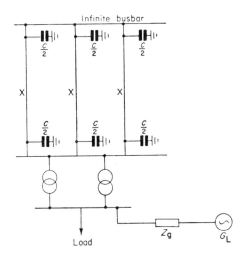

Figure 5.20 Line diagram of three long lines in parallel—effect of the loss of one line. G_L = local generators.

been represented as in a π section. If one of the three lines is removed from the circuit due to a fault, the total system reactance will increase from $X/3$ to $X/2$, and the capacitance, which normally improves the power factor decreases to $2C$ from $3C$. Thus the overall voltage drop is greatly increased and owing to the increased I^2X loss of the lines and the decreased generation of vars by the shunt capacitances the power factor decreases; hence the possibility of voltage instability. The same argument will of course apply to two lines in parallel.

Usually there will be local generation feeding the receiving-end busbars at the end of long lines. If this generation is electrically close to the load busbars, i.e. low connecting impedance Z_g, a fall in voltage will automatically increase the local var generation and this may be sufficient to keep the reactive power transmitted low enough to avoid large voltage drops in the long lines. Often, however, the local generators supply lower voltage networks and are electrically remote from the high-voltage busbar of Figure 5.20 and Z_g is high. The fall in voltage now causes little change in the local generator var output and the use of capacitors, static or rotary, at the load may be required. As Z_g is inversely proportional to the three-phase short-circuit level at the load busbar due to the local generation, the reactive-power contribution of the local machines is proportional to this fault level.

At the time of writing about 40 per cent of the power generated in the British system is fed to load centres less than 32 km in distance. This is changing, however, owing to the construction of the 400 kV network and the greater distance between generation and loads. It has been estimated that the reactive demand in Britain at peak loads is 0.4 kVAr/kW and the net var loss in the lines, etc. 0.1 kVAr/kW, giving a total of 0.5 kVAr/kW. This is supplied almost entirely by generators operating with an average power factor of 0.9 lagging. At minimum load the demand is 0.5 kVAr/kW with a net generation of vars in the network. For modern machines with short-circuit ratios of 0.55, operation at unity power factor is close to the theoretical stability limit with a 20 per cent margin. With an extensive very-high-voltage network, however, the vars absorbed by the network on load will amount to 0.3 kVAr/kW and on light load the network will generate 0.25 kVAr/kW. Hence the voltage regulation problem will be greatly aggravated.

5.9 Voltage Control in Distribution Networks

Single-phase supplies to houses and other small consumers are tapped off from three-phase feeders. Although efforts are made to allocate equal loads to each phase the loads are not applied at the same time and some unbalance occurs. An empirical method for modifying the voltage drop obtained by assuming balanced operation to allow for the average degree of unbalance is reported in reference 6.

In the distribution network (British practice) shown in Figure 5.21 an 11 kV distributor supplies a number of lateral feeders in which the voltage is 420 V and then each phase loaded separately.

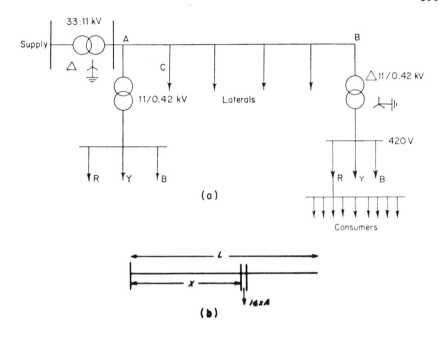

Figure 5.21 Line diagrams of typical radial distribution schemes.

The object of design is to keep the consumers' nominal 415 V supply within the statutory ±6 per cent of the declared voltage. The main 33/11 kV transformer gives a 5 per cent rise in voltage between no load and full load. The distribution transformers have fixed taps and a secondary phase voltage of 250 V which is 4 per cent high on the nominal value of 240 V. A typical distribution of voltage drops would be as follows: main distributor 6 per cent, 11/0.42 kV transformer 3 per cent, 420 V circuit 7 per cent, consumer circuit 1.5 per cent giving a total of 17.5 per cent. On very light load (10 per cent of full load) the corresponding drop may be 1.5 per cent. To offset these drops various voltage boosts are employed as follows: main transformer, +5 per cent (zero on light load); distribution transformer, inherent boost of +4 per cent (i.e. 250 V secondary) plus a fixed 2.5 per cent boost. These add to give a total boost on full load of 11.5 per cent and on light load of 6.5 per cent. Hence the consumers' voltage varies between $(-17.5 + 11.5)$, i.e. -6 per cent and $6.5 - 1.5$, i.e. $+5$ per cent, which is just permissible. There will also be a difference in consumer voltage depending upon the position of the lateral feeder on the main distributor, obviously a consumer supplied from C will have a higher voltage than one from B.

Uniformly loaded feeder fed from one end

In areas with high load densities a large number of tappings are made from feeders and a uniform load along the length of a feeder may be considered to

200

exist. Consider the voltage drop over a length dx of the feeder distant x metres from the supply end. Let iA be the current tapped per unit length and r and x the resistance and reactance per phase per metre. The length of the feeder is L(m), see Figure 5.21(b).

The voltage across d$x = rix$ dx cos $\phi + xix$ dx sin ϕ, where cos ϕ is the power factor (assumed constant) of the uniformly distributed load.

The total voltage drop

$$= r \int_0^L ix \, dx \cos \phi + x \int_0^L ix \, dx \sin \phi$$

$$= ri\frac{L^2}{2}\cos \phi + xi\frac{L^2}{2}\sin \phi$$

$$= \frac{Lr}{2}I \cos \phi + \frac{Lx}{2}I \sin \phi$$

where $I = Li$ the total current load. Hence the uniformly distributed load may be represented by the total load tapped at the centre of the feeder length.

5.10 Long Lines

On light loads the charging volt-amperes of a line exceeds the inductive vars consumed and the voltage rises, causing problems for generators. With very long lines the voltage drop can be massive. A length of 1500 km at 50 Hz corresponds to a quarter-wavelength line. Series capacitors would normally be installed to improve the power capability and these effectively shorten the line

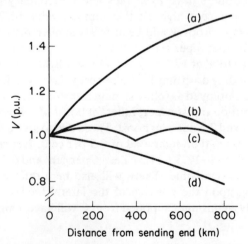

Figure 5.22   Voltage variation along a long line (a) On no load with no compensation (b) On no load with compensation at ends (c) On no load with compensation at ends and at centre. (d) Transmitting natural load, compensation at ends and centre.

electrically. In addition shunt reactors are switched in circuit at times of light
load to absorb the generated vars.

A 500 km line can operate within ±10 per cent voltage variation without
shunt reactors. However, with say, a 800 km line, shunt reactors are essential
and the effects of these are shown in Figure 5.22. For long lines in general it is
usual to divide the system into sections with compensation at the ends of each
section. This controls the voltage profile, helps switching, and reduces short-
circuit currents. Shunt compensation can be varied by switching discrete
amounts of inductance. A typical 500 kV, 1000 km scheme uses compensation
totalling 1200 MVAr.

Improvement in voltage profile may be obtained by compensation at intermediate points, as well as at the ends of the line, as shown in Figure 5.22. If the
natural load is transmitted there is of course constant voltage along the line

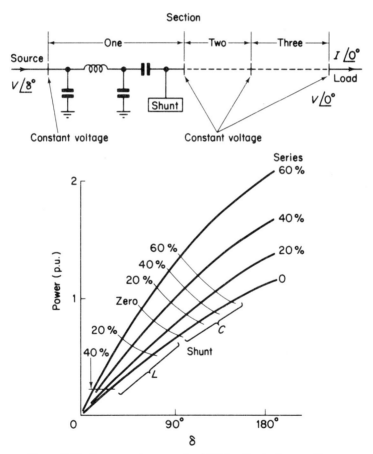

Figure 5.23   Power-angle curves for 1500 km line in three sections.
Voltages at section-busbars maintained constant by variable compen-
sation. Percentage of series and shunt compensation indicated.
(*Permission I.E.E.*).

with no compensation. If the various busbars of a sectioned line can be maintained at *constant voltage* regardless of load each section has a theoretical maximum transmission angle of 90°. Thus for a three-section line a total angle of much greater than 90° would be possible. This is illustrated in Figure 5.23 for a three-section, 1500 km line with a unity power-factor load.

Manual control of shunt units can achieve a reasonably constant voltage profile under steady-state conditions, but cannot cope with transient conditions. Similarly, synchronous compensators are too slow to help materially under transient conditions and also contribute to fault currents. A method of providing variable-shunt compensation lies in the use of saturated reactors which have response times of about 0.03 s. The equivalent circuit and characteristics of a saturable, iron-cored, three-phase reactor are shown in Figure 5.24. It has a high inductance and low var absorption if the voltage is below a prescribed level. Above this voltage the inductance decreases for part of the cycle and more vars are absorbed. Capacitance C_0 changes the slope of the Q–V characteristic. The overall cost is likely to be in the same order as a synchronous compensator.

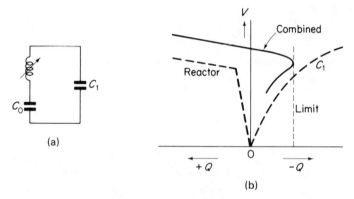

Figure 5.24 Saturated reactor for voltage control (a) equivalent circuit (b) V–Q characteristics.

With the more widespread use of silicon-controlled rectifiers (thyristors) the possibility exists of the thyristor switched reactor. Here the reactive current is controlled by varying the thyristor firing angle. Because of the harmonics generated, filters will be required.

Reactive-power requirements for the voltage control of long lines

It is advantageous to use the generalized line equation,

$$\mathbf{V_S} = \mathbf{A V_r} + \mathbf{B I_r}$$

which takes into account the presence of transformers.

Let the received load current $\mathbf{I}_r$ lag $\mathbf{V}_r$ by ϕ_r where $\mathbf{V}_r$ is the reference phasor

$$\mathbf{A} = A\angle\alpha \quad \text{and} \quad \mathbf{B} = B\angle\beta$$

$$V_s\angle\delta_s = AV_r\angle\alpha + BI_r\angle\beta - \phi_r$$

$$\therefore \quad V_s = AV_r\cos\alpha + jAV_r\sin\alpha + BI_r\cos(\beta - \phi_r) + jBI_r\sin(\beta - \phi_r)$$

As the modulus of the left-hand side is equal to that of the right

$$V_s^2 = A^2V_r^2 + B^2I_r^2 + 2ABV_rI_r\cos(\alpha - \beta + \phi_r)$$
$$= A^2V_r^2 + B^2I_r^2 + 2ABV_rI_r[\cos(\alpha - \beta)\cos\phi_r - \sin(\alpha - \beta)\sin\phi_r] \quad (5.7)$$

$$P_r = V_rI_r\cos\phi_r \quad \text{and} \quad Q_r = V_rI_r\sin\phi_r.$$

Hence equation 5,7 becomes

$$V_s^2 = A^2V_r^2 + B^2I_r^2 + 2ABP_r\cos(\alpha - \beta) - 2ABQ_r\sin(\alpha - \beta)$$

Also,

$$\mathbf{I}_r = I_p - jI_q \quad \text{and} \quad I_r^2 = I_p^2 + I_q^2 \quad \text{and} \quad I_p = \frac{P_r}{V_r} \quad \text{and} \quad I_q = \frac{Q_r}{V_r}$$

$$\therefore \quad V_s^2 = A^2V_r^2 + B^2\left(\frac{P_r^2}{V_r^2} + \frac{Q_r^2}{V_r^2}\right) + 2ABP_r\cos(\alpha - \beta) - 2ABQ_r\sin(\alpha - \beta)$$

$$(5.8)$$

For a given network P_r, $\mathbf{A}$, and $\mathbf{B}$ will be known. The magnitude of Q_r such that V_r is equal to, or a specified ratio of, V_s can be determined by the use of equation (5.8).

Subsynchronous oscillation

The combination of series capacitors and the natural inductance of the line (plus that of the connected systems) creates a resonant circuit of subsynchronous resonant frequency. This resonance can interact with the generator shaft critical torsional frequency, and a mechanical oscillation is superimposed on the rotating generator shaft which may have sufficient magnitude to cause mechanical failure.

Subsynchronous resonance has been reported caused by line-switching in a situation where trouble-free switching was normally carried out with all capacitors in service, but trouble occurred when one capacitor bank was out of service. Although this phenomena may be a rare occurrence the damage resulting is such that at the design stage an analysis of possible resonance effects is required.

5.11 General System Considerations

As voltages and line lengths increase and also the wider use of underground circuits the light load reactive power problem for an interconnected system becomes substantial, particularly with modern generators of limited var

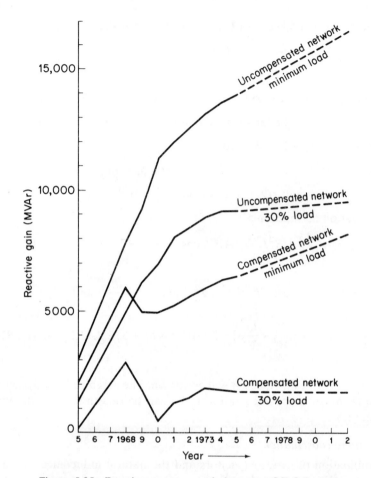

Figure 5.25   Reactive power growth by year—C.E.G.B. system
(British). (*Permission C.E.G.B.*).

absorption capacity. The variation of generated vars for the British C.E.G.B.
system over the years is shown in Figure 5.25 with the system both compensated and uncompensated. To produce var absorption as the load reduces the
following order of measures is advisable:

Switch out capacitor banks.
Switch in synchronous compensators, shunt reactors, and transductors.
Declutch gas turbines and run as compensators.
Reduce system voltage.
Switch out lines and cables (one circuit of double-circuit links).
Tap-stagger transformers.
Minimum-load, high-cost generation and use for var absorption.
Part-load, low-cost generation and use for var absorption.

References

BOOKS

1. *Electrical Transmission and Distribution Reference Book*, Westinghouse Electric. Corp., East Pittsburgh, Pennsylvania, 1964.
2. Miller, J. E. (Ed.), 'Reactive Power Control', In *Electric Systems*, Wiley, New York, 1982.
3. Sullivan, R. L., *Power System Planning*, McGraw-Hill, New York, 1977.
4. Jones, B., *New Aspects in the Design of EHV Plant*, Pergamon Press, London, 1972.

PAPERS

5. Jancke, G., *et al.*, 'Series capacitors in power systems', *I.E.E.E. Trans.*, **PAS-94** (1975), 915.
6. Happ, H. H., and Wirgan, K. A., 'Static and dynamic VAR compensation in system planning', *I.E.E.E. Trans.*, **PAS-97** (1978), 1564.
7. Friedlander, E., 'Transient reactance effects in static shunt reactive compensators for long a.c. lines', *I.E.E.E. Trans.*, **PAS-95** (1976), 1669.
8. Casson, W., and H. J. Sheppard, 'Technical and economic aspects of the supply of reactive power in England and Wales', *Proc. I.E.E.*, **108A** (1961).
9. Booth, E. S., *et al.*, 'The 400 kV Grid System for England and Wales', *Proc. I.E.E.*, **1962**, Pt. A.
10. Csicy, J., 'The influence of the regulating transformer and the excitation of alternators on system voltages and stability', *C.I.G.R.E.*, **1962**, No. 323.
11. Gyugyi, L., *et al.*, 'Principles and application of static thyristor-controlled shunt compensation', *I.E.E.E. Trans.*, **PAS-97** (1978), 1935.
12. Venikov, V. A., 'The influence of power factor correction on load stability', (Russian) *Elektrichestro*, **1960**, No. 6.
13. I.E.E.E. Committee Report, 'Bibliography on power capacitors 1959–1962', *I.E.E.E., P.A.S.*, **83**, (1964), 1110.
14. Breuer, G. D., H. M. Rustebakke, R. A. Gibley, and H. O. Simmons, Jr., 'The use of series capacitors to obtain maximum EHV transmission capability', *I.E.E.E., P.A.S.*, **83** (1964), 1090.
15. Schmill. J. V., 'Optimum size and location of shunt capacitors on distribution feeders', *I.E.E.E., P.A.S.*, **84** (1965), 825.
16. Baum, W. V., and W. A. Frederick, 'A method of applying switched and fixed capacitors for voltage control', *I.E.E.E., P.A.S.*, **84** (1965), 42.
17. Kimbark, E. W., 'Improvement of system stability by switched series capacitors', *I.E.E.E., P.A.S.*, **85** (1966) 180.
18. Duran, H., 'Optimum number, location, and size of shunt capacitors in radial distribution feeders—A dynamic programming approach', *I.E.E.E., P.A.S.*, **87** (1968) 1769.
19. Kumai, K., and K. Ode, 'Power system voltage control using a process control computer', *I.E.E.E., P.A.S.*, **87** (1968) 1985.
20. Chang, N. E., 'Locating shunt capacitors on primary feeder for voltage control and loss reduction', *I.E.E.E., P.A.S.*, **88** (1969), 1574.
21. Boehne, E. W., and S. S. Low, 'Shunt capacitor energization with vacuum interrupters—A possible source of overvoltage', *I.E.E.E., P.A.S.*, **88** (1969) 1424.
22. Kumar, B. S. A., K. Parthasarathy, F. S. Prabhakara, and H. P. Khincha, 'Effectiveness of series capacitors in long distance transmission lines', *I.E.E.E., P.A.S.*, **89** (1970), 941.

23. O'Kelly, D., 'Power transmission systems', *I.E.E. Electronics & Power*, **1977**, 721–726.
24. Iliceto, F., 'Series capacitor compensation compared with shunt reactor compensation', *Trans. I.E.E.E.*, **PAS–96** (1977), 1819–1830.
25. I.E.E.E. Committee (Power Systems Engineering), 'Computer simulation of subsynchronous Resonance', *Trans. I.E.E.E.*, **P.A.S.–96** (1977), 1565–1572.

Problems

5.1. An 11 kV supply busbar is connected to an 11/132 kV, 100 MVA, 10 per cent reactance, transformer. The transformer feeds a 132 kV transmission link consisting of an overhead line of impedance $(0.014 + j0.04)$ p.u. and a cable of impedance $(0.03 + j0.01)$ p.u. in parallel. If the receiving end is to be maintained at 132 kV when delivering 80 MW, 0.9 p.f. lagging calculate the power and reactive power carried by the cable and the line. All p.u. values relate to 100 MVA and 132 kV bases.
(Answer: Line $(23 + j38)$ MVA; cable $(57 + j3.8)$ MVA)

5.2. A three-phase induction motor delivers 500 hp at an efficiency of 0.91, the operating power factor being 0.76 lagging. A loaded synchronous motor with a power consumption of 100 kW is connected in parallel with the induction motor. Calculate the necessary kVA and the operating power factor of the synchronous motor if the overall power factor is to be unity.
(Answer: 365 kVA, 0.274)

5.3. The load at the receiving end of a three-phase, overhead line is 25 MW, power factor 0.8 lagging, at a line voltage of 33 kV. A synchronous compensator is situated at the receiving end and the voltage at both ends of the line is maintained at 33 kV. Calculate the MVAr of the compensator. The line has resistance 5 Ω per phase and inductive reactance (line to neutral) 20 Ω per phase.
(Answer: 25 MVAr.)

5.4. A transformer connects two infinite busbars of equal voltage. The transformer is rated at 500 MVA and has a reactance of 0.15 p.u. Calculate the var flow for a tap setting of (a) 0.85 : 1, (b) 1.1 : 1.
(Answer: 427 MVAr; −367 MVAr)

5.5. A three-phase transmission line has resistance and inductive reactance of 25 Ω and 90 Ω respectively. With no load at the receiving end a synchronous compensator there takes a current lagging by 90°, the voltage at the sending end is 145 kV and 132 kV at the receiving end. Calculate the value of the current taken by the compensator.

When the load at the receiving end is 50 MW, it is found that the line can operate with unchanged voltages at sending and receiving ends, provided that the compensator takes the same current as before but now leading by 90°.

Calculate the reactive power of the load.
(Answer: 83.5 A; Q_L24.2 MVAr)

5.6. Repeat question 5.3 making use of $\partial Q / \partial V$ at the receiving end.

5.7. In Example 5.3 determine the tap ratios if the receiving end voltage is to be maintained at 0.9 p.u. of the sending-end voltage.
(Answer: $1.19t_s$, $0.84t_r$)

5.8. In the transmission system in Example 5.4 a synchronous compensator is connected to the tertiary winding and produces 10 MVAr when a secondary load of 50 MVA 0.8 p.f. lagging is taken. On zero secondary load the compensator is adjusted to absorb 2 MVAr. Calculate the lower limit of voltage at the load busbar if the supply is constant at 132 kV.
(Answer: 127 kV)

5.9. In the system shown in Fig. 5.26 determine the supply voltage necessary at D to maintain a phase voltage of 240 V at the consumer's terminals at C. The following data

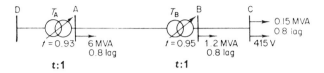

Figure 5.26   Line diagram for system in Problem 5.9

apply:

Line or/ trans- formers	Rated voltage (kV)	Trans. rating	Nominal tap ratio	Impedance (ohms)
BC	0.415			0.0127 + j0.00909
AB	11			1.475 + j2.75
DA	33			1.475 + j2.75
T$_A$	33/11	10 MVA	30.69/11	1.09 + j9.8 referred to 33 kV
T$_B$	11/0.415	2.5 MVA	10.450/0.415	0.24 + j1.95 referred to 33 kV

(Answer: 33 kV)

5.10. A load is supplied through a 275 kV link of total reactance 50 Ω from an infinite busbar at 275 kV. Plot the receiving-end voltage against power graph for a constant load power factor of 0.95 lagging. The system resistance may be neglected.

5.11. In the system shown in Figure 5.20 each line is 160 km long and is rated at 275 kV with 2×113 mm^2 conductors. The lines are carrying their rated (normal weather) MVA at 0.95 power factor lagging. Calculate the voltage at the load busbar, (a) initially and (b) when one line is switched out, taking into account the shunt capacitances of the lines. Examine the effect on the load voltage of generation in the load area which may be represented by a voltage source in series with a reactance of (a) 2 p.u. and (b) 0.5 p.u. connected to the load busbar. The local generator may be assumed to export zero current when all three lines are in service. Base for local generation circuit per unit values is 1290 MVA.

(Answer: When $V_s = 1$ p.u. $V_R = 0.8$ p.u.; $V_R = 0.64, 0.72$ p.u.)

 (*Note*. Use Table 3.2a for line ratings. Initially P_g and Q_g from local generator = 0 and generator e.m.f. (E_g) = received busbar voltage (V_R). After outage, if V'_R is new received voltage and V_s sending voltage,

$$V_s - V'_R = \frac{RP + XQ_r}{V'_R}$$

Also $E_g - V_R = X_g Q_g / V'_R$ as P_g is still zero.)

5.12. Explain the limitations of tap-changing transformers. A transmission link (Figure 5.27a) connects an infinite busbar supply of 400 kV to a load busbar supplying 1000 MW, 400 MVAr. The link consists of lines of effective impedance $(7 + j70)$ Ω feeding the load busbar via a transformer with a maximum tap ratio of 0.9 : 1. Connected to the load busbar is a synchronous compensator. If the maximum overall voltage drop is to be 10 per cent with the transformer taps fully utilized, calculate the reactive power requirement from the compensator.

(Answer: 148 MVAr)

208

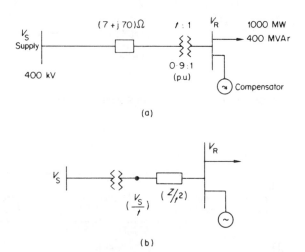

(a)

(b)

Figure 5.27 Circuits for Problem 5.12.

Note. Refer voltage and line Z to load side of transformer in Figure 5.27b.

$$V_R = \frac{V_s}{t} - \left(\frac{\dfrac{RP}{t^2} + \dfrac{X}{t^2}Q}{V_R} \right)$$

5.13. A generating station consists of four 500 MW, 20 kV, 0.95 p.f. (lag) generators, each feeding through a 525 MVA, 0.1 p.u. reactance transformer on to a common busbar. It is required to transmit 2000 MW at 0.95 p.f., lagging, to a substation maintained at 500 kV in a power system at a distance of 500 km from the generating station. Design a suitable transmission link of nominal voltage 500 kV to achieve this, allowing for a reasonable margin of stability and a maximum voltage drop of 10 per cent. Each generator has synchronous and transient reactances of 2 p.u. and 0.3 p.u. respectively, and incorporates a fast-acting automatic voltage regulator. The 500 kV transmission lines have an inductive reactance per phase of 0.4 Ω/km and a shunt capacitive reactance per phase of 0.3×10^6 Ω/km. Both series and shunt capacitors may be used if desired and the number of three-phase lines used should be not more than three—less if feasible. Use approximate methods of calculation, ignore resistance, and state clearly any assumption made. Assume shunt capacitance to be lumped at the receiving end only. (Use two 500 kV lines with series capacitors compensating to 70 per cent of series inductance.)

5.14. It is required to transmit power from a hydroelectric station to a load centre 480 km away using two lines in parallel for security reasons.

Assume sufficient bundle conductors are used such that there are no thermal limitations and that the effective reactance per phase per km is 0.44 Ω and the resistance negligible. The shunt capacitive reactance of each line is 0.44 MΩ per phase per km, and each line may be represented by the nominal π-circuit with half the capacitance at each end. The load is 2000 MW at 0.95 lagging and is independent of voltage over the permissible range.

Investigate the feasibility and performance of the link if the sending-end voltage is 345, 500, and 765 kV from the point of view of stability and voltage drop.

The lines may be compensated up to 70 per cent by series capacitors and at the load end synchronous compensators of 120 MVAr capacity are available. The maximum

209

permissible voltage drop is 10 per cent. As two lines are provided for security reasons, your studies should include the worst operating case of only one line in use.

Calculate the voltage at the receiving end on no-load with the synchronous compensators not in operation. The stability requirement is met by a transmission angle of 30°. (Use 765 kV.)

5.15. A transmission system is shown in Figure 5.28. A constant load of 2000 MW, 0.9 p.f. lagging, is fed by three 500 kV lines each of inductive reactance 0.385 Ω/ph per km and shunt capacitive reactance 0.24 MΩ per km. The line resistance is negligible.

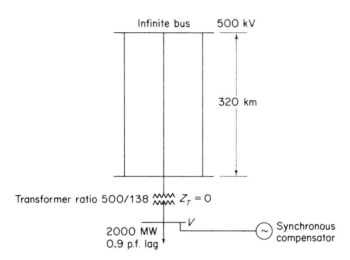

Figure 5.28   System for Problem 5.15.

The synchronous compensator operates as follows:

(a) zero var output with three lines in circuit;
(b) an output of 10 MVAr for each 1 per cent voltage change at the load-bus expressed on the voltage (V) with three lines in circuit. This output is linear with voltage change.

Determine:

(a) the load voltage (V) with three lines in circuit;
(b) the load voltage (V) with one line open circuit.
(Answer: 446 kV; 415 kV)

5.16. Two substations are connected by two lines in parallel of negligible impedance, each containing a transformer of reactance 0.18 p.u. and rated at 120 MVA. Calculate the net absorption of reactive power when the transformer taps are set to 1:1.15 and 1:0.85 respectively (i.e. tap stagger is used). The p.u. voltages are equal at the two ends and are constant in magnitude.
(Answer: 30 MVAr)

6

Load Flows

6.1 Introduction

A load flow is power system parlance for the steady-state solution of a network. This does not essentially differ from the solution of any other type of network except that certain constraints are peculiar to power supply. In previous chapters the manner in which the various components of a power system may be represented by equivalent circuits has been demonstrated. It should be stressed that the simplest representation should always be used consistent with the accuracy of the information available. There is no merit in using very complicated machine and line models when the load and other data are known only to a limited accuracy, for example the long-line representation should only be used where absolutely necessary. Similarly synchronous-machine models of more sophistication than given in this text are needed only for very specialized purposes, for example in some stability studies. Usually the size and complexity of the network itself provides more than sufficient intellectual stimulus without undue refinement of the components. Often resistance may be neglected with little loss of accuracy and an immense saving in computation.

The following combinations of quantities are usually specified at the system busbars for load-flow studies.

(a) *Slack or floating busbar* One node is always specified by a voltage, constant in magnitude and phase. The effective generator at this node supplies the losses to the network; this is necessary because the magnitude of the losses will not be known until the calculation of currents is complete and this cannot be achieved unless one busbar has no power constraint and can feed the required losses into the system. The location of the slack node can influence the complexity of the calculations; the node approaching most closely an infinite busbar should be used.

(b) *Load nodes* The complex power $S = P \pm jQ$ is specified.

(c) *Generation nodes* The voltage magnitude and power are usually specified. Often limits to the value of the reactive power are given depending upon the characteristics of individual machines.

Load-flow studies are performed to investigate the following.

1. Flow of MW and MVAr in the branches of the network.
2. Busbar voltages.
3. Effect of rearranging circuits and incorporating new circuits on system loading.
4. Effect of temporary loss of generation and transmission circuits on system loading.
5. Effect of injecting in-phase and quadrature boost voltages on system loading.
6. Optimum system running conditions and load distribution.
7. Optimum system losses.
8. Optimum rating and tap range of transformers.
9. Improvement from change of conductor size and system voltage.

Studies will normally be performed for minimum load conditions (possibility of instability due to high voltage levels and self-excitation of induction machines) and maximum load conditions (possibility of synchronous instability). Having ascertained that a network behaves reasonably under these conditions further load flows will be performed to attempt to optimize various quantities. The design and operation of a power network to obtain optimum economy is of paramount importance and the furtherance of this ideal will be greatly advanced by the growing use of centralized automatic control of generating stations.

Although the same approach can be used to solve all problems, e.g. the nodal voltage method, the object should be to use the quickest and most efficient method for the particular type of problem. Radial networks will require less sophisticated methods than closed loops. In very large networks the problem of organizing the data is almost as important as the method of solution and the calculation must be carried out on a systematic basis and the nodal-voltage method is the most advantageous. Such methods as network reduction combined with the Thevenin or superposition theorems are at their best with smaller networks. In the nodal method, greater accuracy is required in the computation as the currents in the branches are derived from the voltage differences between the ends. These differences are small in well-designed networks and numerical accuracy of a high order is necessary. Hence this method is ideally suited for computation using digital computers.

6.2 Radial and Simple Loop Networks

In radial networks the phase shifts due to transformer connexions along the circuit are not usually important. The following examples illustrate the solution of this type of network.

Example 6.1 Distribution feeders with several tapped loads.

A distribution feeder with several tapped inductive loads (or laterals) and fed at one end only, is shown in Figure 6.1(a). Determine the total voltage drop.

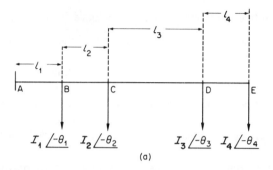

Figure 6.1(a)   Feeder with several load tappings along its length.

Solution   The current in

$$AB = (I_1 \cos \theta_1 + I_2 \cos \theta_2 + I_3 \cos \theta_3 + I_4 \cos \theta_4)$$
$$-j(I_1 \sin \theta_1 + I_2 \sin \theta_2 + I_3 \sin \theta_3 + I_4 \sin \theta_4)$$

Similarly, the currents in the other section of the feeder are obtained. The approximate voltage drop is obtained from $\Delta V = RI \cos \theta + XI \sin \theta$ for each section. That is

$$R_1(I_1 \cos \theta_1 + I_2 \cos \theta_2 + I_3 \cos \theta_3 + I_4 \cos \theta_4)$$
$$+R_2(I_2 \cos \theta_2 + I_3 \cos \theta_3 + I_4 \cos \theta_4)$$
$$+R_3(I_3 \cos \theta_3 + I_4 \cos \theta_4) + R_4(I_4 \cos \theta_4)$$
$$+X_1(I_1 \sin \theta_1 + I_2 \sin \theta_2 + I_3 \sin \theta_3 + I_4 \sin \theta_4) \text{ and so on}$$

Rearranging and letting the resistance per loop metre be r ohms and the reactance per loop metre be x ohms (the term loop metre refers to single-phase circuits and includes the go and return conductors),

$$\Delta V = r[I_1 \cdot l_1 \cos \theta_1 + I_2 \cos \theta_2 \cdot (l_1 + l_2) + I_3 \cos \theta_3(l_1 + l_2 + l_3)$$
$$+I_4 \cos \theta_4(l_1 + l_2 + l_3 + l_4)] + x[I_1 l_1 \sin \theta_1 + I_2 \sin \theta_2(l_1 + l_2)$$
$$+I_3 \sin \theta_3(l_1 + l_2 + l_3) + I_4 \sin \theta_4(l_1 + l_2 + l_3 + l_4)]$$

In the system shown in Figure 6.1(b) calculate the size of cable required if the voltage drop at the end load on the feeder must not exceed 12 V (line value).

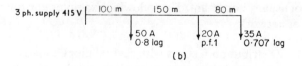

Figure 6.1(b)   Line diagram of feeder in Example 6.1.

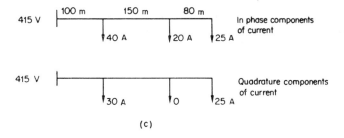

Figure 6.1(c)   In-phase and quadrature quantities.

Let the resistance and reactance per metre per phase be r ohms and x ohms respectively. Then, referring to Figure 6.1(c),

$$r[100 \times 40 + 250 \times 20 + 330 \times 25] + x[100 \times 30 + 250 \times 0 + 330 \times 25] = 12/\sqrt{3}$$

i.e.

$$r(4000 + 5000 + 8250) + x(3000 + 0 + 8250) = 12/\sqrt{3}$$

i.e.

$$17{,}250r + 11{,}250x = 12/\sqrt{3}$$

The procedure now is to consult the appropriate overhead line or cable tables to select a cross-section which gives values of r and x which best fit the above equation. Usually, if underground cables are used the size of cross-section is selected on thermal considerations and then the voltage drop calculated. With overhead lines the volt drop is the prime consideration and the conductor size will be determined accordingly.

Example 6.2   The system shown in Figure 6.2(a) feeds two distinct loads, a group of domestic consumers and a group of induction motors which on

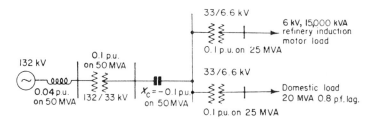

Figure 6.2(a)   Line diagram of system in Example 6.2.

starting takes $5 \times$ full-load current at zero power-factor lagging. The induction motor voltage is nominally 6 kV. Calculate the dip in voltage on the domestic-load busbar when the induction motor is started.

Solution   The problem of the drop in the voltage to other consumers when an abnormally large current is taken for a brief period from an interconnected

214

busbar is often serious. The present problem poses a typical situation. The 132 kV system is not an infinite busbar and is represented by a voltage source in series with a 0.04 p.u. reactance on a 50 MVA base. Using a 50 MVA base the equivalent single-phase circuit is as shown in Figure 6.2(b).

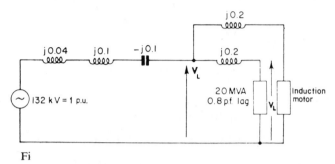

Fi

Figure 6.2(b)   Equivalent circuit of system in Example 6.2.

Starting current of the induction motors

$$= -j\frac{15{,}000 \times 10^3}{\sqrt{3} \times 6000} \times 5 \text{ A}$$

$$= -j7210 \text{ A}$$

Domestic load voltage before the induction-motor load is switched on

$$= 1 - \mathbf{IZ}$$

where

$$\mathbf{Z} = j0.24 \text{ p.u.}$$

and

$$\mathbf{I} = \frac{20 \times 10^6}{\sqrt{3} \times 6000} = 1920 \text{ A, } 0.8 \text{ p.f. lagging}$$

Base current in the 6.6 kV circuit

$$= \frac{50 \times 10^6}{\sqrt{3} \times 6600} = 4380 \text{ A}$$

∴ **I** p.u. = 0.438 p.u. at 0.8 p.f. lagging (i.e. 1920/4380)

$$\mathbf{V_L} = 1 - 0.438(0.8 - j0.6)(j0.24) \text{ p.u.}$$

$$= 0.9366 - j0.0843 \text{ p.u.}$$

and

$$V_L = 0.935 \text{ p.u.}$$

Starting current of the induction motors

$$= \frac{-j7210}{4380} = -j1.645 \text{ p.u.}$$

Voltage $\mathbf{V}'_L$ when the motors start

$$= 1 - j0.04[0.438(0.8 - j0.6) - j1.645] - j0.2[0.438(0.8 - j0.6)]$$

$$= 0.871 - j0.084 \text{ p.u.}$$

and

$$V'_L = 0.874 \text{ p.u.}$$

Hence the voltage dips from 0.935 to 0.87 p.u. or from 6.175 kV to 5.82 kV.

It will be noticed that a series capacitor has been installed partly to neutralize the network reactance. Without this capacitor the dip will be much more serious; it is left to the reader to determine by how much. Often in steel mills large pulses of power are taken at regular intervals and the result is a regular fluctuation on the consumer busbars. This is known as *voltage flicker* owing to the effect on electric lights.

Load flows in closed loops In radial networks any phase shifts due to transformer connexions are not usually of importance as the currents and voltages are shifted by the same amount. In a closed loop to avoid circulating currents the product of the transformer transformation ratios (magnitudes) round the loop should be unity and the sum of the phase shifts in a common direction round the loop should be zero. This will be illustrated by the system shown in Figure 6.3.

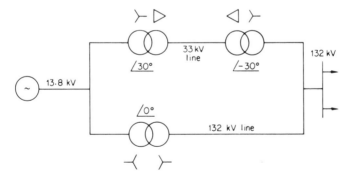

Figure 6.3   Loop with transformer phase shifts.

Neglecting phase shifts due to the impedances of components, for the closed loop formed by the two lines in parallel the total shift and transformation ratio is

$$\left(\frac{33}{13.8}\angle 30°\right)\left(\frac{132}{33}\angle -30°\right)\left(\frac{13.8}{132}\angle 0°\right) = 1\angle 0°$$

In practice the transformation ratios of transformers are frequently changed by the provision of tap-changing equipment. This results in the product of the ratios round a loop being no longer unity, although the phase shifts are still equal to zero. To represent this condition a fictitious auto-transformer is connected as shown in Figure 6.4. The increase in voltage on the 132 kV line by tap changing has the effect, in a largely reactive path, of changing the flow of reactive power and thus the power factor of the current. An undesirable effect is the circulating current set up around the loop.

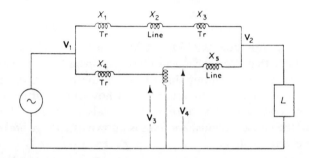

Figure 6.4 Equivalent circuit of network in Figure 6.3 showing autotransformer.

Often the out-of-balance or remnant voltage represented by the auto-transformer can be neglected. If this is not the case the best method of calculation is to determine the circulating current and consequent voltages due to the remnant voltage acting alone, and then superpose these values on those obtained for operation with completely nominal voltage ratios.

Real-power flows

In many network investigations especially in the initial design stages, analysis giving only power flows suffices. The calculations can be considerably reduced by neglecting resistance in the system. In a system with reactance and power flows only it can be shown that round a loop

$$\sum P_{ij}X_{ij} = 0 \tag{6.1}$$

(angle is related to $(PX - RQ)/V$, see equation (2.7)). Also the algebraic sum of the powers meeting at a point is zero. Hence power flow is analogous to current, and the nodal equation for node 1 of a multinode system could be expressed as follows:

$$P_1 = \frac{V_1}{X_{11}} - \frac{V_2}{X_{12}} - \frac{V_3}{X_{13}} \cdots \frac{V_k}{X_{1k}}$$

where X_{11} is the self-reactance of node 1 and X_{1k} the mutual reactance between nodes 1 and k. Generally,

$$V_i = \left[P_i + \sum_{j=2}^{j=n} \frac{V_j}{X_{ij}} \right] \sum X_{ii}$$

The initial voltages at the nodes are assumed as for the complex load flow. An approximate estimation of the I^2R losses can be obtained by evaluating the currents, these losses of course are ignored in the actual flow calculations. Similarly, the reactive power absorption or the I^2X loss can be obtained.

Owing to the relationships between power and reactance in a network being analogous to those for current and resistance a simple resistive analogue can be designed for real power analysis.

If resistance $R = k_1 X$ and current $I = k_2 P$, then

$$\sum X_{ij} P_{ij} = \sum k_1 k_2 I_{ij} R_{ij} = k \sum I_{ij} R_{ij}$$

Example 6.3   In Figure 6.5 is shown a system in which the receiving-end transformer is tapped from its nominal 33/11 kV to 33/11.8 kV. Calculate the circulating current.

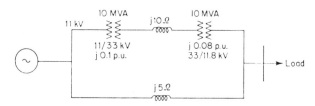

Figure 6.5   Line diagram of system for Example 6.3.

Solution   The equivalent circuit is as shown in Figure 6.4, where $V_1 = 11$ kV (base), $V_2 = 11.8$ kV (base), $X_1 = j0.1$ p.u., $X_2 = j10\ \Omega$, $X_3 = j0.08$ p.u., $X_4 = j5\ \Omega$, $V_3 = 11$ kV and $V_4 = 11.8$ kV.

A base of 10 MVA is taken. Ratio of the autotransformer

$$= \frac{11.8}{11} = 1.073.$$

A first approximation to the circulating current is

$$\frac{1.073 - 1}{\text{loop impedance}} \text{p.u.}$$

Base impedance in the 33 kV circuit

$$= \frac{33^2 \times 10^6}{10 \times 10^6} = 109\ \Omega$$

$$\therefore\ X_2 = j \frac{10}{109} = j0.09 \text{ p.u.}$$

Similarly, in the 11 kV circuit,

$$X_4 = j0.413 \text{ p.u.}$$

The circulating current

$$= \frac{0.073}{j0.683} = -j0.107 \text{ p.u.}$$

or 52 A at zero power-factor lagging.

If this current is significant compared with the normal load current the superposition method already mentioned should be applied.

Example 6.4   Perform a real power load for the network shown in Figure 6.6.

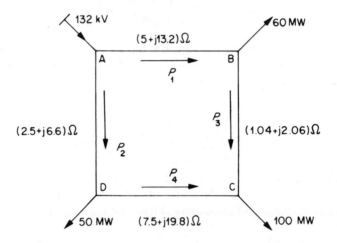

Figure 6.6   Line diagram of loop network for Example 6.4.

Solution   An estimate may be obtained by assuming zero resistance in the network and using the equation

$$\sum P_{ij}X_{ij} = 0 \qquad (6.1)$$

where i and j refer to the branch between nodes i and j of the network. This follows from the fact that $\delta V \propto \angle \delta \doteq (XP/V)$ and the sum of the angular shifts round the loop is zero. Labelling the powers as in Figure 6.6 the following equations hold:

$$13.2P_1 + 2.06P_3 - 19.8P_4 - 6.6P_2 = 0$$

$$P_4 + P_3 = 100 \quad \text{or} \quad P_4 = 100 - P_3$$

$$P_2 - P_4 = 50; \qquad P_1 = 60 + P_3$$

$$P_2 + P_1 = 210 \qquad P_2 = 210 - 60 - P_3 = 150 - P_3$$

$$\therefore \quad 13.2(60 + P_3) + 2.06(P_3) - 19.8(100 - P_3) - 6.6(150 - P_3) = 0$$

and

$$792 + 13.2P_3 + 2.06P_3 - 1980 + 19.8P_3 - 990 + 6.6P_3 = 0.$$

From which $P_3 = 52.2$ MW, which is a reasonably accurate estimate.
Once P in BC is known the flows in the other branches follow readily, i.e.

$$P_{DC} = 100 - 52.2 = 47.8 \text{ MW}$$

$$P_{AB} = 60 + 52.2 = 112.2 \text{ MW}$$

$$P_{AD} = 47.8 + 50 = 97.8 \text{ MW}$$

6.3 Large Systems

The calculations of load flows carried out so far have been on small systems. Similar methods on large networks would be extremely tedious if not virtually impossible. One method of approaching the analyses of large systems is to represent them by model lumped-constant networks at low voltages (<100 V) and currents (few milliamperes). Lines and cables are represented by the π model, generators by the internal voltage (E) in series with the appropriate impedance and transformers by their series impedance. The effects of tap changing are simulated by the use of miniature auto-transformers. These equivalent-circuit models are usually known as 'alternating current analysers' and for several years formed the main approach to large-system studies. They are steady-state models and often operate at higher than mains frequency to reduce the physical size of the components to give the actual system reactances. The many variable resistors, inductors, and capacitors which comprise analysers are scaled in p.u. values so that the actual power-system components (also in p.u.) can be represented directly. One step further towards the physical system is the micro-reseaux system[3] developed in France. In this the generators are represented by actual machines scaled down to laboratory proportions. This arrangement is not economical for large systems and is confined entirely to the study of a few interconnected machines. Special-purpose electronic analogue computers have also been used for the assessment of the effect of prime mover and generator characteristics on synchronous stability.

With the rapid development of digital computers in the 1950s the attention of power engineers has been directed towards the use of numerical methods for analysis. This has advantages over the use of alternating-current analysers in that analysis can be placed on a systematic basis once suitable programs have been developed, thus requiring less trained-engineer time. The various components of the network are uniquely labelled and these designations together with the p.u. values of admittance, power, reactive power, and voltage are fed into the computer which then performs the analysis to prescribed values of accuracy. With digital methods the nodal-voltage method of network solution is preferred. The admittance matrix is formed, the constraints of the network expressed, and then a suitable method of solution used. Usually the voltage magnitudes at the nodes are evaluated and from these the current, watt, and var flows obtained. In a well-designed network the voltage magnitudes at the various nodes will be close and high accuracy required in computation. Before discussing in detail the methods of solution, a simple problem will be solved to illustrate the nodal-voltage method.

Example 6.5 Perform a load flow on the interconnected busbars shown in Figure 6.7.

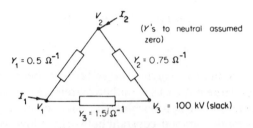

Figure 6.7   Three-busbar system—load flow using matrix inversion.

Solution   The nodal equations may be written directly as follows:

$$\begin{bmatrix} 2 & -0.5 & -1.5 \\ -0.5 & 1.25 & -0.75 \\ -1.5 & -0.75 & +2.25 \end{bmatrix} \begin{bmatrix} V_1 \\ V_2 \\ V_3 \end{bmatrix} = \begin{bmatrix} I_1 \\ I_2 \\ I_3 \end{bmatrix}$$

It is usual to write the admittance matrix direct. If the values in the current column vector are specified it is required to determine the values of V_1, V_2, and V_3. [**Y**] may be inverted or the equations solved by elimination. As

$$[\mathbf{Y}][\mathbf{V}] = [\mathbf{I}] \quad \text{then} \quad [\mathbf{Y}]^{-1}[\mathbf{I}] = [\mathbf{V}]$$

hence it is required to invert the admittance matrix. A simple method will be used in this case,

$$V_1 = 0.5 I_1 + 0.25 V_2 + 0.75 V_3$$

Substitute for V_1 in equations for I_2 and I_3,

$$-0.25 I_1 + 1.125 V_2 - 1.125 V_3 = I_2$$

$$-0.75 I_1 - 1.125 V_2 + 1.125 V_3 = I_3$$

Also from the equation for I_2,

$$V_2 = 0.222 I_1 + 0.888 I_2 + V_3$$

Substitute for V_2 in the other equations,

$$0.555 I_1 + 0.222 I_2 + V_3 = V_1$$

$$0.222 I_1 + 0.888 I_2 + V_3 = V_2$$

$$-I_1 - I_2 - 0 V_3 = I_3$$

i.e.

$$\begin{bmatrix} 0.555+0.222+1.0 \\ 0.222+0.888+1.0 \\ -1 \quad\quad -1 \quad\quad 0.0 \end{bmatrix} \begin{bmatrix} I_1 \\ I_2 \\ V_3 \end{bmatrix} = \begin{bmatrix} V_1 \\ V_2 \\ I_3 \end{bmatrix}$$

The node at which the voltage is specified, the slack busbar, is node (3), hence the manipulation of the matrix into the form shown. Next some loadings at the nodes will be specified.

$$I_1 = 1000 \text{ A (generator node)}$$

$$I_2 = -1500 \text{ A (load node)}$$

$$V_3 = 100 \text{ kV (phase value), slack busbar.}$$

Dealing in kiloamperes and kilovolts,

$$V_1 = 0.555(1)+0.222(-1.5)+(+1)(100)$$

$$= 100.222 \text{ kV}$$

$$V_2 = 0.222(1)+0.888(-1.5)+(1)(100)$$

$$= 98.89 \text{ kV}$$

$$I_3 = -(1)(1)+(1)(1.5)+0(100)$$

$$= +0.5 \text{ kA, i.e. a generator node.}$$

Note that $I_1 + I_2 + I_3 = 0$

In this example the loads have been specified as currents and the solution is obtained directly after the matrix manipulation. In practice, however, power, reactive power, and voltage are specified. The normal combination of quantities at a busbar have already been discussed and it is evident that the following equations require solution:

$$[\mathbf{Y}][\mathbf{V}] = [\mathbf{I}] \tag{6.2}$$

and

$$\mathbf{S} = \mathbf{V}\mathbf{I}^* \tag{6.3}$$

The problem is non-linear and iterative procedures must be used either completely or combined with matrix inversion.

6.4 Methods of Solution for Large Systems

A comprehensive summary of the many numerical methods available is given in reference 4.

(a) DIRECT METHODS These basically invert the admittance matrix, a process consuming both in time and computer storage. Many methods are available and a text on numerical methods should be consulted. These methods include Gaussian elimination.

Triangulation and partitioning

Direct methods solve only linear systems, i.e. $[\mathbf{Y}][\mathbf{V}] = [\mathbf{I}]$ where $[\mathbf{I}]$ is specified. The fact that powers are specified in practice makes the problem nonlinear. A new value for $\mathbf{I}$ must be obtained from $\mathbf{S} = \mathbf{VI}^*$ after each direct solution and this value used to obtain a new one.

Partitioning is useful for handling large problems on small computers, for network reduction, and the elimination of unwanted nodes. A matrix

$$[\mathbf{Y}] = \begin{bmatrix} Y_{11} & Y_{12} & \vdots & Y_{13} \\ Y_{21} & Y_{22} & \vdots & Y_{23} \\ \hline Y_{31} & Y_{32} & \vdots & Y_{33} \end{bmatrix} \tag{6.4}$$

can be partitioned into four submatrices by the dividing lines shown in equation (6.4). Hence, in this typical admittance matrix for a network,

$$[\mathbf{Y}] = \begin{bmatrix} \mathbf{B} & \mathbf{C} \\ \mathbf{C}^t & \mathbf{D} \end{bmatrix}$$

where $\mathbf{C}^t$ is the transpose of $\mathbf{C}$,

$$\mathbf{B} = \begin{bmatrix} Y_{11} & Y_{12} \\ Y_{21} & Y_{22} \end{bmatrix} \quad \text{and so on}$$

A saving in computation time and storage is often obtained by manipulating the submatrices instead of the main one, and in the equation $[\mathbf{I}] = [\mathbf{Y}][\mathbf{V}]$ it is possible to eliminate nodes at which $\mathbf{I}$ is not injected. The matrix $[\mathbf{Y}]$ is partitioned so that the nodes to be eliminated are grouped together in a submatrix, e.g.

$$\begin{bmatrix} \mathbf{I}_w \\ \mathbf{I}_u \end{bmatrix} = \begin{bmatrix} \mathbf{B} & \mathbf{C} \\ \mathbf{C}^t & \mathbf{D} \end{bmatrix} \begin{bmatrix} \mathbf{V}_w \\ \mathbf{V}_u \end{bmatrix} \tag{6.5}$$

where the suffix u indicates unwanted and w wanted nodes. For example if the currents at the unwanted nodes are zero,

$$\mathbf{I}_u = 0 \qquad -\mathbf{C}^t\mathbf{V}_w = \mathbf{D}\mathbf{V}_u \quad \text{and} \quad -\mathbf{D}^{-1}\mathbf{C}^t\mathbf{V}_w = \mathbf{V}_u$$

Substituting this value of $\mathbf{V}_u$ in the expression for $\mathbf{I}_w$

$$\mathbf{I}_w = \mathbf{B}\mathbf{V}_w - \mathbf{C}\mathbf{D}^{-1}\mathbf{C}^t\mathbf{V}_w$$

which gives a new admittance matrix containing only wanted nodes, i.e.

$$[\mathbf{Y}] = \mathbf{B} - \mathbf{C}\mathbf{D}^{-1}\mathbf{C}^t \tag{6.6}$$

Methods are available for modifying the admittance matrix to allow for changes in the configuration of the network, say, due to line outages. These modifications can be performed such that a completely fresh inversion of the new admittance matrix is not necessary.[4]

Most power system matrices are symmetric, i.e. $\mathbf{Y}_{ij} = \mathbf{Y}_{ji}$ and it is only required to store elements on and above the diagonal. Frequently most of the non-zero elements lie within a narrow band about the diagonal and again considerable savings in computer storage may be attained.

Accuracy   When solving the set of equations $[\mathbf{A}][\mathbf{x}] = [\mathbf{b}]$ the residual vector $\mathbf{r} = [\mathbf{b}] - [\mathbf{A}][\mathbf{x}]$ is not zero because of rounding errors. Troubles arise when ill-conditioned equations are obtained in which although the residual is small the solution may be inaccurate. Such matrices are often large and sparse and for most rows the diagonal element is equal to the sum of the non-diagonal elements and of opposite sign.

(b) ITERATIVE METHODS   The two sets of equations (6.2) and (6.3) are solved simultaneously as the iteration proceeds. A number of methods are available, and as two of these have gained wide adoption they will be described in detail. The first is the Gauss–Seidel method which has been widely used for many years and is simple in approach; the second is the Newton–Raphson method which although more complex has certain advantages. The speed of convergence of these methods is of extreme importance, as apart from the cost of computer time the use of these methods in schemes for the automatic control of power systems requires very fast load-flow solutions.

The Gauss–Seidel method

In this method the unknown quantities are initially assumed and the value obtained from the first equation for, say, V_1 is then used when obtaining V_2 from the second equation and so on. Each equation is considered in turn and then the complete set solved again until the values obtained for the unknowns converge to within required limits.

Application of the method to the simple network of Example 6.5. The nodal equations are

$$2V_1 - 0.5V_2 - 1.5V_3 = I_1 = 1$$

$$-0.5V_1 + 1.25V_2 - 0.75V_3 = I_2 = -1.5$$

$$-1.5V_1 - 0.75V_2 + 2.25V_3 = I_3$$

where the voltages are in kV and the currents in kA.

As V_3 is known, i.e. 100 kV (slack busbar voltage) it is necessary only to solve the first two equations. Initially make $V_2 = 100$ kV.

$$\therefore \ ^1V_1 = 0.5(1 + 150 + 50) = 100.5000$$

Using this value to evaluate V_2,

$$^1V_2 = \frac{1}{1.25}(-1.5 + 75 + 50.25) = 99.0000$$

Using this value of V_2 to evaluate V_1,

$$^2V_1 = 0.5(1 + 150 + 49.5) = 100.2500$$

$$^2V_2 = \frac{1}{1.25}(-1.5 + 75 + 50.125) = 98.9000$$

$$^3V_1 = 100.2250 \qquad ^3V_2 = 98.8900$$

$$^4V_1 = 100.2225 \qquad ^4V_2 = 98.8890$$

$$^5V_1 = 100.22225 \qquad ^5V_2 = 98.8888$$

Iterations are now producing changes only in the fourth place of decimals and the process may be stopped.

$$I_3 = -1.5 \times 100.22225 - 0.75 \times 98.8888 + 2.25 \times 100 = 0.5 \text{ kA}.$$

The solution has been obtained with much less computation than the more direct method previously used.

In the three-node system of Example 6.5 the iterative form of the three nodal equations with the nodal constraints that $\mathbf{S} = \mathbf{VI}^*$ is obtained as follows (p indicates the iteration number and is *not* a power). For node 1:

$$\mathbf{I}_1 = \mathbf{V}_1\mathbf{Y}_{11} + \mathbf{V}_2\mathbf{Y}_{12} + \mathbf{V}_3\mathbf{Y}_{13}$$

$$\therefore \ \mathbf{V}_1 = -\mathbf{V}_2\frac{\mathbf{Y}_{12}}{\mathbf{Y}_{11}} - \mathbf{V}_3\frac{\mathbf{Y}_{13}}{\mathbf{Y}_{11}} + \frac{\mathbf{I}_1}{\mathbf{Y}_{11}}$$

$$\therefore \ \mathbf{V}_1^* = -\frac{\mathbf{Y}_{12}^*}{\mathbf{Y}_{11}^*}\mathbf{V}_2^* - \frac{\mathbf{Y}_{13}^*}{\mathbf{Y}_{11}^*}\mathbf{V}_3^* + \frac{\mathbf{I}_1^*}{\mathbf{Y}_{11}^*}$$

Substituting $\mathbf{I}_1^* = \mathbf{S}_1/\mathbf{V}_1$ and writing in the iterative form,

$$\mathbf{V}_1^{p+1} = -\frac{\mathbf{Y}_{12}}{\mathbf{Y}_{11}}\mathbf{V}_2^p - \frac{\mathbf{Y}_{13}}{\mathbf{Y}_{11}}\mathbf{V}_3^p + \frac{\mathbf{S}_1^*}{\mathbf{Y}_{11}}\frac{1}{\mathbf{V}_1^{*p}}$$

Similarly,

$$\mathbf{V}_2^{p+1} = -\frac{\mathbf{Y}_{21}}{\mathbf{Y}_{22}}\mathbf{V}_1^{p+1} - \frac{\mathbf{Y}_{23}}{\mathbf{Y}_{22}}\mathbf{V}_3^p + \frac{\mathbf{S}_2^*}{\mathbf{Y}_{22}}\frac{1}{\mathbf{V}_2^{*p}}$$

and

$$\mathbf{V}_3^{p+1} = -\frac{\mathbf{V}_{31}}{\mathbf{Y}_{33}}\mathbf{V}_1^{p+1} - \frac{\mathbf{V}_{32}}{\mathbf{Y}_{33}}\mathbf{V}_2^{p+1} + \frac{\mathbf{S}_3^*}{\mathbf{Y}_{33}}\frac{1}{\mathbf{V}_3^{*p}}$$

At any node i the already scanned nodes up to i will have new values appropriate to the $(p+1)$ iteration and the nodes yet to be scanned ($j > i$) are appropriate to iteration p. Generally,

$$(\mathbf{V}_i^{p+1}) = +\frac{\mathbf{S}_i^*}{\mathbf{Y}_{ii}} \cdot \frac{1}{\mathbf{V}_i^{*p}} - \sum_{\substack{j>i \\ j \neq i}} \frac{\mathbf{Y}_{ij}}{\mathbf{Y}_{ii}}\mathbf{V}_j^p - \sum_{\substack{j<i \\ j \neq i}} \frac{\mathbf{Y}_{ij}}{\mathbf{Y}_{ii}}\mathbf{V}_j^{p+1} \qquad (6.7)$$

In this particular case node 3 is the slack-bus, and as $\mathbf{V}_i$ is known the equation for it is not required. It is seen in the above equations that the new value of $\mathbf{V}_i$ in the preceding equation is immediately used in the next equation, i.e. $\mathbf{V}_1^{p+1}$ is used in the $\mathbf{V}_2$ equation. Each node is scanned in turn over a complete iteration.

Equation (6.7) refers to a busbar with P and Q specified. At a generator node (i), usually V_i and P_i are specified with perhaps upper and lower limits to Q_i.

The magnitude of $\mathbf{V}_i$ is fixed, but its phase depends on Q_i. The values of $\mathbf{V}_i$ and Q_i from the previous iteration are not related by (6.7) because $\mathbf{V}_i$ has been modified to give a constant value of magnitude. It is necessary at the next iteration to calculate the value of Q_i corresponding to $\mathbf{V}_i$ from equation (6.7), i.e. Q_i = imaginary part of $\mathbf{S}_i$, i.e. of

$$\mathbf{Y}_{ii}^{*}\mathbf{V}_i^{p-1}\left[\mathbf{V}_i^{*p}+\sum_{\substack{j\neq i\\j>i}}\frac{\mathbf{Y}_{ij}^{*}}{\mathbf{Y}_{ii}^{*}}\mathbf{V}_j^{*p-1}+\sum_{\substack{j\neq i\\j<i}}\frac{\mathbf{Y}_{ij}^{*}}{\mathbf{Y}_{ii}^{*}}\mathbf{V}_j^{*p}\right]$$

This value of Q_i holds for the existing value of $\mathbf{V}_i$ and is then substituted into

$$-\sum_{\substack{j\neq i\\j>i}}\frac{\mathbf{Y}_{ij}^{*}}{\mathbf{Y}_{ii}^{*}}\mathbf{V}_j^{*p}+\frac{P_i+\mathrm{j}Q_i}{\mathbf{Y}_{ii}^{*}\mathbf{V}_i^{p}}-\sum_{\substack{j\neq i\\j<i}}\frac{\mathbf{Y}_{ij}^{*}}{\mathbf{Y}_{ii}^{*}}\mathbf{V}_j^{*p+1}$$

to obtain $(\mathbf{V}_i^{p+1})$.*

The real and imaginary components of $\mathbf{V}_i^{p+1}$ are then multiplied by the ratio V_i/V_i^{p+1}, thus complying with the constant V_i constraint. This final step is a slight approximation, but the error involved is small and the saving in computation large. The phase of $\mathbf{V}_i$ is thus found and the iteration can proceed to the next node. The process continues until the value of V^{p+1} at any node differs from V^p by a specified amount, a common figure is 0.0001 p.u. The study is commenced by assuming 1 p.u. voltage at all nodes except one; this exception is necessary in order that current flows may be obtained in the first iteration.

Acceleration factors   The number of iterations required to reach the specified convergence can be greatly reduced by the use of acceleration factors.[5] The correction in voltage from V^p to V^{p+1} is multiplied by such a factor so that the new voltage is brought closer to its final value, i.e.

$$^1\mathbf{V}^{p+1} = \mathbf{V}^p + \omega(\mathbf{V}^{p+1} - \mathbf{V}^p)$$

$$= \mathbf{V}^p + \omega\,\Delta\mathbf{V}^p$$

where $^1\mathbf{V}^{p+1}$ is the accelerated new voltage. It can be shown that a complex value of ω reduces the number of iterations more than a real value. However up to the present real values only seem to be used; the actual value depends on the nature of the system under study but a value of 1.6 is widely used. The Gauss–Seidel method with the use of acceleration factors is known as the method of successive over-relaxation.

Transformer tap-changing

Further changes which must be accommodated in the admittance matrix are those due to transformer tap-changing. When the ratio is at the nominal value the transformer is represented by a single series impedance, but when off-nominal, adjustments have to be made as follows.

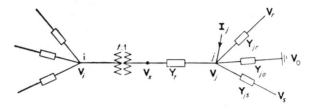

Figure 6.8(a) Equivalent circuit of transformer with off-nominal tap ratio. Transformer series admittance on non-tap side.

Consider a transformer of ratio $t:1$ connected between two nodes i and j; the series admittance of the transformer is $\mathbf{Y}_t$. Referring to Figure 6.8(a) the following nodal-voltage equation holds.

$$\mathbf{I}_j = \mathbf{V}_j(\mathbf{Y}_{jr} + \mathbf{Y}_{j0} + \mathbf{Y}_{js} + \mathbf{Y}_t) - (\mathbf{V}_r\mathbf{Y}_{jr} + \mathbf{V}_x\mathbf{Y}_t + \mathbf{V}_s\mathbf{Y}_{js} + \mathbf{V}_0\mathbf{Y}_{j0})$$

As $\mathbf{V}_0 = 0$ and $\mathbf{V}_x = \mathbf{V}_i/t$,

$$\mathbf{I}_j + \mathbf{V}_s\mathbf{Y}_{js} = \mathbf{V}_j(\mathbf{Y}_{jr} + \mathbf{Y}_{j0} + \mathbf{Y}_{js} + \mathbf{Y}_t) - \left(\mathbf{Y}_{jr}\mathbf{V}_r + \frac{\mathbf{Y}_t\mathbf{V}_i}{t}\right)$$

x is an artificial node between the voltage transforming element and the transformer admittance. From this last equation it is seen that for the node on the off-tap side of the transformer (i.e. the more remote of the two nodes i and j),

when forming $\mathbf{Y}_{jj}$ use $\mathbf{Y}_t$ for the transformer, and when forming $\mathbf{Y}_{ij}$ use $\mathbf{Y}_t/t$ for the transformer.

It can be similarly shown for the tap-side node that,

when forming $\mathbf{Y}_{ii}$ use $\mathbf{Y}_t/t^2$, and when forming $\mathbf{Y}_{ij}$ use $\mathbf{Y}_t/t$.

These conditions may be represented by the π section shown in Figure 6.8(b), although it is probably easier to modify the mutual and self-admittances directly.

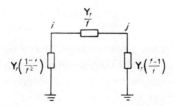

Figure 6.8(b) π-section to represent transformer with off-nominal tap ratio.

The Newton–Raphson method

Although the Gauss–Seidel method is well established, more recently the Newton–Raphson method is being increasingly used. With some systems it gives a greater assurance of convergence and is at the same time economical in computer time.

The basic iterative procedure is as follows:

$$\text{value at new iteration, } x^{p+1} = x^p - \frac{f(x^p)}{f'(x^p)}$$

extending this to a multi-equation system,

$$\mathbf{x}^{p+1} = \mathbf{x}^{(p)} - J^{-1}(\mathbf{x}^p)f(\mathbf{x}^p)$$

where $\mathbf{x}$ and $\mathbf{f}$ are column vectors and $J(\mathbf{x}^p)$ is a matrix known as the Jacobian matrix, of the form

$$\begin{bmatrix} \dfrac{\partial f_1}{\partial x_1} & \cdot & \cdot & \cdot & \dfrac{\partial f_1}{\partial x_n} \\ \cdot & & \cdot & & \cdot \\ \cdot & \cdot & \dfrac{f_k}{x_k} & \cdot & \cdot \\ \cdot & & \cdot & & \cdot \\ \dfrac{\partial f_n}{\partial x_1} & \cdot & \cdot & \cdot & \dfrac{\partial f_n}{\partial x_n} \end{bmatrix} \quad p\text{th iteration.}$$

Consider now the application to an n-node power system, for a link connecting nodes k and j of admittance $\mathbf{Y}_{kj}$,

$$P_k + jQ_k = \mathbf{V}_k \mathbf{I}_k^* = \mathbf{V}_k \sum_{j=1}^{n-1} (\mathbf{Y}_{kj}\mathbf{V}_j)^*$$

Let

$$\mathbf{V}_k = a_k + jb_k \quad \text{and} \quad \mathbf{Y}_{kj} = G_{kj} - jB_{kj}$$

Then,

$$P_k + jQ_k = (a_k + jb_k) \sum_{1}^{n-1} [(G_{kj} - jB_{kj})(a_j + jb_j)]^* \tag{6.8}$$

from which,

$$P_k = \sum_{j=1}^{n-1} [a_k(a_jG_{kj} + b_jB_{kj}) + b_k(b_jG_{kj} - a_jB_{kj})] \tag{6.9}$$

$$Q_k = \sum_{j=1}^{n-1} [b_k(a_jG_{kj} + b_jB_{kj}) - a_k(b_jG_{kj} - a_jB_{kj})] \tag{6.10}$$

Hence, there are two non-linear simultaneous equations for each node. Note that $(n-1)$ nodes are considered because the slack node n is completely specified.

Changes in P and Q are related to changes in a and b by equations (6.9) and (6.10), e.g.

$$\Delta P_1 = \frac{\partial P_1}{\partial a_1} \Delta a_1 + \frac{\partial P_1}{\partial a_2} \Delta a_2 + \cdots \frac{\partial P_1}{\partial a_{n-1}} \Delta a_{n-1}$$

Similar equations hold in terms of ΔP and Δb, and ΔQ in terms of Δa and Δb.

The equations may be expressed generally in the following manner.

$$
\begin{bmatrix}
\Delta P_1 \\
\cdot \\
\cdot \\
\cdot \\
\Delta P_{n-1} \\
\Delta Q_1 \\
\cdot \\
\cdot \\
\cdot \\
\Delta Q_{n-1}
\end{bmatrix}
=
\begin{bmatrix}
\dfrac{\partial P_1}{\partial a_1} & \cdots & \dfrac{\partial P_1}{\partial a_{n-1}} & \dfrac{\partial P_1}{\partial b_1} & \cdots & \dfrac{\partial P_1}{\partial b_{n-1}} \\
\cdots & & \cdots & & \cdots & \\
\cdots & & \cdots & & \cdots & \\
\cdots & & \cdots & & \cdots & \\
\dfrac{\partial P_{n-1}}{\partial a_1} & \cdots & \dfrac{\partial P_{n-1}}{\partial a_{n-1}} & \dfrac{\partial P_{n-1}}{\partial b_1} & \cdots & \dfrac{\partial P_{n-1}}{\partial b_{n-1}} \\
\dfrac{\partial Q_1}{\partial a_1} & \cdots & \dfrac{\partial Q_1}{\partial a_{n-1}} & \dfrac{\partial Q_1}{\partial b_1} & \cdots & \dfrac{\partial Q_1}{\partial b_{n-1}} \\
\cdots & & \cdots & & \cdots & \\
\cdots & & \cdots & & \cdots & \\
\cdots & & \cdots & & \cdots & \\
\dfrac{\partial Q_{n-1}}{\partial a_1} & \cdots & \dfrac{\partial Q_{n-1}}{\partial a_{n-1}} & \dfrac{\partial Q_{n-1}}{\partial b_1} & \cdots & \dfrac{\partial Q_{n-1}}{\partial b_{n-1}}
\end{bmatrix}
\begin{bmatrix}
\Delta a_1 \\
\cdot \\
\cdot \\
\cdot \\
\Delta a_{n-1} \\
\Delta b_1 \\
\cdot \\
\cdot \\
\cdot \\
\Delta b_{n-1}
\end{bmatrix}
$$

$$(6.11)$$

For convenience denote the Jacobian matrix by $\begin{array}{|c|c|} \hline J_A & J_B \\ \hline J_C & J_D \\ \hline \end{array}$

The elements of the matrix are evaluated for the values of P, Q, and V at each iteration as follows.

For the submatrix J_A and from equation (6.9),

$$\frac{\partial P_k}{\partial a_j} = a_k G_{kj} - b_k B_{kj} \tag{6.12}$$

where $k \neq j$, i.e. off-diagonal elements.

Diagonal elements,

$$\frac{\partial P_k}{\partial a_k} = 2a_k G_{kk} + b_k B_{kk} - b_k B_{kk} + \sum_{\substack{j=1 \\ j \neq k}}^{n-1} (a_j G_{kj} + b_j B_{kj}) \tag{6.13}$$

This element may be more readily obtained by expressing some of the quantities in terms of the current at node k, $\mathbf{I}_k$ which can be determined separately at each iteration.

Let

$$\mathbf{I}_k = c_k + jd_k = (G_{kk} - jB_{kk})(a_k + jb_k) + \sum_{\substack{j=1 \\ j \neq k}}^{n-1} (G_{kj} - jB_{jk})(a_j + jb_j)$$

from which,

$$c_k = a_k G_{kk} + b_k B_{kk} + \sum_{\substack{j=1 \\ j \neq k}}^{n-1} (a_j G_{kj} + b_j B_{kj})$$

and

$$d_k = b_k G_{kk} - a_k B_{kk} + \sum_{\substack{i=1 \\ j \neq k}}^{n-1} (b_j G_{kj} - a_j B_{jk})$$

So that,

$$\frac{\partial P_k}{\partial a_k} = a_k G_{kk} - b_k B_{kk} + c_k$$

For J_{B},

$$\frac{\partial P_k}{\partial b_k} = a_k B_{kk} + b_k G_{kk} + d_k$$

and

$$\frac{\partial P_k}{\partial b_j} = a_k B_{kj} + b_k G_{kj} \qquad (k \neq j)$$

For J_{C},

$$\frac{\partial Q_k}{\partial a_k} = a_k B_{kk} + b_k G_{kk} - d_k$$

and

$$\frac{\partial Q_k}{\partial a_j} = a_k B_{kj} + b_k G_{kj} \qquad (k \neq j)$$

For J_{D},

$$\frac{\partial Q_k}{\partial b_j} = -a_k G_{kj} + b_k B_{kj} \qquad (k \neq j)$$

and

$$\frac{\partial Q_k}{\partial b_k} = -a_k G_{kk} + b_k B_{kk} + c_k$$

The process commences with the iteration counter 'p' set to zero and all the nodes except the slack bus being assigned voltages, usually 1 p.u.

From these voltages, P and Q are calculated from equations (6.9) and (6.10). The changes are then calculated,

$$\Delta P_k^p = P_k \text{ (specified)} - P_k^p \text{ and } \Delta Q_k^p = Q_k \text{ (specified)} - Q_k^p,$$

where p is the iteration number.

Next the node currents are computed,

$$I_k^p = \left(\frac{P_k^p + jQ_k^p}{V^p}\right)^* = c_k^p + jd_k^p$$

The elements of the Jacobian matrix are then formed and from (6.11),

$$\begin{array}{|c|} \hline \Delta a \\ \hline \Delta b \\ \hline \end{array} = \begin{array}{|c|c|} \hline J_A & J_B \\ \hline J_C & J_D \\ \hline \end{array}^{-1} \begin{array}{|c|} \hline \Delta P \\ \hline \Delta Q \\ \hline \end{array}$$

(6.14)

Hence, a and b are determined and the new values, $a_k^{p+1} = a_k^p + \Delta a_k^p$ and $b_k^{p+1} = b_k^p + \Delta b_k^p$ obtained. The process is repeated ($p = p + 1$) until ΔP and ΔQ are less than a prescribed tolerance.

The Newton–Raphson method is reported to have better convergence characteristics and for many systems to be faster than the Gauss–Seidel, it has a much larger time per iteration but requires very few iterations (four is general), whereas the Gauss–Siedel requires at least 30 iterations, the number increasing with the size of system.

Acceleration factors may be used for the Newton–Raphson method. The quantities are frequently expressed in polar form.

The polar form of the equations has advantages and the equations are:

$$\mathbf{P}_k = \mathbf{P}(V, \theta)$$
$$Q_k = Q(V, \theta)$$

(6.15)

The power at a bus is

$$\mathbf{S}_k = P_k + jQ_k = \mathbf{V}_k \mathbf{I}_k^*$$

$$= V_k \sum_{m \neq k} Y_{km}^* V_m^*$$

(6.16)

$$P_k = \sum_{m \neq k} V_k V_m (G_{km} \cos \theta_{km} + B_{km} \sin \theta_{km})$$

(6.17)

$$Q_k = \sum_{m \neq k} V_k V_m (G_{km} \sin \theta_{km} - B_{km} \cos \theta_{km}) \qquad (6.18)$$

where $\theta_{km} = \theta_k - \theta_m$.

For a load bus,

$$\Delta P_k = \sum_{m \neq k} \frac{\partial P_k}{\partial \theta_m} \Delta \theta_m + \sum_{m \neq k} \frac{\partial P_k}{\partial V_m} \Delta V_m \qquad (6.19)$$

$$\Delta Q_k = \sum \frac{\partial Q_k}{\partial \theta_m} \Delta \theta_m + \sum \frac{\partial Q_k}{\partial V_m} \Delta V_m$$

For a generator (P, V) busbar only the ΔP_k equation is used as Q_k is not specified. The mismatch equation is:

$$
\begin{array}{|c|}
\hline
\mathbf{\Delta P}^{P-1} \\
\hline
\mathbf{\Delta Q}^{P-1} \\
\hline
\end{array}
=
\begin{array}{|c|c|}
\hline
\mathbf{H}^{P-1} & \mathbf{N}^{P-1} \\
\hline
\mathbf{J}^{P-1} & \mathbf{L}^{P-1} \\
\hline
\end{array}
\begin{array}{|c|}
\hline
\mathbf{\Delta \theta}^{P} \\
\hline
\left(\dfrac{\mathbf{\Delta V}^{P}}{\mathbf{V}^{P-1}} \right) \\
\hline
\end{array}
\qquad (6.20)
$$

$\Delta \theta^P$ is the correction to P, Q and PV buses and $\Delta V^P/V^{P-1}$ is the correction to P, Q buses.

For buses k and m,

$$H_{km} = \frac{\partial P_k}{\partial \theta_m} = V_k V_m (G_{km} \sin \theta_{km} - B_{km} \cos \theta_{km})$$

$$N_{km} = V_m \frac{\partial P_k}{\partial V_m} = V_k V_m (G_{km} \cos \theta_{km} + B_{km} \sin \theta_{km})$$

$$J_{km} = \frac{\partial Q_k}{\partial \theta_m} = -V_k V_m (G_{km} \cos \theta_{km} + B_{km} \sin \theta_{km})$$

$$L_{km} = V_m \frac{\partial Q_m}{\partial V_m} = V_k V_m (G_{km} \sin \theta_{km} - B_{km} \cos \theta_{km})$$

Also,

$$H_{kk} = -Q_k - B_{kk} V_k^2$$

$$N_{kk} = P_k + G_{kk} V_k^2$$

$$J_{kk} = P_k - G_{kk} V_k^2$$

$$L_{kk} = Q_k - B_{kk} V_k^2$$

In the above the admittance of the link km, $Y_{km} = G_{km} + jB_{km}$. The computational process can be enhanced by the following:

Preordering in which nodes are numbered in sequence of increasing number of connections.

Dynamic Ordering in which at each step in the elimination the next row to be operated on has the fewest non-zero terms.

Decoupled load flow

The coupling between the P–θ and Q–V components is weak. Hence the equations can be reduced to

$$[P] = [T][\theta]$$
$$[Q] = [U][V - V_0]$$

(6.21)

At reference node $\theta_0 = 0$ and $V_k = V_0$

$$T_{km} = -\frac{V_k V_m}{Z_{km}^2/X_{km}}; \qquad U_{km} = -\frac{1}{Z_{km}^2/X_{km}}$$

$$T_{kk} = -\sum_{m \neq k} T_{km}; \qquad U_{kk} = -\sum_{m \neq k} U_{km}$$

where Z_{km} and X_{km} are the branch impedance and reactance and $[U]$ is constant valued.

$$[\Delta P] = [T][\Delta \theta]$$
$$[\Delta Q/V] = [U][\Delta V]$$

These are solved alternatively. Advantage is obtained by using the following:

$$[\Delta P/V] = [A][\Delta \theta]$$
$$[\Delta Q/V] = [C][\Delta V]$$

(6.22)

Fact decoupled load flow

This makes the Jacobians of the decoupled method constant in value throughout the iteration. The following assumptions are made:

$$E_k = E_m = 1 \text{p.u.}$$

$G_{km} \ll B_{km}$ and can be ignored; this is reasonable for lines and cables

$$\cos(\theta_k - \theta_m) \doteq 1$$
$$\sin(\theta_k - \theta_m) \doteq 0$$

Hence,

$$[\Delta P] = [B][\Delta \theta]$$
$$[\Delta Q] = [B][\Delta V]$$

where $B_{km} = -B_{km}$ for $m \neq k$

$$\bar{B}_{kk} = \sum_{m \neq k} B_{km}$$

Further assumptions yield:

$$\left[\frac{\Delta P}{V}\right] = [B'][\Delta\theta]$$

$$\left[\frac{\Delta Q}{V}\right] = [B''][V]$$

(6.23)

Where

$$B'_{km} = -\frac{1}{X_{km}} \, (m \neq k)$$

$$B'_{kk} = \sum_{m \neq k} \frac{1}{X_{km}}$$

$$B''_{km} = -B_{km}(m \neq k)$$

$$B''_{kk} = \sum_{m \neq k} B_{km}$$

Matrices B' and B'' are real and constant in value and need to be triangulated only once.

Example 6.6 Using the fast decoupled method calculate the angles and voltages after the first iteration for the 3-node network described by the following admittance matrix (From Figure 6.9). V_1 is 230 kV, initially $V_2 = 220$ kV, $\theta_2 = 0$, and V_3 is 228 kV, $\theta_3 = 0$. Node 2 is a load node consuming 200 MW, 120 MVAr, node 3 is a generator node set at 70 MW and 228 kV. V_1 is an infinite busbar.

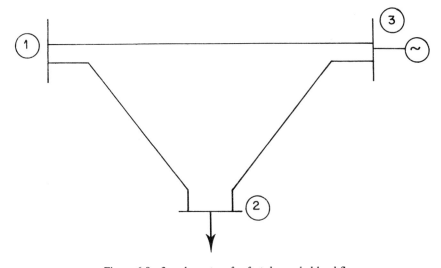

Figure 6.9   3-node system for fast decoupled load flow.

1	$0.00819 - j0.049099$	$-0.003196 + j0.019272$	$-0.004994 + j0.030112$
$\mathbf{Y} = 2$	$-0.003196 + j0.019272$	$0.007191 - j0.043099$	$-0.003995 + j0.02409$
3	$-0.004994 + j0.030112$	$-0.003995 + j0.02409$	$0.008989 - j0.053952$

$$\begin{bmatrix} \dfrac{\Delta p_2}{V_2} \\[2mm] \dfrac{\Delta P_3}{V_3} \end{bmatrix} = \begin{bmatrix} \mathbf{B}'_{22} & \mathbf{B}'_{23} \\ \mathbf{B}'_{32} & \mathbf{B}'_{33} \end{bmatrix} \begin{bmatrix} \Delta\theta_2 \\ \Delta\theta_3 \end{bmatrix}$$

$V_1 = 230\,\text{kV}, \; V_2 = 220\,\text{kV}, \; \theta_2 = 0, \; V_3 = 220\,\text{kV}, \; \theta_3 = 0$

$$B'_{22} = \frac{V_1}{X_{12}} + \frac{V_3}{X_{23}} = \frac{220}{50.5} + \frac{220}{40.4} = 9.80198$$

$$B'_{23} = -\frac{V_3}{X_{23}} = -\frac{220}{40.4} = -5.4455$$

$$B'_{32} = -\frac{V_2}{X_{23}} = -\frac{220}{40.4} = -5.4455$$

$$B'_{33} = \frac{V_1}{X_{13}} + \frac{V_2}{X_{23}} = \frac{220}{32.32} + \frac{220}{40.4} = 12.25247$$

$$C_{32} = L_{32} = \frac{B'_{32}}{B'_{22}} = -\frac{5.4455}{9.80198} = -0.5555$$

$$D_{32} = -B'_{22} = 9.80198$$

$$D_{33} = B'_{33} - \frac{B'_{32}B'_{23}}{B'_{22}} = 12.25247 - \frac{5.4455^2}{9.80198} = 9.2272$$

$$\Delta P_2^0 = -200 - 220(-0.003196 \times 230 + 0.00719 \times 220 - 0.003995 \times 228)$$
$$= -185.937\,\text{MW}$$

$$\Delta P_3^0 = 70 - 228\,(-0.004994 \times 230 - 0.003995 \times 220 + 0.008989 \times 228)$$
$$= 64.99\,\text{MW}$$

$$\begin{bmatrix} D_{22} & \\ & D_{33} \end{bmatrix} \begin{bmatrix} 1 & L_{23} \\ 0 & 1 \end{bmatrix} \begin{bmatrix} \Delta\theta_2 \\ \Delta\theta_3 \end{bmatrix} = \begin{bmatrix} \Delta\theta'_2 \\ \Delta\theta'_3 \end{bmatrix}$$

$$\begin{bmatrix} 1 & 0 \\ C_{32} & 1 \end{bmatrix} \begin{bmatrix} \Delta\theta'_2 \\ \Delta\theta'_3 \end{bmatrix} = \begin{bmatrix} \dfrac{\Delta P_2}{V_2} \\[2mm] \dfrac{\Delta P_3}{V_3} \end{bmatrix}$$

$$\Delta\theta_i' = \frac{\Delta P_i}{V_i} - \sum_{k=2}^{i-1} C_{ik}\Delta\theta_k' \quad (i = 2, 3)$$

$$\Delta\theta_i = \frac{\Delta\theta_i'}{D_{ii}} - \sum_{k=3}^{i+1} L_{ik}\Delta\theta_k \quad (i = 3, 2)$$

$$\Delta\theta_2' = \frac{\Delta P_2^{(0)}}{V_2} = \frac{-185.937}{220} = -0.845168$$

$$\Delta'\theta_3 = \left(\frac{\dfrac{64.99}{228} - 0.5555 \times 0.84516}{9.2272}\right) = \frac{-0.1844}{9.2272} = -0.1998 \text{ rad}$$

$$\Delta\theta_2' = \frac{-0.845168}{9.80198} - 0.5555 \times 0.01988 = -0.09732 \text{ rad}$$

Hence,

$$\theta_2' = \theta_2^{(0)} + \Delta\theta_2^{(0)} = -0.09732 \text{ rads} = -5.576°$$

$$\theta_3' = \theta_3^0 + \Delta\theta_3' = -0.012 \text{ rads} = -1.14477°$$

Considering reactive power at node 2,

$$\begin{aligned}\Delta Q_2^0 &= Q_2 - V_2^0\{V_1[G_{21}\sin(\theta_2' - \theta_1) - B_{21}\cos(\theta_2' - \theta_1)] - V_2^0 B_{22} \\ &\quad + V_3^0[G_{23}\sin(\theta_2' - \theta_3') - B_{23}\cos(\theta_2' - \theta_3')]\} \\ &= -120 - 220\{230\,[-0.00319\sin(-5.576 - 0) \\ &\quad -0.019272\cos(-5.576 - 0) + 220 \times 0.043049 \\ &\quad + 228[-0.003995\sin(-5.576 + 1.14477) \\ &\quad -0.02409\cos(-5.576 + 1.14477)]\} = -59.476 \text{ MVAr}\end{aligned}$$

Hence

$$\Delta V_2' = \frac{\Delta Q_2^0}{B_{22}'' V_2^0} = \frac{-59.476}{0.043049 \times 220} = -6.28 \text{ kV}$$

and

$$V_1 = 230 \text{ kV}$$

$$V_2' = V_2^0 + \Delta V_2' = 220 - 6.28 = 213.72 \text{ kV}$$

$$V_3' = V_3^0 = 228 \text{ kV}$$

The process is repeated for the next iteration and so on until convergence is reached when,

$$P_1 = 134.389 \text{ MW}$$

$$Q_1 = 56.77 \text{ MVAr}$$

236

6.5 Example of a Complex Load Flow

The line diagram of a system is shown in Figure 6.10 along with details of line and transformer impedances and loads. This represents a slightly simplified arrangement of the network used in reference 6. All quantities are expressed as per unit and the nodes are numbered as indicated. They are not ordered consecutively to show in the solution that the numbering system is not important, although in more sophisticated studies on larger systems it has been shown that certain orderings of nodes can produce faster convergence and solutions.

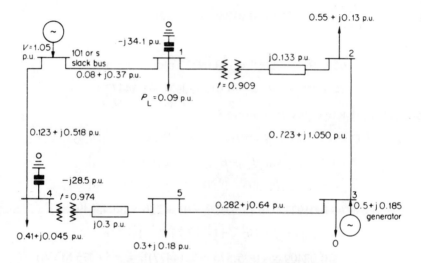

Figure 6.10   System for complex load flow using iterative method.

The solution of this problem has been carried out by digital computer, and a brief description of the arrangement of essential data will be given and the first iteration performed on a desk calculating machine. This program was developed for instructional purposes and is not as refined or sophisticated as commercial programs; it is hoped, however, that the beginner will find it easier to understand than the latter. In view of the nature of equation (6.7) the basic system data will be modified to allow less computation during the actual solution of the equations; this will be obvious as the method is described. The system admittance-matrix $[\mathbf{Y}]$ is stored by specifying the following: order of matrix (n), number of off-diagonal elements, number of off-diagonal elements per row, column numbers of the off-diagonal elements in each row, a list of off-diagonal elements (Y_{ij}/Y_{ii}). This input data is arranged in two tables in the following manner.

Table A

l = number of links
n = number of non-slack nodes.

i Sending end note	j – Receiving end note	R Resistance per unit	X Reactance per unit	B Susceptance per unit	t Off-nominal transformer ratio

$n = 1$ to 99.
slack nodes are numbered starting at 101,
earth connexions are numbered 0.
When representing transformers, node i must be the
tap-side node and R and X refer to the non-tap side

Table B

l = number of links
n = number of non-slack nodes
s = number of slack nodes
(suffix G → generation, suffix L → loads)

Node number	V_{real} per unit	V_{imag} per unit	P_G per unit	Q_G per unit	P_L per unit	Q_L per unit

P and Q are considered positive for watts and lagging vars generated or supplied and negative when received. It should be noted that the $\mathbf{Y}_{ij}$'s are in fact negative although they are shown as positive in the tables; this is useful as the $\mathbf{Y}_{ij}$ terms appear as negative in the nodal equations. The connexions to the slack busbar (V_s) are shown separately, no equation is necessary for the slack nodes as the voltages are fully specified. The input data are shown in Table 6.1 and identification of the various quantities will be made easier by reference to the following system matrix:

$$\begin{bmatrix} \mathbf{Y}_{11} & \mathbf{Y}_{12} & \mathbf{Y}_{13} & \mathbf{Y}_{14} & \mathbf{Y}_{15} & \mathbf{Y}_{1s} \\ \mathbf{Y}_{21} & \mathbf{Y}_{22} & \mathbf{Y}_{23} & \mathbf{Y}_{24} & \mathbf{Y}_{25} & \mathbf{Y}_{2s} \\ \mathbf{Y}_{31} & \mathbf{Y}_{32} & \mathbf{Y}_{33} & \mathbf{Y}_{34} & \mathbf{Y}_{35} & \mathbf{Y}_{3s} \\ \mathbf{Y}_{41} & \mathbf{Y}_{42} & \mathbf{Y}_{43} & \mathbf{Y}_{44} & \mathbf{Y}_{45} & \mathbf{Y}_{4s} \\ \mathbf{Y}_{51} & \mathbf{Y}_{52} & \mathbf{Y}_{53} & \mathbf{Y}_{54} & \mathbf{Y}_{55} & \mathbf{Y}_{5s} \end{bmatrix}$$

The initial part of the computer flow diagram is shown in Figure 6.11.

Before proceeding with the first iteration the calculation of the admittances associated with one of the transformers will be shown. Consider the transformer 1–2 of impedance j0.133 per unit and having $t = 0.909$. For the off-tap side node 2, $\mathbf{Y}_{22}$ is formed using $\mathbf{Y}_t$ only, i.e.

$$\mathbf{Y}_{22} = \frac{1}{0.723 + j1.05} + \frac{1}{j0.133}$$

$$= 0.444860 - j8.164860$$

$$\mathbf{Y}_{21} = \frac{1}{j0.133} \cdot \frac{1}{0.909} = -j8.278000.$$

The formation of the admittance matrix is necessary for both the Gauss–Seidel and Newton–Raphson methods. At this point the methods differ and first the application of Gauss–Seidel method will be given with a complete computer solution and then an outline of the Newton-Raphson method applied to the problem. For node 1:

$$\mathbf{Y}_{11} = \frac{1}{0.08 + j0.37} + \frac{1}{-j34.1} + \frac{1}{j0.133} \cdot \frac{1}{0.909^2} = 0.558269 - j11.652234$$

$$\mathbf{Y}_{12} = \mathbf{Y}_{21}$$

First iteration. Node 1:

$$\mathbf{V}_1^{1*} = -\left(\frac{\mathbf{Y}_{12}}{\mathbf{Y}_{11}}\right)^* \mathbf{V}_2^* - \left(\frac{\mathbf{Y}_{1s}}{\mathbf{Y}_{11}}\right)^* \mathbf{V}_s^* + \frac{\mathbf{S}_1}{\mathbf{Y}_{11}^*} \frac{1}{\mathbf{V}_1}$$

$$= (0.708238 - j0.033932)^* 1^* + (0.234539 + j0.039069)^* 1.05^*$$

$$+ (-0.000369 + j0.007706)\frac{1}{1}$$

$$\therefore \ \mathbf{V}_1^{1*} = 0.942409 + j0.002569 \ \text{p.u.}$$

Node 2:

$$\mathbf{V}_2^{1*} = -\left(\frac{\mathbf{Y}_{21}}{\mathbf{Y}_{22}}\right)^* \mathbf{V}_1^{1*} - \left(\frac{\mathbf{Y}_{23}}{\mathbf{Y}_{22}}\right)^* \mathbf{V}_3^* + \left(\frac{\mathbf{S}_2}{\mathbf{Y}_{22}^*}\right) \frac{1}{\mathbf{V}_2}$$

$$= (1.010063 + j0.055033)(0.942409 + j0.002569)$$

$$+ (0.081853 - j0.050025)1 + (-0.019534 + j0.066298)\frac{1}{1}$$

$$\therefore \ \mathbf{V}_2^{1*} = 1.014073 + j0.070671 \ \text{p.u.}$$

Node 3:

$$\mathbf{V}_3^{1*} = -\left(\frac{\mathbf{Y}_{32}}{\mathbf{Y}_{33}}\right)^* \mathbf{V}_2^{1*} - \left(\frac{\mathbf{Y}_{35}}{\mathbf{Y}_{33}}\right)^* \mathbf{V}_5^* + \left(\frac{\mathbf{S}_3}{\mathbf{Y}_{33}^*}\right) \frac{1}{\mathbf{V}_3}$$

$$= (0.353069 - j0.043097)(1.014073 + j0.070671)$$

$$+ (0.646931 + j0.043097)1 + (0.179357 - j0.162088)\frac{1}{1}$$

$$\mathbf{V}_3^{1*} = 1.187371 - j0.137743 \ \text{p.u.}$$

Table 6.1

+5- n			
+8- l			

+1		
+2		
+2		} Off-diagonal elements per row
+1		
+2		

+2		
+3	+1	
+5	+2	} Column numbers
+5		
+4	+3	

+0.708238	−0.033932	Y_{12}/Y_{11}
+0.081853	+0.050025	Y_{23}/Y_{22}
+1.010063	−0.055033	Y_{21}/Y_{22}
+0.646931	−0.043097	Y_{35}/Y_{33}
+0.353069	+0.043097	Y_{32}/Y_{33} } Off-diagonal elements $\mathbf{Y}_{ij}/\mathbf{Y}_{ii}$
+0.640699	−0.052397	Y_{45}/Y_{44}
+0.726081	−0.090184	Y_{54}/Y_{55}
+0.292797	+0.087839	Y_{53}/Y_{55}

Generation Table 1

+1.000000	+0.000000	
+1.000000	+0.000000	
+1.000000	+0.000000	} Voltage estimates, v_i
+1.000000	+0.000000	
+1.000000	+0.000000	

−0.000369	+0.007706	
−0.019534	+0.066298	
+0.179357	−0.162088	} $P+\mathrm{j}Q/Y_{ii}^{*}$'s
−0.014702	+0.076068	
−0.046094	+0.058905	

+0.234539	+0.039069	
+0	+0	
+0	+0	} $V_sY_{is}/\mathbf{Y}_{ii}$'s slack bus connex
+0.366206	+0.055921	ions
+0	+0	

+0.558269	−11.652234	
+0.444860	−8.164860	
+1.021401	−1.954524	} $\mathbf{Y}_{ii}$'s
+0.433934	−5.306045	
+0.576541	−4.641795	

+0.223371	+0.037209	+101.000000
+0.000000	+0.000000	+0.000000
+0.000000	+0.000000	+0.000000 } $\mathbf{Y}_{is}/Y_{ii}$'s + slack nos.
+0.348767	+0.053259	+101.000000
+0.000000	+0.000000	+0.000000

Node 4:

$$\mathbf{V}_4^{1*} = -\left(\frac{\mathbf{Y}_{45}}{\mathbf{Y}_{44}}\right)^* \mathbf{V}_5^* - \left(\frac{\mathbf{Y}_{4s}}{\mathbf{Y}_{44}}\right)^* \mathbf{V}_s^* + \left(\frac{\mathbf{S}}{\mathbf{Y}_{44}^*}\right)\frac{1}{\mathbf{V}_4}$$

$$= (0.640699 + j0.052397)(1) + (0.366206 - j0.055921)$$

$$+ (-0.014702 + j0.076068)\frac{1}{1}$$

$$\mathbf{V}_4^{1*} = 0.992202 + j0.072543 \text{ p.u.}$$

Node 5:

$$\mathbf{V}_5^{1*} = -\left(\frac{\mathbf{Y}_{54}}{\mathbf{Y}_{55}}\right)^* \mathbf{V}_4^{1*} - \left(\frac{\mathbf{Y}_{53}}{\mathbf{Y}_{55}}\right)^* \mathbf{V}_3^{1*} + \left(\frac{\mathbf{S}}{\mathbf{Y}_{55}^*}\right)\frac{1}{\mathbf{V}_5}$$

$$= (0.726081 + j0.090184)(1.010513 + j0.069748)$$

$$+ (0.292797 - j0.087839)(1.187371 - j0.137743)$$

$$+ (-0.046094 + j0.058905)\frac{1}{1}$$

$$\therefore \ \mathbf{V}_5^{1*} = 1.003342 - j0.056430 \text{ p.u.}$$

$\mathbf{V}_s$ or $\mathbf{V}_{101} = 1.05000 + j0.000000$ p.u.

It will be noticed that the latest value of each nodal voltage is used. In this iteration no acceleration factor has been used (i.e. the factor = 1); if a factor of 1.6 is used, $\mathbf{V}_1^1$ becomes

$$\mathbf{V}_1 + 1.6(\mathbf{V}_1^1 - \mathbf{V}_1)$$

i.e.

$$1 + 1.6(-0.057591 - j0.002569)$$

or

$$0.907854 - j0.004110 \text{ p.u.}$$

This modified value of $\mathbf{V}_1^1$ should then be used to evaluate $\mathbf{V}_2^1$ when $\mathbf{V}_2^1$ is modified to $\mathbf{V}_2 + 1.6(\mathbf{V}_2^1 - \mathbf{V}_2)$ and so on. The busbar voltages after 30 iterations are

$$\mathbf{V}_1 = 0.918345 - j0.159312$$

$$\mathbf{V}_2 = 0.978674 - j0.221811$$

$$\mathbf{V}_3 = 1.101718 - j0.065242$$

$$\mathbf{V}_4 = 0.901468 - j0.194617$$

$$\mathbf{V}_5 = 0.903003 - j0.196604$$

Evaluation of line currents and power flows

Knowing the nodal voltages and admittances, the current, power, and var flows between nodes are readily obtained. Links with transformers, however, need special attention. Consider a transformer with its impedance referred to the non-tap side (Figure 6.8),

$$\mathbf{I}_i = \frac{\mathbf{I}_j}{t} \quad \text{and} \quad \mathbf{V}_i = t\mathbf{V}_x$$

$$\therefore \mathbf{I}_j = (\mathbf{V}_x - \mathbf{V}_j)\mathbf{Y}_t$$

$$= \left(\frac{\mathbf{V}_i}{t} - \mathbf{V}_j\right)\mathbf{Y}_t$$

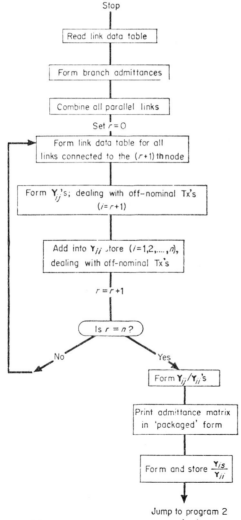

Figure 6.11(a)   Initial part of flow diagram for data processing of load-flow program.

242

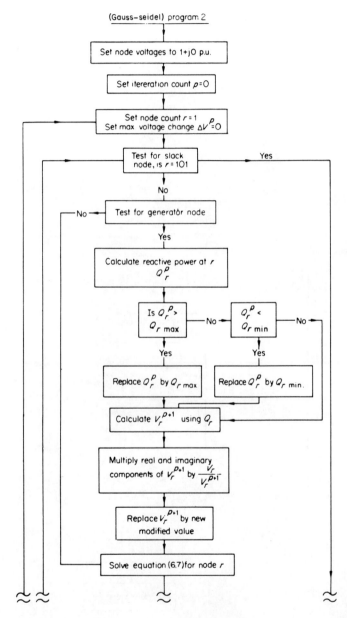

Figure 6.11(b) (*continued opposite*)

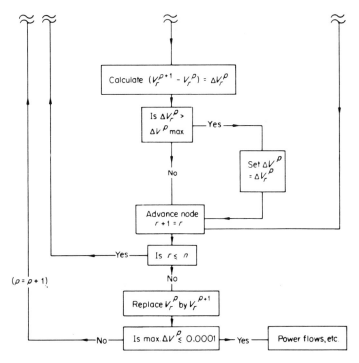

Figure 6.11(b) Remainder of flow diagram for Gauss–Seidel method.

Power at j,

$$P_j = \mathbf{V}_j\left[\left(\frac{\mathbf{V}_i}{t} - \mathbf{V}_j\right)\mathbf{Y}_t\right]^*$$

Also,

$$\mathbf{I}_i = \left(\frac{\mathbf{V}_i}{t^2} - \frac{\mathbf{V}_j}{t}\right)\mathbf{Y}_t$$

$$\therefore\ P_i = \mathbf{V}_i\left[\left(\frac{\mathbf{V}_i}{t^2} - \frac{\mathbf{V}_j}{t}\right)\mathbf{Y}_t\right]^*$$

Therefore power transferred i to j

$$= P_i - P_j$$

Application of the Newton–Raphson method to the system in Figure 6.10

The elements of the admittance matrix are determined as in the previous method, and it should be noted that for off-diagonal elements both G and B will be (-1) times the values derived from the network.

Let all the busbar voltages be assigned a voltage of $(1 + j0)$ p.u.

At node 3 $\qquad P_3 = 0.5 \qquad Q_3 = 0.185$ (generated)

$$\mathbf{Y}_{32} = 0.444860 - j0.646063 \text{ p.u.}$$

$$\mathbf{Y}_{35} = 0.576541 - j1.308461 \text{ p.u.}$$

$$\mathbf{Y}_{33} = 1.021401 - j1.954524 \text{ p.u.}$$

$$a_3 = 1, b_3 = 0 \qquad a_2 = 1, b_2 = 0 \qquad a_5 = 1, b_5 = 0$$

First iteration ($p = 0$)

$$P_3^0 = 1 \times 1 \times (-0.444860) + 1 \times 0 \times (-0.646063)$$

$$+ 0 \times 0 \times (-0.444860) - 0 \times 1 \times (-0.646063)$$

$$+ 1 \times 1 \times (+1.021401) + 1 \times 0 \times (1.954524)$$

$$+ 0 \times 0 \times (1.021401) - 0 \times 1 \times (1.954524)$$

$$+ 1 \times 1 \times (-0.576541) + 1 \times 0 \times (-1.308461)$$

$$+ 0 \times 0 \times (-0.576541) - 0 \times 1 \times (-1.308461)$$

$$= 0.0 \text{ p.u.}$$

Note that $\mathbf{Y}_{31}$, $\mathbf{Y}_{34}$ do not exist. (The above result would be expected in the initial case as all the involved voltages are equal to 1 p.u.) Similarly,

$$Q_3^0 = 0.0 \text{ p.u.}$$

Therefore,

$$\Delta P_3^0 = 0.5 - 0 = 0.5 \text{ p.u.}$$

and

$$\Delta Q_3^0 = 0.185 - 0 = 0.185 \text{ p.u.}$$

ΔP and ΔQ for the remaining non-slack nodes are similarly obtained.

$$\mathbf{I}_k^p = \frac{P_k^p - jQ_k^p}{(V_k^p)^*} \quad \text{and} \quad \mathbf{I}_3^0 = \frac{0.5 - j0.185}{1 - j0}$$

hence

$$c_3^0 = 0.5 \quad \text{and} \quad d_3^0 = -0.185$$

The elements of the Jacobian are determined next.

$$\frac{\partial P_3}{\partial a_3} = a_3^0 G_{33} - b_3^0 B_{33} + C_3^0$$

$$= 1(1.021401) - 0.0 + 0.5 = 1.521401$$

$$\frac{\partial P_3}{\partial a_2} = a_3^0 G_{32} - b_3^0 B_{32}$$

$$= 1(-0.444860) + 0.0$$

$$\frac{\partial P_3}{\partial a_5} = -0.576541$$

$$\frac{\partial P_3}{\partial b_3} = a_3^0 B_{33} + b_3^0 G_{33} + d_3^0$$

$$= 1(1.954520) + 0.0 + (-0.185) = 1.76952$$

$$\frac{\partial P_3}{\partial b_2} = a_3^0 B_{32} + b_3^0 G_{32} = 1(-0.646063) + 0.0$$

$$\frac{\partial P_3}{\partial b_5} = 1(-1.308461) + 0.0$$

Similarly, obtain

$$\frac{\partial Q_3}{\partial a_3}, \ \frac{\partial Q_3}{\partial a_2}, \ \frac{\partial Q_3}{\partial a_5} \quad \frac{\partial Q_3}{\partial b_3}, \ \frac{\partial Q_3}{\partial b_2}, \ \frac{\partial Q_3}{\partial b_5}$$

The Jacobian matrix for the first iteration is thus formed and inverted and then Δa_k^0 and $\Delta b_k^0 (k = 1 \ldots 5)$ are evaluated. Hence, $V_k^{0+1} = V_k^0 + \Delta a^0 + j\Delta b^0$. The process is repeated until changes in real and reactive power at each bus are less than a prescribed amount, say 0.01 p.u.

Summary

The direct method involving matrix inversion and a final iterative procedure to deal with the restraints at the nodes has advantages for smaller networks. For large networks the completely iterative methods are preferable, especially using accelerating factors. System information, such as specified generation and loads, line and transformer series, and shunt admittances are fed into the computer on tape or punched cards. The data are expressed in per unit on arbitrary MVA and voltage bases. The computer formulates the self- and mutual-admittances for each node.

If the system is very sensitive to reactive power flows, i.e. the voltages change considerably with change in load and network configuration, the computer program may diverge. It is preferable to allow the reactive power outputs of generators to be initially without limits to ensure an initial convergence. Convergence having been attained the computer evaluates the real and reactive power flows in each branch of the system along with losses, absorption of vars, and any other information that may be required. Programs are available which automatically adjust the tap settings of transformers to optimum values. Also, facilities exist for the printing out only of information regarding overloaded and underloaded lines; this is very useful when carrying out a series of load flows investigating the outages of plant and lines.

6.6 Design Considerations

The entire chapter so far has been devoted to the analysis of existing systems, so also is the following chapter on faults. Before analysis can take place a network must be designed and this will be based on the following considerations: standards of security, degree of utilization of plant, the existing system if any, standardized sizes of plant and voltage levels, amenity, location geographically. To meet a number of requirements several schemes may be produced and then the optimum, on grounds of cost, number of transmission circuits, and other factors chosen. Loads will have to be predicted (the system will come into existence some time after the initial-design stage) and rough checks such as generation meeting load requirements made.

Of great importance is the reinforcement of an existing network to meet the growing load. As the latter is a continual process the examination of the

Table 6.2 Flow diagram for design–analysis operation

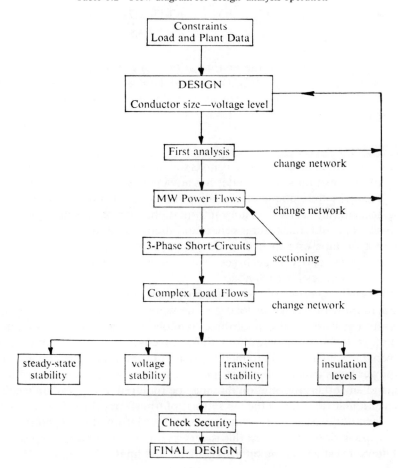

existing network must also be frequently performed. The delay of adding an extra link may save considerable expense, even if delayed for only a year, and this must be borne in mind.

The process of obtaining an optimum design is thus one of design–analysis–design–analysis and so on until the design constraints are met. Future development here, as with many design operations, e.g. electrical machines, lies in the realm of automatic computation. The data regarding load growth and the other considerations already mentioned must be stored, along with the methods of design and analysis traditionally used by engineers. A flow diagram illustrating the sequence of operations is shown in Table 6.2. The methods of logical design[7] and linear programming are being used to express the design constraints in mathematical terms.

References

BOOKS
1. *Electrical Transmission and Distribution Reference Book*, Westinghouse Electric. Corp., East Pittsburgh, Pennsylvania, 1964.
2. El Abiad, A. H., and G. W. Stagg, *Computer Methods in Power System Analysis*, McGraw-Hill, New York, 1968.

PAPERS
3. Robert, R., 'Micromachines and microreseaux', *C.I.G.R.E.*, **1950,** paper 338, Paris.
4. Laughton, M. A., and M. W. H. Davies, 'Numerical techniques in solution of power system load-flow problems', *Proc. I.E.E.*, **111,** No. 9 (1964).
5. Carré, B. A., 'The determination of optimum accelerating factor for successive over-relaxation', *Computer Jour.*, **4** (1962).
6. Ward, J. B., and H. W. Hale, 'Digital computer solution of power flow problem', *Trans. A.I.E.E.*, **75** (1956), 111.
7. Knight, U. G. W., 'The logical design of electrical networks using linear programming methods', *Proc. I.E.E.*, **107,** Pt. A, (1960).
8. Braess, D., *et al.*, 'A numerical analysis of load-flow calculation methods', *I.E.E.E. Trans.*, **PAS-101** (1982), 3642.
9. Brown, H. E., G. K. Carter, H. H. Happ, and C. E. Person, 'Power flow solution by impedance matrix iterative method', *Trans. I.E.E.E.*, *P.A.S.*, **82** (1963), 1.
10. Brown, R. J., and W. F. Tinney, 'Digital solutions for large power networks', *Trans. A.I.E.E.*, **76,** Pt. III (1957), 347.
11. Chamorel, P. A., *et al.*, 'An efficient constrained power flow technique based on active–reactive decoupling', *I.E.E.E. Trans.*, **PAS-101** (1982), 158.
12. Glimn, A. F., and G. W. Stagg, 'Automatic calculation of load flows', *Trans. A.I.E.E.*, **76,** Pt. III, (1957), 817.
13. Tripathy, S. C., *et al.*, 'Load-flow solutions for ill-conditioned power systems', *I.E.E.E. Trans.*, **PAS-101** (1982), 3648.
14. Xia, D., *et al.*, 'Harmonic power flow studies—I Formulation and Solution', *I.E.E.E. Trans.*, **PAS-101** (1982), 1257.
15. Launay, M., 'Use of computer graphics for distribution network planning in "Electricité de France" ', *I.E.E.E. Trans.*, **PAS-101** (1982), 276.
16. Wu, F. F., 'Fast decoupled load flow analysis', *Trans. I.E.E.E.*, **PAS–96,** (1977) 268–275.

Problems

6.1. A single-phase distributor has the following loads at the stated distances from the supply end: 10 kW at 10 m, 10 kW at 0.9 p.f. lagging at 16 m 5 kW at 0.8 p.f. lagging at 91.5 m, and 20 kW at 0.95 p.f. lagging at 137 m. The loads may be assumed to be constant current at their nominal voltage values (240 V). If the supply voltage is 250 V and the maximum voltage drop is 5 per cent of the nominal value determine the nearest commercially available conductor size.
(Answer: $r + 0.376x = 0.7 \times 10^{-3}$.)

6.2. In the d.c. network shown in Figure 6.12 calculate the voltage at node B by inverting the admittance matrix. Check the answer by Thevenin's theorem.
(Answer: 247.7 V.)

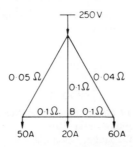

Figure 6.12   Network for Problem 6.2.

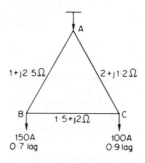

Figure 6.13   Network for Problem 6.3.

6.3. In the interconnected network shown in Figure 6.13 calculate the current in feeder BC.
(Answer: 26.75 A.)

6.4. In the network shown in Figure 6.14 the loads are represented by constant impedances $1 + j1$ p.u. Determine the current distributions in the network, (a) when the transformer has its nominal ratio, (b) when the transformer is tapped up 10 per cent. (Note: determine the distribution with the off-nominal voltage alone and use superposition.)
(Answer: Transformer branch, (a) 0, (b) $(0.0735 - j0.075$ (p.u.).).)

6.5. For the system shown in Figure 6.10 calculate the busbar voltages after the second iteration without the use of an accelerating factor.

6.6. For the system shown in Figure 6.10 calculate the power, reactive power, and current flows in the lines for the completed load flow. (Voltages after 30 iterations are given in the text.)

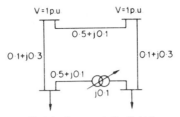

Each load represented by Z=1+j1p.u.

Figure 6.14   Network for Problem 6.4.

6.7. For the system shown in Figure 6.10 recalculate the nodal voltages and power and reactive power flows at the end of the first iteration using an accelerating factor of 1.6.
6.8. In the system shown in Figure 6.10 the busbar quantities are to remain the same except at busbar 3 where the generation is now to produce a constant voltage of magnitude 1.15 p.u. and a power output of 0.5 p.u. with no restriction on the var output. Determine the voltages after the first iteration without the use of an accelerating factor.
6.9. For the system of Figure 6.10 perform a real power load flow (ignoring var loading and component resistance) and compare the power flows with those obtained for the complex study given in the text. Design a resistance network operating from a 50 V supply to simulate the system for real power flows only.
6.10. Enumerate the information which may be obtained from a load-flow study. Part of a power system is shown in Figure 6.15. The line to neutral reactances and values of

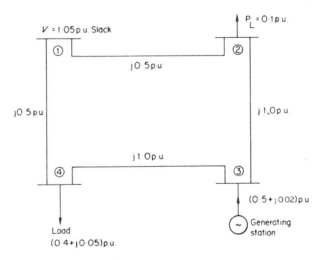

Figure 6.15   System for Problem 6.10.

real and reactive power (in the form $P \pm jQ$) at the various stations are expressed as per unit values on a common MVA base. Resistance may be neglected. By the use of an iterative method suitable for a digital computer, calculate the voltages at the stations after the first iteration without the use of an accelerating factor.
(Answer: V_2 1.03333 − j0.03333 p.u.; V_3 1.11666 + j0.23333 p.u.; V_4 1.05556 + j0.00277 p.u.)

250

6.11. A 400 kV interconnected system is supplied from bus A which may be considered to be an infinite busbar. The loads and line reactances are as indicated in Figure 6.16.

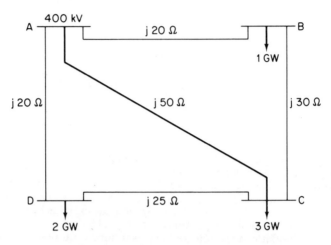

Figure 6.16 System for Problem 6.11.

Determine the flow of power in each line. In line AC a phase shift transformer is installed which gives a phase advance of 10° at C with respect to A. Calculate the new power flow in line AC.
(Answer: P_{AC} 1.35 GW; P_{AC} 1 GW.)

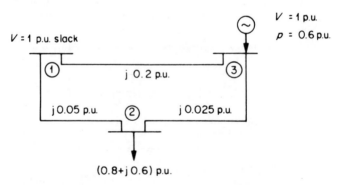

Figure 6.17 System for Problem 6.12.

6.12. Determine the voltage at bus (2) and the reactive power at bus (3) in Figure 6.17 after the first iteration of a Gauss–Seidel load flow method. Assume the initial voltage to be $1\angle0°$ p.u. All the quantities are in per unit on a common base.
(Answer: $0.99 - j0.0133$)

7

Fault Analysis

7.1 Introduction

An essential part of the design of a power supply network is the calculation of the currents which flow in the components when faults of various types occur. In a fault survey, faults are applied at various points in the network and the resulting currents obtained by hand calculation, or, more likely now on large networks, by analogue or digital processes. The magnitude of the fault currents give the engineer the current settings for the protection to be used and the ratings of the circuit breakers.

The types of fault commonly occurring in practice are illustrated in Figure 7.1 and the most common of these is the short circuit of a single conductor to earth. Often the path to earth contains resistance in the form of an arc as shown in Figure 7.1(f). Although the single line to earth fault is the most common, calculations are frequently performed with the three-line, balanced short circuit (Figure 7.1(d) and (e)). This is the most severe fault and also the most amenable to calculation. The causes of faults are summarized in Table 7.1

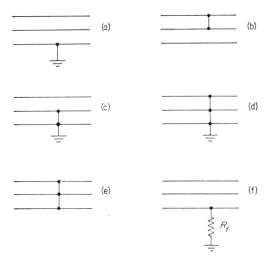

Figure 7.1 Common types of fault.

Table 7.1   Causes of overhead-line faults, British system
66 kV and above

	Faults/160 km of line
Lightning	1.59
Dew, Fog, Frost	0.15
Snow, ice	0.01
Gales	0.24
Salt spray	0.01
Total	2 faults per 160 km giving a total of 232 faults on system

which gives the distribution of faults due to various causes on the British Central Electricity Generating Board system in a typical year. Table 7.2 shows the components affected. In tropical countries the incidence of lightning is much greater than in Britain, resulting in larger numbers of faults.

Table 7.2   Distribution of faults, British system

Type	Number of faults
Overhead lines	289
Cables	67
Switchgear	56
Transformers	59
Total	471

As well as fault current, fault MVA is frequently considered; this is obtained from the expression $\sqrt{(3)}\,V_L I_F \times 10^{-6}$ fault, where V_L is the nominal line voltage of the faulted part. The fault MVA is often referred to as the *fault level*. The calculation of fault currents can be divided into the following two main types.

(a) Faults short-circuiting all three phases when the network remains balanced electrically. For these calculations normal single-phase equivalent circuits may be used as in ordinary load-flow calculations.

(b) Faults other than three-phase short circuits when the network is electrically unbalanced. To facilitate these calculations a special method for dealing with unbalanced networks is used known as the method of *symmetrical components*.

The main objects of fault analysis may be enumerated as follows:

1. To determine maximum and minimum three-phase short-circuit currents.
2. To determine the unsymmetrical fault current for single and double line-to-earth, line-to-line faults, and open-circuit faults.
3. Investigation of the operation of protective relays.

4. Determination of rated rupturing capacity of breakers.
5. To determine fault-current distribution and busbar-voltage levels during faults.

7.2 Calculation of Three-phase Balanced Fault Currents

The action of synchronous generators on three-phase short circuits has been described in Chapter 3. There it was seen that dependent on the time elapsing from the incidence of the fault, either the transient or the subtransient reactance should be used to represent the generator. For specifying switchgear the value of the current flowing at the instant at which the circuit breakers open is required, It has been seen, however, that the initially high fault current associated with the subtransient reactance decays with the passage of time. Modern air-blast circuit breakers usually operate in 2.5 cycles of 60 Hz alternating current and are associated with extremely fast protection. Older circuit breakers and those on lower voltage networks usually associated with relatively cruder protection can take in the order of 8 cycles or more to operate. In calculations it is usual to use the subtransient reactance of generators and to ignore the effects of induction motors. The calculation of fault currents ignores the direct-current component, the magnitude of which depends on the instant in the cycle that the short circuit occurs. If the circuit breaker opens a reasonable time after the incidence of the fault the direct-current component will have decayed considerably. With fast-acting circuit breakers the actual current to be interrupted is increased by the direct-current component and it must be taken into account. To allow for the direct-current component of the fault current the symmetrical r.m.s. value is modified by the use of multiplying factors such as the following:

8 cycle circuit breaker opening time, multiply by 1
3 cycle circuit breaker opening time, multiply by 1.2
2 cycle circuit breaker opening time, multiply by 1.4

Consider an initially unloaded generator with a short circuit across the three terminals as in Figure 7.2. The generated voltage per phase is E and therefore the short-circuit current is $[E/Z(\Omega)]A$, where Z is either the transient or

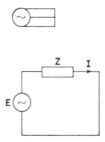

Figure 7.2   Voltage source with short circuit and equivalent circuit.

subtransient impedance. If Z is expressed in per unit notation,

$$Z \text{ (p.u)} = \frac{I_{\text{FL}} Z \text{ }(\Omega)}{E}$$

$$\therefore \text{ } Z \text{ }(\Omega) = \frac{EZ \text{ (p.u.)}}{I_{\text{FL}}}$$

Taking I_{FL} (full load or rated current) and E as base values of voltage and current, the short-circuit current

$$I_{\text{s.c.}} = \frac{E}{Z(\Omega)}$$

$$= \frac{E \cdot I_{\text{FL}}}{E \cdot Z \text{ (p.u.)}} = \frac{I_{\text{FL}}}{Z \text{ (p.u.)}} \tag{7.1}$$

Also the three-phase short-circuit volt-amperes

$$= \sqrt{(3)} \, V I_{\text{s.c.}} = \frac{\sqrt{(3)} \, V I_{\text{FL}}}{Z \text{ (p.u.)}}$$

$$= \frac{\text{Base volt-amperes}}{Z \text{ (p.u.)}} \tag{7.2}$$

Hence the short-circuit level is immediately obtained if the impedance from the source of the voltage to the point of the fault is known.

Example 7.1 An 11.8 kV busbar is fed from three synchronous generators having the following ratings and reactances,

20 MVA, X' 0.08 p.u.;   60 MVA, X' O.1 p.u.;   20 MVA, X' 0.09 p.u.

Calculate the fault current and MVA if a three-phase symmetrical fault occurs on the busbars. Resistance may be neglected. The voltage base will be taken as 11.8 kV and the VA base as 60 MVA.

Solution The transient reactance of the 20 MVA machine on the above bases is (60/20) × 0.08, i.e. 0.24 p.u. and of the 20 MVA machine (60/20) × 0.09, i.e. 0.27 p.u. These values are shown in the equivalent circuit in Figure 7.3. As the generator e.m.f.s are equal one source may be used (Figure 7.3(c)). The equivalent reactance,

$$X_{\text{e}} = \frac{1}{1/0.24 + 1/0.27 + 1/0.1}$$

$$= 0.054 \text{ p.u.}$$

Therefore fault MVA

$$= \frac{60}{0.054} = 1111 \text{ MVA}$$

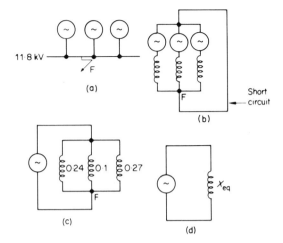

Figure 7.3   Line diagram and equivalent circuits for Example 7.1.

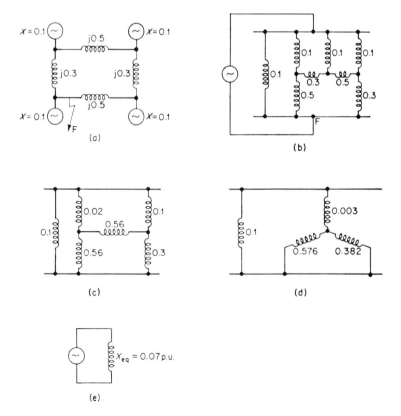

Figure 7.4   Line diagram and equivalent circuits for Example 7.2.

and fault current

$$= \frac{1111 \times 10^6}{\sqrt{(3)} \times 11,800}$$

$$= 54,300 \text{ A}$$

Example 7.2   In the network shown in Figure 7.4 a three-phase fault occurs at point F. Calculate the fault MVA at F. The per unit values of reactance given, all refer to a base of 100 MVA. Resistance may be neglected.

Solution   The equivalent single-phase network of generator and line reactances is shown in Figure 7.4(b). This is replaced by the network in Figure 7.4(c) by the use of the delta–star transformation. A further transformation is carried out on the network in Figure 7.4(d) to give the final single equivalent reactance,

$$X_e = 0.07 \text{ p.u.}$$

The fault level at point F

$$= \frac{100}{0.07} = 1430 \text{ MVA}$$

This value is based on the assumption of symmetrical fault current. Allowing for a multiplying factor of 1.4, the fault level is 2000 MVA.

Current limiting reactors   The impedances presented to fault currents by transformers and machines when faults occur on substation or generating station busbars are low. To reduce the high fault current which would do considerable damage mechanically and thermally, artificial reactances are sometimes connected between bus sections. These current limiting reactors usually consist of insulated copper strip embedded in concrete formers; this is necessary to withstand the high mechanical forces produced by the current in neighbouring conductors. The position in the circuit occupied by the reactor is a matter peculiar to individual designs and installations (see Figure 7.5(a) and (b)).

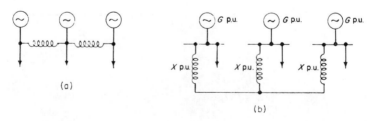

Figure 7.5   Connexion of artificial reactors: (a) ring system; (b) tie-bar system. Transient reactance of machines, G p.u.; reactance of artificial reactors X p.u.

Example 7.3   A 400 kV power system contains three substations A, B, and C having fault levels (GVA) of 20, 20, and 30 respectively.

The system is to be reinforced by three lines each of reactance j5 Ω connecting together the three substations. Calculate the new fault level at C (three-phase symmetrical fault). Neglect resistance.

Solution   The equivalent circuits for the new system is shown in Figure 7.6(a) and (b) The substations are represented by a voltage source (400 kV) in series

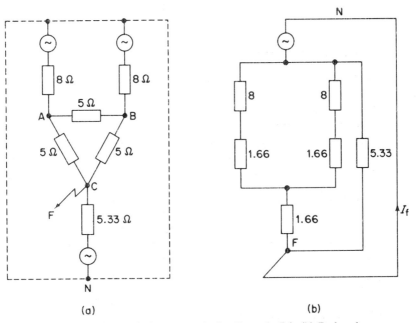

(a)   (b)

Figure 7.6(a)   Equivalent networks for Example 7.3. (b) Reduced network.

with the value of reactance to give the specified initial fault level, e.g. for A the fault current (before reinforcement).

$$= \frac{20 \times 10^9}{\sqrt{3} \times 400,000} \text{ A}$$

Hence the effective reactance 'behind' A when subject to a 3-phase short circuit

$$= \frac{400,000}{\sqrt{3}} \times \frac{\sqrt{(3)} \times 400,000}{20 \times 10^9} = 8 \ \Omega$$

Similarly, the effective reactance at B = 8 Ω and at C = 5.33 Ω.

The 5 Ω mesh is transformed into a wye with arms of value 1.66 Ω and the equivalent circuit with the fault on C after reinforcement is shown in Figure 7.6(b). Fault current at C

$$= \frac{400,000}{\sqrt{3}} \times \frac{1}{j2.9} = -j78.5 \text{ kA}$$

Fault level at C

$$= \sqrt{3} \times 400 \times 78.5 \text{ MVA} = 54,870 \text{ MVA}$$

Note that this value is very high and that the result of reinforcement is always to increase the fault levels. One great advantage of using HVDC links is that this does not happen.

7.3 Method of Symmetrical Components

This method formulates a system of three separate phasor systems which when superposed give the true unbalanced conditions in the circuit. It should be stressed that the systems to be discussed are essentially artificial and used merely as an aid to calculation. The various sequence-component voltages and currents do not exist as physical entities in the network.

The method postulates that a three-phase unbalanced system of voltages and currents may be presented by the following three separate systems of phasors,

(a) a balanced three-phase system in the normal a–b–c (red–yellow–blue) sequence, called the *positive phases-sequence* system;
(b) a balanced three-phase system of reversed sequence, i.e. a–c–b (red–blue–yellow), called the *negative phase-sequence* system;
(c) three phasors equal in magnitude and phase revolving in the positive phase rotation, called the *zero phase-sequence* system.

In Figure 7.7 an unbalanced system of currents is shown with the corresponding system of symmetrical components. If each of the red-phase phasors are added, i.e. $\mathbf{I}_{1R} + \mathbf{I}_{2R} + \mathbf{I}_{R0}$ the resultant phasor will be I_R in magnitude and direction; similar reasoning holds for the other two phases.

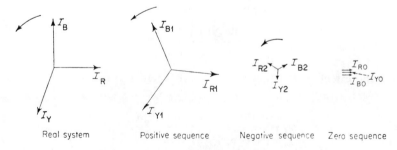

Figure 7.7   Real system and corresponding symmetrical components.

To express the phasors algebraically, use is made of the complex operator 'a' (sometimes denoted by λ or h) which denotes a phase-shift operation of $+120°$ and a multiplication of unit magnitude; i.e.

$$V\angle\phi \times \mathbf{a} = V\angle\phi \times 1\angle120° = V\angle\phi + 120°, \ \mathbf{a} = e^{j2\pi/3} \text{ and } \mathbf{a}^3 = e^{j3\times2\pi/3} = 1$$

Also,

$$\mathbf{a}^2 + \mathbf{a} = (-0.5 - j0.866) + (-0.5 + j0.866) = -1$$

$$\therefore \ \mathbf{a}^3 + \mathbf{a}^2 + \mathbf{a} = 0$$

and

$$1 + \mathbf{a} + \mathbf{a}^2 = 0$$

For positive-sequence phasors, taking the red phasor as reference;

$$\mathbf{I}_{1R} = \mathbf{I}_{1R} e^{j0} = \text{reference phasor (Figure 7.8)}$$

$$\mathbf{I}_{1Y} = \mathbf{I}_{1R}(-0.5 - j0.866) = \mathbf{a}^2\mathbf{I}_{1R}$$

and

$$\mathbf{I}_{1B} = \mathbf{I}_{1R}(-0.5 + j0.866) = \mathbf{a}\mathbf{I}_{1R}$$

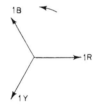

Figure 7.8   Positive-sequence phasors.

For negative sequence quantities,

$$\mathbf{I}_{2R} = \mathbf{I}_{2R}(1 + j0) \text{ (Figure 7.9)}.$$

$$\mathbf{I}_{2Y} = \mathbf{I}_{2R}(-0.5 + j0.866) = \mathbf{a}\mathbf{I}_{2R}$$

$$\mathbf{I}_{2B} = \mathbf{I}_{2R}(-0.5 - j0.866) = \mathbf{a}^2\mathbf{I}_{2R}$$

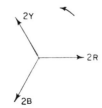

Figure 7.9   Negative-sequence phasors.

Returning to the original unbalanced system of currents, I_R, I_Y and I_B;

$$I_R = I_{1R} + I_{2R} + I_{0R}$$

$$I_Y = I_{1Y} + I_{2Y} + I_{0Y} = a^2 I_{1R} + a I_{2R} + I_{0R}$$

$$I_B = I_{1B} + I_{2B} + I_{0B} = a I_{1R} + a^2 I_{2R} + I_{0R}$$

Hence in matrix form,

$$\begin{bmatrix} I_R \\ I_Y \\ I_B \end{bmatrix} = \begin{bmatrix} 1 & 1 & 1 \\ 1 & a^2 & a \\ 1 & a & a^2 \end{bmatrix} \begin{bmatrix} I_{0R} \\ I_{1R} \\ I_{2R} \end{bmatrix} \tag{7.3}$$

Inverting the matrix,

$$\begin{bmatrix} I_{0R} \\ I_{1R} \\ I_{2R} \end{bmatrix} = \tfrac{1}{3} \begin{bmatrix} 1 & 1 & 1 \\ 1 & a & a^2 \\ 1 & a^2 & a \end{bmatrix} \begin{bmatrix} I_R \\ I_Y \\ I_B \end{bmatrix} \tag{7.4}$$

The above also holds for voltages, i.e.

$$[E_{actual}] = [T_s][E_{1,2,0}]$$

where $[T_s]$ is the symmetrical component transformation matrix,

$$\begin{bmatrix} 1 & 1 & 1 \\ 1 & a^2 & a \\ 1 & a & a^2 \end{bmatrix}$$

In a three-wire system the instantaneous voltages and currents add to zero and therefore no single-phase component is required. A fourth wire or connexion to earth must be provided for single-phase currents to flow. In a three-wire system the zero phase-sequence components are replaced by zero in equations (7.3) and (7.4). Also,

$$I_{0R} = \frac{I_R + I_Y + I_B}{3} = \frac{I_N}{3}$$

where I_N is the neutral current.

$$\therefore \quad I_N = 3I_{R0} = 3I_{Y0} = 3I_{B0}$$

In the application of this method it is necessary to calculate the symmetrical components of the current in each line of the network and then combine them to obtain the actual values. The various phase-sequence values are obtained by considering a network derived from the actual network in which only a particular sequence current flows; for example, in a zero-sequence network only zero-sequence currents and voltages exist. The positive-sequence network is identical with the real, balanced equivalent network, i.e. it is the same as used for three-phase symmetrical short-circuit studies. The negative-sequence network is almost the same as the real one except that the values of impedance used for rotating machines are different and there are no generated voltages. The zero-sequence network is considerably different from the real one.

The above treatment assumes that the respective sequence impedances in each of the phases are equal, i.e. $Z_{1R} = Z_{1Y} = Z_{1B}$, etc. Although this covers most cases met in practice, unequal values may occur in certain circumstances, e.g. an open circuit on one phase. The following equation applies for the voltage drops across the phase impedances:

$$\begin{bmatrix} V_R \\ V_Y \\ V_B \end{bmatrix} = [T_S] \begin{bmatrix} V_{0R} \\ V_{1R} \\ V_{2R} \end{bmatrix} = \begin{bmatrix} Z_R & & \\ & Z_Y & \\ & & Z_B \end{bmatrix} [T_S] \begin{bmatrix} I_{0R} \\ I_{1R} \\ I_{2R} \end{bmatrix}$$

From which,

$$V_{0R} = \tfrac{1}{3}I_{1R}(Z_R + a^2 Z_Y + a Z_B) + \tfrac{1}{3}I_{2R}(Z_R + a Z_Y + a^2 Z_B) + \tfrac{1}{3}I_{0R}(Z_R + Z_Y + Z_B)$$

and similarly expressions V_{1R} and V_{2R} may be obtained.

It is seen that the voltage drop in each sequence is influenced by the impedances in all three phases. If as previously assumed, $Z_R = Z_Y = Z_B = Z$, then the voltage drops become $V_{1R} = I_{1R}Z$, etc. as before.

7.4 Representation of Plant in the Phase-sequence Networks

(a) *The synchronous machine* (Table 3.1) The positive-sequence impedance Z_1 is the normal transient or subtransient value. Negative-sequence currents set up a rotating magnetic field in the opposite direction to that of the positive-sequence currents and which rotates round the rotor surface at twice the synchronous speed; hence the effective impedance (Z_2) is different from Z_1. The zero-sequence impedance Z_0 depends upon the nature of the connexion between the star point of the windings and the earth and the single-phase impedance of the stator windings in series. Resistors or reactors are frequently connected between the star point and earth for reasons usually connected with protective gear and the limitation of over-voltages. Normally, the only voltage sources appearing in the networks are in the positive sequence one, as the generators only generate positive-sequence e.m.f.s.

(b) *Lines and cables* The positive- and negative-sequence impedances are the normal balanced values. The zero-sequence impedance depends upon the nature of the return path through the earth if no fourth wire is provided. It is also modified by the presence of an earth wire on the towers which protects the lines against lightning surges. In the absence of detailed information the following rough guide to the value of Z_0 may be used. For a single-circuit line $(Z_0/Z_1) = 3.5$ with no earth wire and 2 with one. For a double-circuit line $(Z_0/Z_1) = 5.5$. For underground cables, $(Z_0/Z_1) = 1$–1.25 for single-core and 3–5 for three-cored cables.

(c) *Transformers*　The positive- and negative-sequence impedances are the normal balanced ones. The zero-sequence connexion of transformers is, however, complicated and depends on the nature of the connexion of the windings. In Table 7.3 is listed the zero-sequence representation of transformers for various winding arrangements. Zero-sequence currents in the

Table 7.3

Connexions of windings Primary　　Secondary	Representation per phase	Comments
	Z_0	Zero-sequence currents free to flow in both primary and secondary circuits
	Z_0	No path for zero-sequence currents in primary circuits
	Z_0	Single-phase currents can circulate in the delta but not outside it
	Z_0	No flow of zero-sequence currents possible
	Z_0	No flow of zero-sequence currents possible
		Tertiary winding provides path for zero-sequence currents

windings on one side of a transformer must produce the corresponding ampere-turns in the other. But three in-phase currents cannot flow in a star connexion without a connexion to earth. They can circulate round a delta winding, but not in the lines outside it. Owing to the mutual impedance between the phases, $Z_0 \neq Z_1$.

An example showing the nature of the three sequence networks for a small transmission link is shown in Figure 7.10 and Table 7.3 shows typical transformer representations for zero-sequence conditions.

7.5　Types of Fault

In the following a single voltage source in series with an impedance is used to represent the power network as seen from the point of the fault. This is an

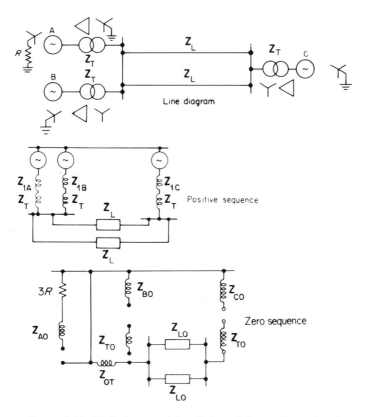

Figure 7.10 Typical transmission link and form of associated sequence networks. Negative-sequence diagram as for positive, but voltage sources omitted and generator impedances are Z_{2A}, Z_{2B}, and Z_{2C}.

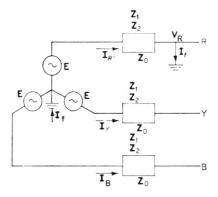

Figure 7.11 Single line-to-earth fault—Thevenin equivalent of system at point of fault.

264

extension of Thevenin's theorem to three-phase systems. It represents the general method used for manual calculation, i.e. the successive reduction of the network to a single impedance and voltage or current source. The network is assumed to be initially on no load before the occurrence of the fault.

(a) *Single-phase to earth fault* The three-phase circuit diagram is shown in Figure 7.11.

Let I_1, I_2, and I_0 be the symmetrical components of I_R and V_1, V_2, and V_0 the components of V_R. For this condition $V_R = 0$, $I_B = 0$, and $I_Y = 0$.

From equation (7.4)

$$I_0 = \tfrac{1}{3}(I_R + I_B + I_Y)$$

$$I_1 = \tfrac{1}{3}(I_R + aI_Y + a^2 I_B)$$

$$I_2 = \tfrac{1}{3}(I_R + a^2 I_Y + aI_B)$$

Hence,

$$I_0 = \frac{I_R}{3} = I_1 = I_2$$

Also,

$$V_R = E - I_1 Z_1 - I_2 Z_2 - I_0 Z_0 = 0$$

Eliminating I_0 and I_2,

$$E - I_1(Z_1 + Z_2 + Z_0) = 0$$

and

$$I_1 = \frac{E}{Z_1 + Z_2 + Z_0} \qquad (7.5)$$

The fault current,

$$I_f = I_R = 3I_1$$

$$= \frac{3E}{Z_1 + Z_2 + Z_0} \qquad (7.6)$$

The e.m.f. of the Y-phase $= a^2 E$, and

$$I_Y = I_0 + a^2 I_1 + aI_2$$

$$\therefore \ V_Y = a^2 E - I_0 Z_0 - a^2 I_1 Z_1 - aI_2 Z_2$$

The pre-fault and post-fault phasor diagrams are shown in Figure 7.12.

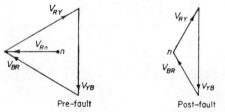

Pre-fault   Post-fault

Figure 7.12   Pre- and post-fault phasor diagrams—single line-to-
earth fault.

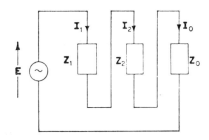

Figure 7.13 Interconnexion of positive-, negative-, and zero-sequence networks for single line-to-earth faults.

It is usual to form an equivalent circuit to represent the equation (7.5) and this can be obtained from an inspection of the equations. The circuit is shown in Figure 7.13 and it will be seen that $I_1 = I_2 = I_0$, and

$$I_1 = \frac{E}{Z_1 + Z_2 + Z_0}$$

(b) *Phase to phase fault* In Figure 7.14, E = e.m.f. per phase and the R-phase is again taken as the reference phasor. In this case, $I_R = 0$, $I_Y = -I_B$ and $V_Y = V_B$.

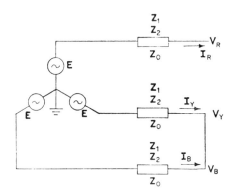

Figure 7.14 Line-to-line fault.

From equation 7.4,

$$I_0 = 0$$

$$I_1 = \tfrac{1}{3}I_Y(a - a^2)$$

and

$$I_2 = \tfrac{1}{3}I_Y(a^2 - a)$$

$$\therefore \; I_1 = -I_2$$

As $\mathbf{V_Y} = \mathbf{Y_B}$

$$\mathbf{a^2E - a^2I_1Z_1 - aI_2Z_2 = aE - aI_1Z_1 - a^2I_2Z_2}$$

$$\therefore \ \mathbf{E(a^2 - a) = I_1[Z_1(a^2 - a) + Z_2(a^2 - a)]}$$

$$\mathbf{I_1 = \frac{E}{Z_1 + Z_2}} \tag{7.7}$$

This can be represented by the equivalent circuit in Figure 7.15 in which of

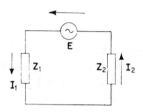

Figure 7.15   Interconnexion of sequence networks for a line-to-line fault.

course there is no zero-sequence network. If the connexion between the two lines has an impedance $\mathbf{Z_f}$ (the fault impedance) this is connected in series in the equivalent circuit.

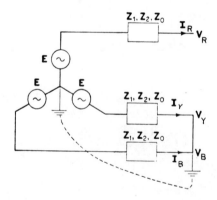

Figure 7.16   Line-to-line to earth fault.

(c) *Double phase to earth fault*   (Figure 7.16)

$$\mathbf{I_R = 0} \qquad \mathbf{V_Y = Y_B = 0}$$

and

$$\mathbf{I_R = I_1 + I_2 + I_0 = 0}$$

$$\therefore \ \mathbf{a^2E - a^2I_1Z_1 - aI_2Z_2 - I_0Z_0 = V_Y = 0}$$

and

$$aE - aI_1 Z_1 - a^2 I_2 Z_2 - I_0 Z_0 = V_B = 0$$

Hence,

$$I_1 = \frac{E}{Z_1 + (Z_2 Z_0 / Z_2 + Z_0)} \tag{7.8}$$

$$I_2 = -I_1 \frac{Z_0}{Z_2 + Z_0} \tag{7.9}$$

and

$$I_0 = -I_1 \cdot \frac{Z_2}{Z_2 + Z_0} \tag{7.10}$$

These can be represented by the equivalent circuit shown in Figure 7.17.

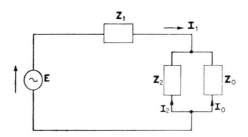

Figure 7.17 Interconnexion of sequence networks—double line-to-earth fault.

The inclusion of impedances in the earth path, such as the star point to earth connexion in a generator or transformer, modifies the sequence diagrams. For a phase-to-earth fault an impedance Z_g in the earth path is represented by an impedance of $3Z_g$ in the zero-sequence network. Z_g can include the impedance of the fault itself, usually the resistance of the arc. As $I_1 = I_2 = I_0$ and $3I_1$ flows through Z_g in the physical system it is necessary to use $3Z_g$ to obtain the required effect. Hence,

$$I_f = \frac{3E}{Z_1 + 3Z_g + Z_2 + Z_0}$$

Again, for a double-phase to earth fault an impedance $3Z_g$ is connected as shown in Figure 7.18. Z_g includes both machine neutral impedances and fault impedances.

The phase shift introduced by star–delta transformers has no effect on the magnitude of the fault currents, although it will affect the voltages at various points. It is shown in Chapter 3 that the positive-sequence voltages and currents are advanced by a certain angle and the negative-sequence quantities retarded by the same angle for a given connexion.

268

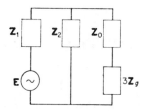

Figure 7.18   Modification of network in Figure 7.17 to account for neutral impedance.

7.6 Fault Levels in a Typical System

In Figure 7.19 a section of a typical system is shown. At each voltage level the fault level can be ascertained from the reactances given, It should be noted that the short-circuit level will change with network conditions, and there will normally be two extreme values, that with all plant connected and that with the

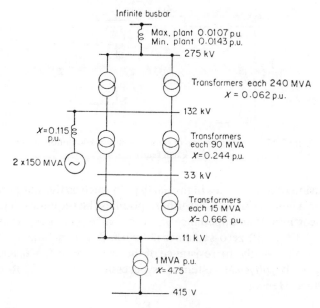

Figure 7.19 Typical transmission system. All reactances on a 100 MVA base

minimum plant normally connected. The short-circuit MVA at 275 kV busbars in Britain is normally 10,000 MVA, but drops to 7000 MVA with minimum plant connected. Maximum short-circuit (three-phase) levels experienced in the British system are as follows: 275 kV, 15,000 MVA; 132 kV, 3500 MVA; 33 kV, 750/1000 MVA; 11 kV, 150/250.MVA; 415 V, 25 MVA.

As the transmission voltages increase, the short-circuit currents also increase and for the 400 kV system, circuit breakers of 35,000 MVA breaking capacity are required. In order to reduce the fault level the number of parallel paths is reduced by sectionalizing. This is usually achieved by opening the circuit breaker connecting two sections of a substation or generating station busbar. One great advantage of direct-current transmission links in parallel with the alternating-current system is that no increase in the short-circuit currents results. The relation between system voltage and fault level is shown in Figure 7.20.

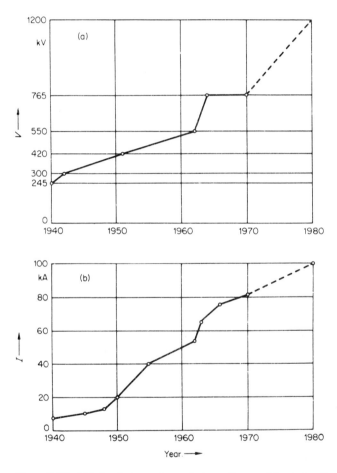

Figure 7.20   Relation between transmission voltage and short-circuit current (European systems). (*Permission of Brown Boveri Co.*)

Circuit parameters with faults

Fault resistance   The resistance of the fault is normally that of the arc and may be approximated by

$$R_a(\Omega) = 44 \,(V \text{ in kV})/I_f \quad \text{for} \quad V < 110 \text{ kV}$$

and

$$R_a(\Omega) = 22 \ (V \text{ in kV})/I_f \quad \text{for} \quad V > 110 \text{ kV}$$

For example, a 735 kV line would have an arc resistance of 4Ω with a fault current (zero fault resistance) of 4 kA.

The overall grounding resistance depends on the footing resistance (resistance of tower metalwork to ground) of the towers (R_T) and also on the resistance per section of the ground wires (R_s) where present. The situation is summarized in Fig. 7.21 where it is seen that the effective grounding resistance is smaller than the individual tower footing resistance. There is normally a spread in the values of R_T, but normally R_T should not exceed 10 Ω (ground wires present).

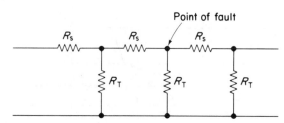

Figure 7.21   Equivalent network of towers with footing resistance (R_T) and ground wire sections (resistance R_S)

Typical values of fault resistance with fault location are as follows: at source, $0\ \Omega$; on line with ground wires, 15 Ω; on line without ground wires, 50 Ω.

X/R ratio   The range of X/R values for typical voltage class (Canadian) are as follows: 735 kV 18.9–20.4; 500 kV, 13.6–16.5; 220 kV, 2–25; 110 kV, 3–26. The X/R value decreases with separation of the fault point from the source and can be substantially decreased by the fault resistance.

For distribution circuits (X/R) is lower, and although data is limited, typical values are 10 (at source point) and 4 on the line.

Example 7.4
A synchronous machine A generating 1 p.u. voltage is connected through a star–star transformer, reactance 0.12 p.u., to two lines in parallel. The other ends of the lines are connected through a star–star transformer of reactance 0.1 p.u. to a second machine B, also generating 1 p.u. voltage. For both transformers $X_1 = X_2 = X_0$.

Calculate the current fed into a double line-to-earth fault on the line-side terminals of the transformer fed from A.

The relevant per unit reactances of the plant all referred to the same base are as follows:

A	X_1 0.3	X_2 0.2	X_0 0.05
B	X_1 0.25	X_2 0.15	X_0 0.03

For each line $X_1 = X_2 = 0.3$. $X_0 = 0.7$.

The star points of machine A and of the two transformers are solidly earthed.

Solution The positive-, negative-, and zero-sequence networks are shown in Figure 7.22. All per unit reactances are on the same base. From these diagrams the following equivalent reactances up to the point of the fault are obtained: $Z_1 = j0.23$ p.u., $Z_2 = j0.18$ p.u. and $Z_0 = j0.17$ p.u.

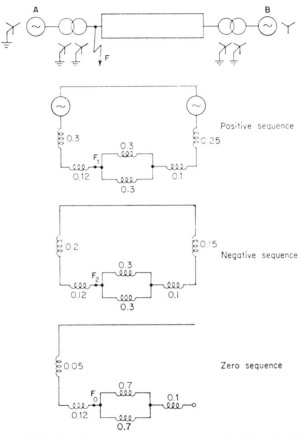

Figure 7.22 Line diagram and sequence networks for Example 7.4.

The red phase is taken as reference phasor and the blue and yellow phases assumed to be shorted at the fault point. From the equivalent circuit for a phase-to-phase fault,

$$I_1 = \frac{1}{Z_1 + (Z_0 Z_2 / Z_0 + Z_2)} \text{ p.u.}$$

$$= \frac{1}{j(0.23 + (0.17 \times 0.18 / 0.17 + 0.18))} \text{ p.u.}$$

$$= -j3.18 \text{ p.u.}$$

$$I_2 = -I_1 \frac{Z_0}{Z_0 + Z_2}$$

$$= j3.18 \times \frac{0.17}{0.17 + 0.18}$$

$$= j1.55 \text{ p.u.}$$

$$I_0 = -I_1 \frac{Z_2}{Z_0 + Z_2}$$

$$= j3.18 \times \frac{0.18}{0.18 + 0.17}$$

$$= j1.63 \text{ p.u.}$$

$$I_Y = I_0 + a^2 I_{R1} + a I_{R2}$$

$$= j1.63 + (-0.5 - 0.866j)(-j3.18) + (-0.5 + j0.866)(j1.55)$$

$$= -4.09 + j2.45 \text{ p.u.}$$

$$I_B = j1.63 + (-0.5 + j0.866)(-j3.18) + (-0.5 - j0.866)(j1.55)$$

$$= 4.09 + j2.45.$$

$$I_Y = I_B = 4.78. \text{p.u.}$$

The correctness of the first part of the solution can be checked as

$$I_R = I_1 + I_2 + I_0 = 0.$$

Example 7.6 An 11 kV synchronous generator is connected to a 11/66 kV transformer which feeds a 66/11/3.3 kV three-winding transformer through a short feeder of negligible impedance. Calculate the fault current when a single-phase-to-earth fault occurs on a terminal of the 11 kV winding of the three-winding transformer. The relevant data for the system are as follows:

Generator $X_1 = j0.15$ p.u , $X_2 = j0.1$ p.u., $X_0 j0.03$ p.u. all on a 10 MVA base; star point of winding earthed through a 3 Ω resistor.

11/66 kV *Transformer* $X_1 = X_2 = X_0 = j0.1$ p.u. on a 10 MVA base; 11 kV winding delta connected and the 66 kV winding star connected with the star point solidly earthed.

Three-winding Transformer 66 kV winding, star connected, star point solidly earthed; 11 kV winding, star connected, star point earthed through a 3 Ω resistor; 3.3 kV winding, delta connected; the three windings of an equivalent star connexion to represent the transformer have sequence impedances, 66 kV winding $X_1 = X_2 = X_0 = j0.04$ p.u., 11 kV winding $X_1 = X_2 = X_0 = j0.03$ p.u., 3.3 kV winding $X_1 = X_2 = X_0 = j0.05$ p.u., all on a 10 MVA base.

Resistance may be neglected throughout.

Solution The line diagram and the corresponding positive, negative-, and zero-sequence networks are shown in Figure 7.23. A 10 MVA base will be

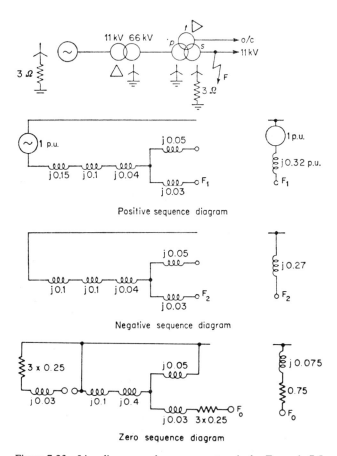

Figure 7.23 Line diagram and sequence networks for Example 7.5.

used. The 3 Ω earthing resistor has the following p.u. value:

$$\frac{3 \times 10{,}000}{(11)^2 \times 1000} \quad \text{or} \quad 0.25 \text{ p.u.}$$

Much care is needed with the zero-sequence network owing to the transformer connexions. For a phase-to-earth fault the equivalent circuit shown in Figure

274

7.13 is used, from which

$$\mathbf{I}_1 = \mathbf{I}_2 = \mathbf{I}_0 \quad \text{and} \quad \mathbf{I}_f = \mathbf{I}_1 + \mathbf{I}_2 + \mathbf{I}_0$$

$$= \frac{3 \times 1}{\mathbf{Z}_1 + \mathbf{Z}_2 + \mathbf{Z}_0 + 3\mathbf{Z}_g}$$

$$= \frac{3}{j0.32 + j0.27 + j0.075 + 0.75}$$

$$= \frac{3}{0.75 + j0.66}$$

$$\therefore \ \mathbf{I}_f = 3 \text{ p.u.} = \frac{3 \times 10 \times 10^6}{\sqrt{(3)} \times 11,000}$$

$$= 1575 \text{ A}$$

An alternative approach using branch and connexion matrices is given in the next example.

Example 7.6 Determine the fault currents in the system shown in Figure 7.24 in which phase (a) has a combined open circuit and earth fault.

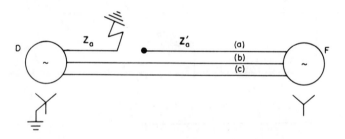

Figure 7.24 Power system with combined ground and open-circuit faults for Example 7.6. Reactances—per unit on a common base—resistance neglected.

Generator D	Impedances of line D to fault $\mathbf{Z}_a$
Z_1 j0.3 p.u.	Z_1 j0.1 p.u.
Z_2 j0.2 p.u.	Z_2 j0.1 p.u.
Z_0 j0.05 p.u.	Z_0 j0.2 p.u.
Generator F	Impedances of line F to fault $\mathbf{Z}_a$
Z_1 j0.25 p.u.	Z_1 j0.05 p.u.
Z_2 j0.15 p.u.	Z_2 j0.05 p.u.
Z_0 j0.03 p.u.	Z_0 j0.1 p.u.

Solution In this solution a branch matrix will be formed and a connexion matrix for currents used. All sequence values refer to the faulted phase.

On the 'D' side, $\mathbf{I}_{1D} = \mathbf{I}_{2D} = \mathbf{I}_{0D}$. (In an earth fault the sequence components are equal.)

On the 'F' side, $I_{aF} = 0$ (open circuit). Also,

$$I_{0F} = \tfrac{1}{3}(I_B + I_C)$$

$$I_{1F} = \tfrac{1}{3}(aI_B + a^2 I_C)$$

$$I_{2F} = \tfrac{1}{3}(a^2 I_B + a I_C)$$

Hence,

$$I_{0F} + I_{2F} + I_{1F} = I_{aF} = 0$$

But $I_{0F} = 0$ as end F is not grounded,

$$\therefore \; I_{2F} = -I_{1F}$$

Hence, connexion matrix $[C]$ is

$$\begin{array}{|c|} I_{0D} \\ I_{1D} \\ I_{1F} \\ I_{2D} \\ I_{2F} \end{array} = \begin{array}{|cc|} 1 & 0 \\ 1 & 0 \\ 0 & 1 \\ 1 & 0 \\ 0 & -1 \end{array} \begin{array}{|c|} I_{0D} \\ I_{1F} \end{array} \quad \text{and} \quad [C_t] = \begin{array}{|ccccc|} 1 & 1 & 0 & 1 & 0 \\ 0 & 0 & 1 & 0 & -1 \end{array}$$

$$v = [C_t]\,[V] = [C_t] \begin{array}{|c|} 0 \\ E_D \\ E_F \\ 0 \\ 0 \end{array} = \begin{array}{|c|} E_D \\ E_F \end{array}$$

$$\text{Branch matrix, } Z_{br} = \begin{bmatrix} Z_{0D} & 0 & 0 & 0 & 0 \\ 0 & Z_{1D} & 0 & 0 & 0 \\ 0 & 0 & Z_{1F} & 0 & 0 \\ 0 & 0 & 0 & Z_{2D} & 0 \\ 0 & 0 & 0 & 0 & Z_{2F} \end{bmatrix}$$

and $[C_t]\,[Z_{br}]\,[C] = [Z] = \begin{bmatrix} Z_{0D} + Z_{1D} + Z_{2D} & 0 \\ 0 & Z_{1F} + Z_{2F} \end{bmatrix}$

As, $[v] = [Z]\,[i]$

$$\begin{array}{|c|} E_D \\ E_F \end{array} = \begin{bmatrix} (Z_{0D} + Z_{1D} + Z_{2D}) & 0 \\ 0 & (Z_{1F} + Z_{2F}) \end{bmatrix} \cdot \begin{array}{|c|} I_{0D} \\ I_{1F} \end{array}$$

$$\begin{array}{|c|} 1.4 \\ 1.4 \end{array} = \begin{bmatrix} j0.95 & 0 \\ 0 & j0.5 \end{bmatrix} \cdot \begin{array}{|c|} I_{0D} \\ I_{1F} \end{array}$$

$$I_{1D} = I_{2D} = I_{0D} = \frac{1.4}{j0.95} = -1.475j \text{ p.u.}$$

Current magnitude into fault from end $D = 3 \times 1.475 = 4.425$ p.u.
Current into fault from end $F \qquad = I_{1F} - I_{2F} = 0$

Note This method can be used for any type of fault, and although it does not use interconnected sequence networks as such it uses the constraints which are necessary for the formation of such networks. For example, in the system shown in Figure 7.26 a connexion matrix $[C]$ can be formed knowing that for an earth-fault current $I_1 = I_2 = I_0$.

7.7 Power in Symmetrical Components

The total power in a three-phase network

$$= V_a I_a^* + V_b I_b^* + V_c I_c^*$$

where V_a, V_b and V_c are phase voltages and I_a, I_b, and I_c are line currents. In phase (a),

$$P_a + jQ_a = (V_{a0} + V_{a1} + V_{a2})(I_{a0}^* + I_{a1}^* + I_{a2}^*)$$
$$= (V_{a0} I_{a0}^* + V_{a1} I_{a1}^* + V_{a2} I_{a2}^*)$$
$$+ (V_{a0} I_{a1}^* + V_{a1} I_{a2}^* + V_{a2} I_{a0}^*)$$
$$+ (V_{a0} I_{a2}^* + V_{a1} I_{a0}^* + V_{a2} I_{a0}^*$$

with similar expressions for phases (b) and (c).

In extending this to cover the total three-phase power it should be noted that

$$I_{b1}^* = (a^2 I_{a1})^* = (a^2)^* I_{a1}^* = a I_{a1}^*.$$

Similarly, I_{b2}^*, I_{c1}^* and I_{c2}^* may be replaced.

$$\text{The total power} \quad = 3(V_{a0} I_{a0}^* + V_{a1} I_{a1}^* + V_{a2} I_{a2}^*) \tag{7.11}$$

i.e. 3 (the sum of the individual sequence powers in any phase).

7.8 Systematic Methods for Fault Analysis in Large Networks

The methods described so far become unwieldy when applied to large networks and a systematic approach using digital computers is used. The digital methods used for load flows are also suitable for three-phase symmetrical fault calculations which after all are merely load flows under certain network conditions. The input information must be modified so that machines are represented by the appropriate reactances. The generators are represented by their no-load voltages in series with the subtransient reactances. If loads are to be taken into account they are represented by the equivalent shunt admittance to neutral. This will involve some modification to the original load-flow admittance matrix, but the general form will be as in the normal load flow.

A current of 1 p.u. is injected at the fault point and removed at the neutral. If the voltage at the fault vecomes V_f then $1/V_f$ is the fault level (i.e. short-circuit admittance) at node f. Also the flow along branch kf into the fault is given by $\dfrac{(\mathbf{V}_f - \mathbf{V}_k)}{\mathbf{V}_f}\mathbf{Y}_{fk}$. The voltages due to the injected currents can be obtained by the same numerical techniques as used for load flows. As loads are represented by admittance the problem is linear and iterative methods are not required.

Three-phase fault studied are performed in conjunction with load flows. For example, if the fault level on a solid busbar is too high the busbar will be sectioned, i.e. split into two or more sections by opening switches, and new load flows required.

The nodal or admittance method may be applied to large networks. It is preferable on grounds of storage and time not to invert the matrix but to use Gauss elimination methods. The computation efficiency may also be improved by utilizing the sparsity of the $\mathbf{Y}$ matrix. The mesh or loop (impedance matrix) method may be used, although the matrix is not so easily formed.

The following example illustrates a method suitable for determination of balanced three-phase fault currents in a large system by means of a digital computer.

Example 7.7 Determine the fault current in the system shown in Figure 7.4 for the balanced fault shown.

As already stated, generators are represented by their voltages behind the transient reactance and normally the system is assumed on no load before the occurrence of the three-phase balanced fault. The voltage sources and transient reactances are converted into current sources and the admittance matrix formed (including the transient reactance admittances). The basic equation $[Y]$ $[V] = [I]$ is formed and solved with the constraint that the voltage at the fault node is zero. A current may be injected to the faulted node, as explained in the previous section, or the voltage made zero and the remaining nodal voltages calculated.

The system in Figure 7.4 is replaced by the equivalent circuit shown in Figure 7.25 in which $\mathbf{V}_D = 0.0$ p.u.

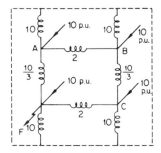

Figure 7.25 Equivalent circuit of system of Figure 7.4. All values are admittances (i.e. $-jY$). The generators, i.e. 1 p.u. voltage behind $-j10$ p.u. admittance transform to $-j10$ p.u. current sources in parallel with $-j10$ p.u. admittance.

$$-j \begin{bmatrix} 15.33333 & -2.00000 & & -3.33333 \\ -2.00000 & 15.33333 & -3.33333 & \\ & -3.33333 & 15.33333 & -2.00000 \\ -3.33333 & & -2.00000 & 15.33333 \end{bmatrix} \begin{bmatrix} V_A \\ V_B \\ V_C \\ 0 \end{bmatrix} = -j \begin{bmatrix} 10 \\ 10 \\ 10 \\ I_F \end{bmatrix}$$

As all the Y's are reactive the j's will be omitted for simplicity.
Here, V_A, and V_C are eliminated, i.e.

$$\begin{bmatrix} 23 & -3 & 0 \\ -3 & 23 & -5 \\ 0 & -5 & 23 \end{bmatrix} \cdot \begin{bmatrix} V_A \\ V_B \\ V_C \end{bmatrix} = \begin{bmatrix} 15 \\ 15 \\ 15 \end{bmatrix}$$

from which

$$\begin{bmatrix} 520 & -115 \\ -5 & 23 \end{bmatrix} \cdot \begin{bmatrix} V_B \\ V_C \end{bmatrix} = \begin{bmatrix} 385 \\ 15 \end{bmatrix}$$

Thus,

$$V_B = 0.92950 \qquad V_C = 0.85510 \quad \text{and} \quad V_A = 0.77495 \text{ p.u.}$$

and

$$I_F = -j\left(10 + \frac{0.77495}{0.3} + \frac{0.88510}{0.5}\right) \text{ p.u.}$$
$$= -j\,14.2985 \text{ p.u.}$$

Fault MVA $= 14.2985 \times 100 = 1429.85$ MVA

Consider the system in Figure 7.26(a) in which a phase-to-earth fault exists. The manner in which the positive-, negative- and zero-sequence networks are interconnected to represent such a fault has been derived and for the system in Figure 7.26(a) is shown in Figure 7.26(b).

In Figure 7.26(b) the generator and transformer reactances have been lumped together. Resistance is neglected and all values are in per unit on a common base. Loop currents are assigned to the network, and the mesh impedance matrix is assembled either by inspection or by the use of the connexion matrix (C). Using the former method the following is obtained.

		F	A₁	B₁	A₂	B₂	A₀	B₀	
E_2		1.40	-0.35	-0.15	-0.25	-0.15	-0.15	-0.35	I_F
$E_1 - E_2$		-0.35	1.07	-0.30	0	0	0	0	I_{A1}
0		-0.15	-0.30	0.60	0	0	0	0	I_{B1}
0	$=j$	-0.25	0	0	0.82	-0.30	0	0	I_{A2}
0		-0.15	0	0	-0.30	0.60	0	0	I_{B2}
0		-0.15	0	0	0	0	1.07	-0.70	I_{A0}
0		-0.35	0	0	0	0	-0.70	1.4	I_{B0}

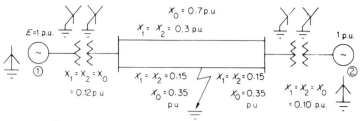

For machine ①, $X_1 = 0.3$, $X_2 = 0.15$ $X_0 = 0.10$

For machine ②, $X_1 = 0.25$, $X_2 = 0.15$ $X_0 = 0.05$

(a)

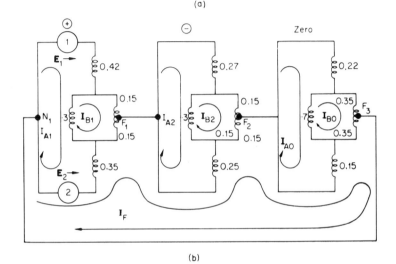

(b)

Figure 7.26(a) Line diagram of faulted system. For machine (1): $X_1 = 0.3$, $X_2 = 0.15$, $X_0 = 0.10$. For machine (2): $X_1 = 0.25$, $X_2 = 0.15$, $X_0 = 0.05$. All values are per unit on a common base. (b) Equivalent symmetrical component circuit diagram for the faulted system of (a).

By inversion of the above matrix (e.g. by successive elimination) the loop currents I_F, $I_A \ldots I_{B0}$ are obtained and from these the sequence currents in each individual network are determined.

For the unfaulted line,

$$I_1 = I_{A1} - I_{B1} \qquad I_2 = I_{A2} - I_{B2} \quad \text{and} \quad I_0 = I_{A0} - I_{B0}$$

For machine (2),

$$I_1 = I_F - I_{A1} \qquad I_2 = I_F = I_{A2}$$

and

$$I_0 = I_F - I_{A0}$$

Hence a connexion matrix relating branch and loop sequence currents may be assembled.

7.9 Bus Impedance (Short circuit Matrix) Method

This method is used for computer fault analysis and has the following advantages:

(a) Matrix inversion is avoided, resulting in savings in computer storage and time.
(b) The matrices for the sequences quantities are determined only once and retained for later use; they are readily modified for system changes.
(c) Mutual impedances between lines are readily handled.
(d) Subdivisions of the main system may be incorporated.

The system is represented by the usual symmetrical component sequence networks and frequently the positive and negative impedances are assumed to be identical. Balanced phase impedances for all items of plant are assumed and equal voltages for all generators. For simplicity in this treatment mutual coupling between lines will be neglected and load currents assumed negligible. The method will be explained using the positive-sequence network as this will be most familiar to the reader.

In the system shown in Figure (7.27) the voltage source supplies a common bus and four busbars of interest in the passive network are identified. The network loop matrix, i.e. $[e] = [Z][i]$, is set up in terms of the various loop currents

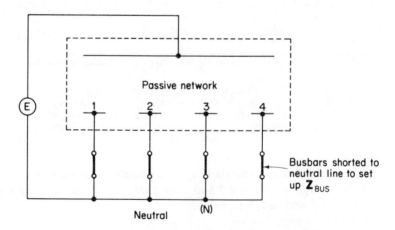

Figure 7.27   Identification of busbars of interest in a large network—
single-phase equivalent network.

which must include those flowing through busbars (1)–(4) to the neutral bus, i.e. these busbars are short-circuited to N. In the matrix equation so formed only nodes 1–4 are of interest and by the process of partitioning described in the section on load flows a new equation is obtained involving only currents from these busbars. $E = 1$ p.u.

$$
\begin{bmatrix} 1.0 \\ 1.0 \\ 1.0 \\ 1.0 \\ \hline 0 \\ \cdot \\ \cdot \\ \cdot \\ 0 \end{bmatrix}
=
\begin{bmatrix}
Z_{11} & Z_{12} & Z_{13} & Z_{14} & \mid & \cdots Z_{1n} \\
Z_{21} & Z_{22} & Z_{23} & Z_{24} & \mid & \cdots Z_{2n} \\
Z_{31} & Z_{32} & Z_{33} & Z_{34} & \mid & \cdots Z_{3n} \\
Z_{41} & Z_{42} & Z_{43} & Z_{44} & \mid & \cdots Z_{4n} \\
\hline
\cdot & \cdot & \cdot & \cdot & \mid & \cdots \\
\cdot & \cdot & \cdot & \cdot & \mid & \cdots \\
\cdot & \cdot & \cdot & \cdot & \mid & \cdots \\
Z_{n1} & Z_{n2} & Z_{n3} & Z_{n4} & \mid & \cdots Z_n
\end{bmatrix}
\cdot
\begin{bmatrix} i_1 \\ i_2 \\ i_3 \\ i_4 \\ \hline \cdot \\ \cdot \\ \cdot \\ i_n \end{bmatrix}
$$

$(\mathbf{Z_A})$ $(\mathbf{Z_B})$

$(\mathbf{Z_c})$ $(\mathbf{Z_D})$

$$[\mathbf{e}] = [\mathbf{Z_{sc}}]\,[\mathbf{i}]$$

where

$$\mathbf{Z_{sc}} = \mathbf{Z_A} - \mathbf{Z_B}\,\mathbf{Z_D}^{-1}\,\mathbf{Z_C}$$

The new matrix is called the *bus impedance* or short-circuit matrix and the equation is given by

$$
\begin{bmatrix} 1.0 \\ 1.0 \\ 1.0 \\ 1.0 \end{bmatrix}
=
\begin{bmatrix}
z_{11} & z_{12} & z_{13} & z_{14} \\
z_{21} & z_{22} & z_{23} & z_{24} \\
z_{31} & z_{32} & z_{33} & z_{34} \\
z_{41} & z_{42} & z_{43} & z_{44}
\end{bmatrix}
\cdot
\begin{bmatrix} I_1 \\ I_2 \\ I_3 \\ I_4 \end{bmatrix}
\tag{7.12}
$$

This matrix may be represented by the simple network shown in Figure 7.28 in

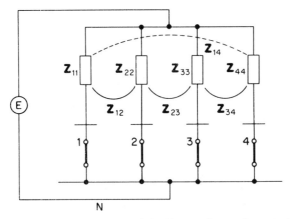

Figure 7.28 Equivalent circuit of the Z_{bus} matrix equation—single-phase equivalent, i.e. balanced-fault conditions.

which the buses of interest are shown short-circuited to the neutral. Consider a fault on bus (1) only, currents $\mathbf{I}_2$, $\mathbf{I}_3$, and $\mathbf{I}_4$ will be zero and, from equation (1), $1.0 = \mathbf{z}_{11} \mathbf{I}_1$, where $\mathbf{I}_1$ = fault current with three-phase symmetrical fault on 1. Similarly, the currents with balanced faults on each of the other buses may be easily determined.

The voltages at the remaining busbars with a balanced fault on bus (1) may also be readily determined, e.g.

$$1.0 - \mathbf{V}_{n2} = \mathbf{Z}_{21} \mathbf{I}_1 = \mathbf{Z}_{21} \frac{1.0}{\mathbf{Z}_{11}}$$

where $\mathbf{V}_{n2}$ is the voltage on bus 2 with a short circuit on (1) and so on.

For unbalanced faults on the busbars the three sequence networks are used in the normal way. In Figure 7.29 the connection of the three reduced sequence

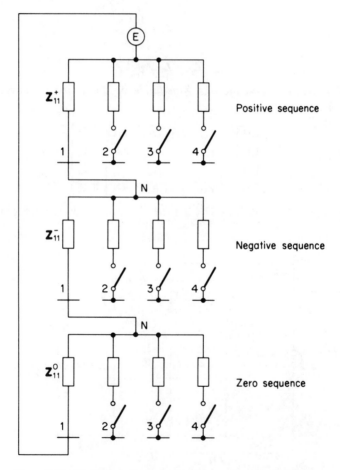

Figure 7.29 Equivalent network using Z_{bus} method for single-phase to ground fault on busbar (1) of network in Figure 7.27.

networks for a line-to-ground fault on (1) is illustrated. As usual the sequence networks refer to phase (a) of the real network.

For a L–G fault on bus (1),

$$\mathbf{I}_1 = 3/(2\mathbf{z}_{11}^+ + \mathbf{z}_{11}^0)$$

and

$$\mathbf{I}_1^+ = \mathbf{I}_1^- = \mathbf{I}_1^0 = 1/(2\mathbf{z}_{11}^+ + \mathbf{z}_{11}^0)$$

The assumption is made that the positive- and negative-sequence impedances are equal. This aids computation and does not produce major error.

Also the voltage on bus (3),

$$\mathbf{V}_3^+ = 1 - \frac{\mathbf{z}_{31}^+}{(2\mathbf{z}_{11}^+ + \mathbf{z}_{11}^0)}$$

$$\mathbf{V}_3^- = -\frac{\mathbf{z}_{31}^-}{2\mathbf{z}_{11}^+ + \mathbf{z}_{11}^0}$$

$$\mathbf{V}_3^0 = -\frac{\mathbf{z}_{31}^0}{(2\mathbf{z}_{11}^+ + \mathbf{z}_{11}^0)}$$

and

$$\mathbf{V}_3 = \mathbf{V}_3^+ + \mathbf{V}_3^- + \mathbf{V}_3^0 = 1 - \left(\frac{2\mathbf{z}_{12}^+ + \mathbf{z}_{13}^0}{2\mathbf{z}_{11}^+ + \mathbf{z}_{11}^0}\right)$$

The bus impedance matrix has so far been derived from the full loop impedance matrix obtained either by inspection or by systematic methods. An alternative approach which is more economical in computer usage is to form the matrix one step at a time as each component of the network is added. These additions fall into four categories as follows:

(a) A new generator bus or a new bus with a direct connexion to the neutral.
(b) A generator (or connexion to neutral) connected to an existing bus.
(c) Connexion from an existing bus to a new bus.
(d) Connexion between two existing buses.

To illustrate the addition of buses a 2×2 matrix describing buses (1) and (2) will be assumed to have already been assembled.

(a) New bus with connexion to neutral through $\mathbf{z}_L$ (generator) (see Figure 7.30).

$$\text{New } \mathbf{z}_B = \begin{array}{c} \text{OLD} \\ \begin{bmatrix} \mathbf{z}_{11} & \mathbf{z}_{12} & 0 \\ \mathbf{z}_{21} & \mathbf{z}_{22} & 0 \\ 0 & 0 & \mathbf{z}_L \end{bmatrix} \end{array}$$

Note that the currents through (1) and (2) have no interaction in (3).

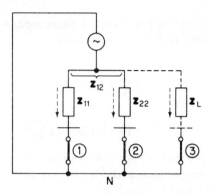

Figure 7.30   New busbar with connection to neutral through Z_L, e.g. generator.

(b) Connection to neutral of generator from an existing bus. $\mathbf{Z}_L$ is connected to bus (2) and is associated with loop current $\mathbf{I}_3$ as in Figure 7.31. The new matrix is

$$
\begin{array}{cc}
& \text{(A)} \qquad\qquad \text{(B)} \\
& \overset{①}{} \ \overset{②}{} \qquad \overset{③}{} \\
\left[\begin{array}{cc|c}
\mathbf{z}_{11} & \mathbf{z}_{12} & \mathbf{z}_{12} \\
\mathbf{z}_{21} & \mathbf{z}_{22} & \mathbf{z}_{22} \\
\hline
\mathbf{z}_{21} & \mathbf{z}_{22} & \mathbf{z}_{22} + \mathbf{z}_L
\end{array}\right] &
\begin{array}{c}
(\mathbf{I}_1) \\[1.2em]
(\mathbf{I}_2) \\[1.8em]
(\mathbf{I}_3)
\end{array}
\end{array}
$$

$$
\text{(C)} \qquad\quad \text{(D)}
$$

Note the $\mathbf{I}_1$ influences loop $\mathbf{I}_3$ via $\mathbf{z}_{12}$. The internal loop is eliminated by applying, $\mathbf{z} = \mathbf{z}_A - \mathbf{z}_B\,\mathbf{z}_D^{-1}\,\mathbf{z}_C$ to the partitioned matrix giving a 2×2 matrix.

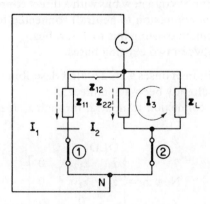

Figure 7.31   Connexion to neutral of generator from an existing bus (2).

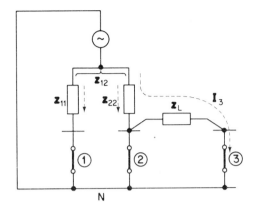

Figure 7.32 Connexion of impedance (Z_L) between an existing bus
(2) and a new bus (3).

(c) Connection of impedance between an existing bus and a new bus (3).   New line z_L connected to bus (2). Consider loop current I_3 as shown in Figure 7.32.

$$\mathbf{z}_{bus} = \begin{array}{c} \quad\; \textcircled{1} \quad\; \textcircled{2} \qquad\;\; \textcircled{3} \\ \left[\begin{array}{ccc|c} \mathbf{z}_{11} & \mathbf{z}_{12} & \mathbf{z}_{12} \\ \mathbf{z}_{21} & \mathbf{z}_{22} & \mathbf{z}_{22} \\ \hline \mathbf{z}_{21} & \mathbf{z}_{22} & \mathbf{z}_{22} + \mathbf{z}_L \end{array} \right] \end{array}$$

(d) Connection of impedance (line, transformer, etc.) between two existing buses.   Consider the new link z_L between existing buses (2) and (3) in Figure 7.33 and loop current I_L.

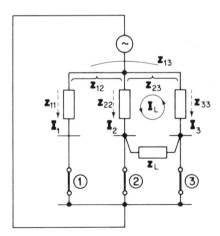

Figure 7.33 Connexion of impedance (Z_L) between two existing buses, (2) and (3).

$$\begin{array}{cccc} \textcircled{1} & \textcircled{2} & \textcircled{3} & \textcircled{L} \\ \end{array}$$

$$\begin{bmatrix} z_{11} & z_{12} & z_{13} & \vdots & (z_{1L} = z_{12} - z_{13}) \\ z_{21} & z_{22} & z_{23} & \vdots & (z_{2L} = z_{22} - z_{23}) \\ z_{31} & z_{32} & z_{33} & \vdots & (z_{3L} = z_{23} - z_{33}) \\ \hdashline z_{L1} & z_{L2} & z_{L3} & \vdots & z_L + z_{33} + z_{22} - 2z_{23} \end{bmatrix}$$

Note I_L in z_{22} induces a voltage in the z_{33} branch due to z_{23}. Similarly, it induces in z_{22} when flowing in z_{33} giving an opposing double voltage $2I_L \times z_{23}$.

As I_L traverses the loop it induces into z_{11} due to z_{12} and z_{13}. Hence $z_{L1} = z_{12} - z_{13}$ or generally the mutual between the loop current I_L and node i with a line added between k and m is

$$z_{Li} = z_{iL} = z_{ki} - z_{mi} \qquad (i \neq L)$$

e.g. for node 3,

$$z_{L3} = z_{23} - z_{33}$$

Again the internal loop is eliminated by partitioning, leaving a 3×3 matrix.

It can be shown that a *single loop* elimination may be accomplished by operation on the individual elements as follows. Let z'_{mn} be an element in the new bus impedance (reduced) matrix (i.e. column n, row m) and z_{mn} an element in the original loop matrix. Then,

$$z'_{mn} = z_{mn} - z_{mL} z_{LL}^{-1} z_{Ln} \qquad (7.13)$$

For example in the above 4×4 matrix the new *term*, $z'_{23} = z_{23} - z_{2L} z_{LL}^{-1} z_{L3}$

$$= z_{23} - (z_{22} - z_{23})(z_L + z_{33} + z_{22} - 2z_{23})^{-1}(z_{23} - z_{33})$$

Similarly, the other eight terms are obtained.

If in existing networks lines or generators are to be removed this can be achieved as outlined above, but with the use of negative impedances. The same general approach is made to systems with coupling between parallel lines and also the splitting of large networks into several smaller networks. For details of these more complex situations the reader is referred to reference 11.

Example 7.8 For the system shown in Figure 7.34 assemble the Z_{bus} matrix, branch by branch.

Solution First consider the j0.2 branch (Figure 7.34(b)) and then add the j0.1 branch (Figure 7.34(c)) giving the matrix

$$\begin{array}{cc} \textcircled{1} & \textcircled{2} \end{array}$$

j.2	0
0	j.1

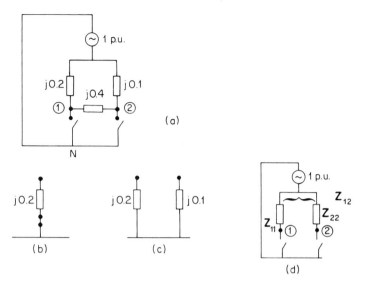

Figure 7.34 Network for Example 7.8.

Finally the j0.4 branch connects nodes 1 and 2 and the matrix is

$$
\begin{array}{c}
 & & n & \\
 & \begin{array}{ccc} ① & ② & ⓛ \end{array} \\
m\downarrow & \begin{bmatrix} j0.2 & 0 & j0.2 \\ 0 & j0.1 & -j0.1 \\ j0.2 & -j0.1 & j0.7 \end{bmatrix}
\end{array}
$$

Eliminate the internal loop, using equation (7.13).

$$\mathbf{Z}_{11}^1 = j0.2 - (j0.2)\,(j0.7)^{-1}\,(j0.2) = 0.145$$

$$\mathbf{Z}_{12}^1 = j0.1 - (-j0.1)\,(j0.7)^{-1}\,(-j0.1) = j0.085$$

$$\mathbf{Z}_{12}^1 = 0 - (j0.2)\,(j0.7)^{-1}\,(-j0.1) = j0.028$$

$$\mathbf{Z}_{21}^1 = j0.028$$

Therefore

$$
\mathbf{Z}_{\text{bus}} = j
\begin{array}{c}
\begin{array}{cc} ① & ② \end{array} \\
\begin{bmatrix} 0.145 & 0.028 \\ 0.028 & 0.085 \end{bmatrix}
\end{array}
$$

and

$$|\mathbf{E}| = |\mathbf{Z}_{\text{bus}}||\mathbf{I}|$$

(see Figure 7.34(d)).

With a three-phase short circuit on bus (1) above

$$\mathbf{I}_1 = \frac{1.0}{\mathbf{Z}_{11}} = \frac{1.0}{j0.145} = 6.9 \text{ p.u.}$$

$$(\mathbf{I}_2 = 0)$$

Similarly, for a fault on (2) above

$$\mathbf{I}_2 = \frac{1}{j0.085} = 17.5 \text{ p.u.}$$

7.10 Neutral Grounding

Introduction

From the analysis of unbalanced fault conditions it has been seen that the connexion of the transformer and generator neutrals greatly influences the fault currents and voltages. In most high-voltage systems the neutrals are solidly grounded, i.e. connected directly to the ground, with the exception of generators which are grounded through a resistance to limit stator fault currents. The advantages of such grounding are as follows:

(a) Voltages to ground are limited to the phase voltage.
(b) Intermittent ground faults and high voltages due to arcing faults are eliminated.
(c) Sensitive protective relays operated by ground fault currents clear these faults at an early stage.

The main advantage in operating with neutrals isolated is the possibility of maintaining a supply with a ground fault on one line which places the remaining conductors at line voltage above ground. Also interference with telephone circuits is reduced because of the absence of zero-sequence currents. With normal balanced operation the neutrals of an ungrounded or isolated system are held at ground potential because of the presence of the system capacitance to earth. For the general case shown in Figure 7.35 the following analysis

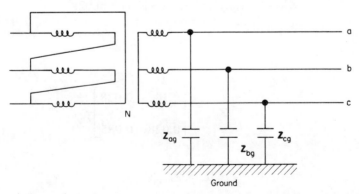

Figure 7.35 Line-to-ground capacitances in an ungrounded system.

applies:

$$\frac{\mathbf{V}_{ag}}{\mathbf{Z}_{ag}} + \frac{\mathbf{V}_{bg}}{\mathbf{Z}_{bg}} + \frac{\mathbf{V}_{cg}}{\mathbf{Z}_{cg}} = 0 \qquad (7.14)$$

Also,

$$\mathbf{V}_{ag} = \mathbf{V}_{an} + \mathbf{V}_{ng}$$

where

$\mathbf{V}_{an}$ = voltage of line a to neutral and

$\mathbf{V}_{ng}$ = voltage of neutral to ground

Similarly

$$\mathbf{V}_{bg} = \mathbf{V}_{bn} + \mathbf{V}_{ng}$$

and

$$\mathbf{V}_{cg} = \mathbf{V}_{cn} + \mathbf{V}_{ng}$$

Substituting in equation (7.14) and separating terms,

$$\frac{\mathbf{V}_{an}}{\mathbf{Z}_{ag}} + \frac{\mathbf{V}_{bn}}{\mathbf{Z}_{bg}} + \frac{\mathbf{V}_{cn}}{\mathbf{Z}_{cg}} + \mathbf{V}_{ng}\left(\frac{1}{\mathbf{Z}_{ag}} + \frac{1}{\mathbf{Z}_{bg}} + \frac{1}{\mathbf{Z}_{cg}}\right) = 0 \qquad (7.15)$$

the coefficient of $\mathbf{V}_{ng}$ gives the ground capacitance admittance of the system.

Arcing faults

Consider the single-phase system in Figure 7.36 at the instant when the instantaneous voltages are v on line (a) and $-v$ on line (b), where v is the maximum instantaneous voltage. The sudden occurrence of a fault to ground causes line (b) to assume a potential of $-2v$ and (a) to become zero. Because of the presence of both L and C in the circuit the sudden change in voltages by v produces a high-frequency oscillation of peak magnitude $2v$ superimposed on

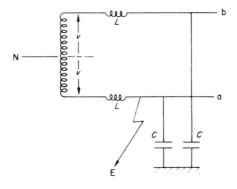

Figure 7.36   Single-phase system with arcing fault to ground.

the power frequency voltages (see Chapter 10) and line (a) reaches $-v$ and (b) $-3v$, as shown in Figure 7.37. These oscillatory voltages attenuate quickly due to the resistance present. The current in the arc to earth on line (a) is approximately 90° ahead of the fundamental voltage and when it is zero the voltage will be at a maximum. Hence, if the arc extinguishes at the first current zero the lines remain charged at $-v$ for (a) and $-3v$ for (b) The line potentials

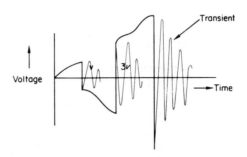

Figure 7.37   Voltage on line 'a' of Figure 7.36.

now change at power frequency until (a) reaches $-3v$ when the arc could restrike causing a voltage change of $-3v$ to 0, resulting in a transient overvoltage of $+3v$ in (a) and $+5v$ in (b) This process could continue and the voltage build up further, but the resistance present usually limits the peak voltage to approximately $4v$. A similar analysis may be made for a three-phase circuit again showing that serious overvoltages may occur with arcing faults.

This condition may be overcome in an isolated neutral system by means of an *arc suppression* or Petersen coil. The reactance of this coil which is connected between the neutral and ground is made in the range 90–110 per cent of the value required to neutralize the capacitance current. The phasor diagram for the network of Figure 7.38(a) is shown in Figure 7.38(b) if the voltage drop across the arc is neglected.

$$\mathbf{I_a} = \mathbf{I_b} = \surd(3)\mathbf{V}\omega C$$

and

$$\mathbf{I_a} + \mathbf{I_b} = \surd(3) \times \surd(3)\mathbf{V}\omega C$$

also,

$$\mathbf{I_L} = \frac{\mathbf{V}}{\omega L}$$

For compensation of the arc current,

$$\frac{\mathbf{V}}{\omega L} = 3\mathbf{V}\omega C$$

$$\therefore \ L = \frac{1}{3\omega^2 C}$$

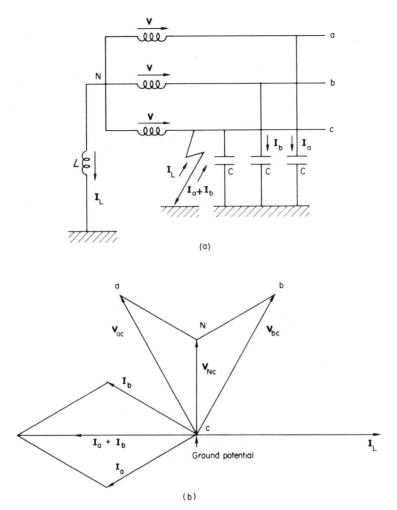

Figure 7.38(a)   System with arc suppression coil. (b) Phasor diagram
of voltages and currents in Figure 7.38(a).

and

$$X_{\mathrm{L}} = \frac{1}{3\omega C} \tag{7.16}$$

This result may be obtained by analysis of the ground fault by means of
symmetrical components.

  Generally isolated neutral systems give rise to serious arcing-fault voltages if
the arc current exceeds the region 5–10 A which covers most systems operating
above 33 kV. If such systems are to be operated with isolated neutrals,
arc-suppression coils should be used. Most systems at normal transmission
voltages have grounded neutrals.

7.11 Interference with Communication Circuits

When power and telephone lines run in parallel under certain conditions voltages sufficient to cause high noise levels may be induced into the communication circuits. This may be caused by electromagnetic and electrostatic unbalance in the power lines, especially if harmonics are present. The major problem, however, is due to ground faults producing large zero-sequence currents in the power line which inductively induce voltage into the neighbouring circuit. The value of induced voltage depends on the spacing, resistivity of the earth immediately below, and the frequency. The induced voltage is induced into each of the communication wires so that if the latter are perfectly transposed no voltage would exist between them. However, the voltage would exist between the wire and earth. This voltage is kept low by the use of a 'drainage' coil connected between the wires and earthed at its electrical midpoint and which shunts to earth the longitudinal induced voltages but gives small attenuation at the communication frequencies.

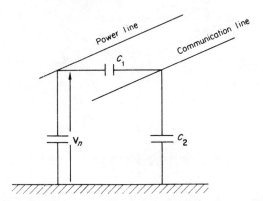

Figure 7.39   Induced voltage between power and communication circuits—induced longitudinal voltage = $V_n[C_1/(C_1 + C_2)]$.

There are two main groups of communication circuits, (a) telegraph circuits with signal fequencies up to 300 Hz and usually single line with ground return, and (b) telephone circuits with signal frequencies of 100–400 Hz and always double-line circuits. Interference from underground cable circuits is much less (10 per cent) than that from overhead lines.

Because of right-of-way constraints telephone and power distribution lines run parallel along the same street in urban areas. However the interference in rural areas is often greater because communication lines and plant may be unshielded or have higher shield resistances, and unlike urban areas there is no extensive network of water and gas pipes to share the ground return currents.

Resistive coupling between power and communication circuits can exist:

(a) because of physical contact between them;
(b) via paths through the soil between telephone and power grounds.

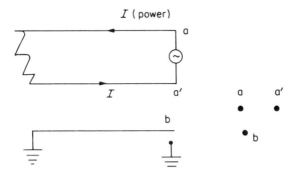

Figure 7.40   Coupling between communication line (b) and power
lines a and a'.

Another important mode of coupling is magnetic. In Figure 7.40 e.m.f.s are induced in the communication line b by the power lines a and a'. If the physical distances a to b and a' to b are equal the inductions cancel, but this is never the case in practice, especially as most of the power fault-current is through the earth.

Various formulae exist to calculate the value of mutual inductance— in H/m between circuits with earth return. These assume that ε_r (soil) is unity, displacement currents are much less than conduction currents, and the length of conductor is infinite. A typical formula is that due to Carson.[22]

During line-to-ground faults induced voltages into communication circuits may be sufficient to be a shock hazard to personnel. Although in transmission circuits equal current loading may be assumed in the phases this is not the case in the lower voltage distribution circuits where significant residual currents may flow.

References

BOOKS

1. *Electrical Transmission and Distribution Reference Book*, Westinghouse Electric Corp., East Pittsburgh, Pennsylvania, 1964.
2. El-Abiad, A. H., and G. W. Stagg, *Computer Methods in Power Systems Analysis*, McGraw-Hill, New York, 1968.
3. Clarke, E., *Circuit Analysis of Alternating Current Power Systems*, Vol. 1, Wiley, New York, 1943.
4. Wagner, C. F. and R. D. Evans, *Symmetrical Components*, McGraw-Hill, New York, 1933.
5. Austin Stigant, S., *Master Equations and Tables for Symmetrical Component Fault Studies*, Macdonald, London, 1964.
6. Klewe, H. R., *Interference Between Power Systems and Telecommunication Lines*, Edward Arnold, London, 1958.

PAPERS

7. Fortescue, C. L., 'Method of symmetrical co-ordinates applied to the solution of polyphase networks', *Trans. A.I.E.E.*, **37**, Pt. II (1918).

8. Lantz, M. J., 'Digital short-circuit solution of power system networks including mutual impedance', *Trans. A.I.E.E.*, **76**, Pt. III (1957), 1230.
9. Siegel, J. C., and G. W. Bills, 'Nodal representation of large complex-element networks including mutual reactances', *Trans. A.I.E.E.*, **77**, Pt. III (1958), 1226.
10. Brown, H. E., C. E. Person, L. E. Kirchmayer, and G. W. Stagg, 'Digital calculation of three-phase short circuits by matrix method', *Trans. A.I.E.E.*, **79**, Pt. III (1960), 1277.
11. El-Abiad, A. H., 'Digital calculation of line-to-ground short circuits by matrix method', *Trans. A.I.E.E.*, **79**, Pt. III (1960), 323.
12. Taylor, G. E., G. K. Carter, and E. H. MacDonald, 'New digital computer short-circuit program for relay studies,' *Trans A.I.E.E.*, **79**, Pt. III (1960), 1257.
13. El-Abiad, A. H., Ruth Guidone, and G. W. Stagg, 'Calculation of short circuits using a high-speed digital computer,' *Trans. A.I.E.E.*, **80**, Pt. III (1961), 702.
14. Graves, J. D., *et al.*, 'Interactive short circuit system', *I.E.E.E. Trans.*, **PAS-101** (1982), 421.
15. Sato, N., and W. F. Tinney, 'Techniques for exploiting the sparcity of the network admittance matrix', *Trans. I.E.E.E., P.A. & S.*, **82**, (1963), 944.
16. Brown, H. E., and C. E. Person, 'Short circuit studies of large systems by the impedance matrix method,' *I.E.E.E. Conference Record, P.I.C.A.* (1967), 335.
17. Proulx, R., and D. Crevier, 'New interactive short circuit calculation algorithm', *I.E.E.E. Trans.*, **PAS-101** (1982), 2681.
18. Sebo, S. A., 'Zero sequence current distribution along transmission lines,' *Trans. I.E.E.E., PA. & S.*, **88**, (1969), 910.
19. Tarsi, D. C. 'Simultaneous solution of line-out and open-end line-to-ground short circuits', *Trans. I.E.E.E., P.A. & S.* **89**, (1970), 1220–1225.
20. Dy Liacco, T. E., and K. A. Ramarao, 'Short-circuit calculations for multiline switching and end faults', *Trans. I.E.E.E., P.A. & S.*, **89** (1970), 1226–1237.
21. Smith, D. R. 'Digital, simulation of simultaneous unbalances involving open and faulted conductors, *Trans. I.E.E.E., P.A. & S.*, (1970), 1826–1835.
22. Carson, J. R., 'Wave propagation in overhead wires with ground returns', *Bell System Tech, J.*, **5** (1926), 539.
23. Meyer, W. S., and H. W. Dommel, 'Telephone-interface calculations for multi-conductor power', *Trans. I.E.E.E.*, **PAS-88** (1969), 35–41.

Problems

7.1 Four 11 kV generators designated A, B, C, and D each have a subtransient reactance of 0.1 p.u. and a rating of 50 MVA. They are connected in parallel by means of three 100 MVA reactors which join A to B, B to C, and C to D; these reactors have per unit reactances of 0.2, 0.4 and 0.2 respectively. Calculate the voltamperes and the current flowing into a three-phase symmetrical fault on the terminals of machine B. Use a 50 MVA base.

Design a resistance fault-analyser to represent the above system using a supply of 50 V and taking 1 mA to represent a 100 MVA fault level.

(Answer: 940 MVA; 49,300 A)

7.2 Two 100 MVA, 20 kV turbo-generators (each of transient reactance 0.2 p.u.) are connected, each through its own 100 MVA, 0.1 p.u. reactance transformer, to a common 132 kV busbar. From this busbar, a 132 kV feeder, 40 km in length, supplies an 11 kV load through a 132/11 kV transformer of 200 MVA rating and reactance 0.1 p.u. If a balanced three-phase short circuit occurs on the low voltage terminals of the load transformer, determine, using a 100 MVA base, the fault current in the feeder and the rating of a suitable circuit breaker at the load end of the feeder. The feeder impedance per phase is $(0.035 + j0.14)$ Ω/km.

(Answer: 482 MVA)

7.3. Two 60 MVA generators of transient reactance 0.15 p.u. are connected to a busbar designated 'A'. Two identical machines are connected to another busbar 'B'. A feeder is supplied from A through a step-up transformer rated at 30 MVA with 10 per cent reactance.

Calculate the reactance of a reactor to connect A and B if the fault level due to a three-phase fault on the feeder side of the transformer is to be limited to 240 MVA. Calculate also the voltage on A under this condition if the generator voltage is 13 kV (line).

(Answer: $X = 0.075$ p.u.; $V_A = 10.4$ kV)

7.4. A 132 kV supply feeds a line of reactance 13 Ω which is connected to a 100 MVA, 132/33 kV transformer of 0.1 p.u. reactance. The transformer feeds a 33 kV line of reactance 6 Ω which in turn is connected to an 80 MVA, 33/11 kV transformer of 0.1 p.u. reactance. This transformer supplies an 11 kV substation from which a local 11 kV feeder of 3 Ω reactance is supplied. This feeder energizes a protective overcurrent relay through 100 A/1 A current transformers. The relay has a true inverse-time characteristic and operates in 10 s with a coil current of 10 A.

If a three-phase fault occurs at the load end of the 11 kV feeder calculate the fault current and time of operation of the relay.

(Answer: 1575 A; 6.35 s.)

7.5. A ring-main system consists of a number of substations designated A, B, C, D, and E connected by transmission lines having the following impedances per phase (ohms): AB (1.5 + j2); BC (1.5 + j2); CD (1 + j1.5); DE (3 + j4); EA (1 + j1).

The system is fed at A at 33 kV from a source of negligible impedance. At each substation except A the circuit breakers are controlled by relays fed from 1500/5 A current transformers. At A the current transformer ratio is 4000/5. The characteristics of the relays are as follows:

Current (A)		7	9	11	15	20
Operating time(s)	relays at A, D, and C	3.1	1.95	1.37	0.97	0.78
	relays at B and E	4	2.55	1.8	1.27	1.02

Examine the sequence of operation of the protective gear for a three-phase symmetrical fault at the mid-point of line CD.

Assume that the primary current of the current transformer at A is the total fault current to the ring and that each circuit breaker opens 0.3 s after the closing of the trip-coil circuit. Comment on the disadvantages of this system.

7.6. The following currents were recorded under fault conditions in a three-phase system:

$$I_R = 1500\angle 45° \text{ A}, \qquad I_Y = 2500\angle 150° \text{A}, \qquad I_B = 1000\angle 300° \text{ A}.$$

If the phase sequence is R–Y–B, calculate the values of the positive, negative, and zero phase sequence components for each line.

(Answer: $I_0 = (-200 + j480)$A.)

7.7. A single line-to-earth fault occurs on the red phase at the load end of a 66 kV transmission line. The line is fed via a transformer by 11 kV generators connected to a common busbar. The line-side of the transformer is connected in star with the star point earthed and the generator side is in delta. The positive sequence reactances of the transformer and line are j10.9 Ω and j44 Ω respectively, and the equivalent positive and negative sequence reactances of the generators, referred to the line voltage, are j18 Ω and j14.5 Ω respectively. Measured up to the fault the total effective zero-sequence reactance is j150 Ω. Calculate the fault current in the lines if resistance may be neglected. If a two-phase to earth fault occurs between the blue and yellow lines, calculate the current in the yellow phase.

(Answer: 392 A and 488 A.)

296

7.8. A single line-to-earth fault occurs in a radial transmission system. The following sequence impedances exist between the source of supply (an infinite busbar) of voltage 1 p.u. to the point of the fault: $Z_1 = 0.3 + j0.6$ p.u., $Z_2 = 0.3 + j0.55$ p.u., $Z_0 = 1 + j0.78$ p.u. The fault path to earth has a resistance of 0.66 p.u. Determine the fault current and the voltage at the point of the fault.
(Answer: $I_f = 0.732$ p.u.; $V_f = (0.43 - j0.23)$p.u.)

7.9. In the system described in Example 7.5 determine the sequence components of the current fed into a double-line to earth fault when the star point of generator B is also connected to earth. Check the result using the fact that $I_R = 0$.

7.10. In the system shown in Figure 7.19 the generation connected to the 132 kV busbar is only used under maximum load conditions. Calculate the fault levels for three-phase faults at each busbar with maximum-plant infeed reactance and with and without the generation connected to the 132 kV busbars.

7.11. Develop an expression, in terms of the generated e.m.f. and the sequence impedances, for the fault current when an earth fault occurs on phase 'a' of a three-phase generator, with an earthed star point. Show also that the voltage to earth of the sound phase 'B' at the point of fault is given by

$$V_B = \frac{-j\sqrt{3}E_A|Z_2 - aZ_0|}{Z_1 + Z_2 + Z_0}$$

Two 30 MVA, 6.6 kV synchronous generators are connected in parallel and supply a 6.6 kV feeder. One generator has its star point earthed through a resistor of 0.4 Ω and the other has its star point isolated. Determine: (1) the fault current and the power dissipated in the earthing resistor, when an earth fault occurs at the far end of the feeder on phase 'A' and (2) the voltage to earth of phase 'B'. The generator phase sequence is ABC and the impedances are

	Generator p.u./ph	Feeder Ω/ph
To positive sequence currents	j0.2	j0.6
To negative sequence currents	j0.16	j0.6
To zero sequence currents	j0.06	j0.4

Use a base of 30 MVA.
(Answer: $5459\angle -52.6°$ A; 11.92 MW; $(-1402 + j2770)$ V)

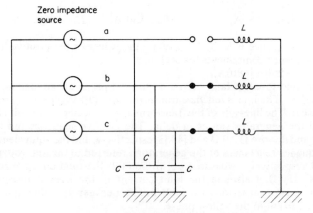

Figure 7.41   System for Problem 7.13.

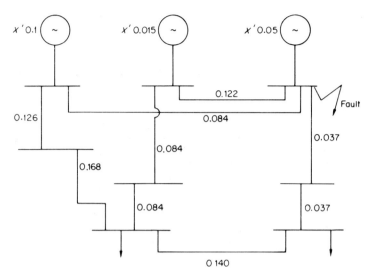

Figure 7.42   System for Problem 7.14.

7.12. A 33 kV, 60 Hz overhead line has a capacitance to ground per phase of approximately 0.7 μF. Calculate the reactance of a suitable arc-suppression coil. (Answer: 1315 Ω)

7.13. The system shown in Figure 7.41 has an open circuit on line (a). Determine the potential between the source neutral and ground. The phase voltage is V and angular

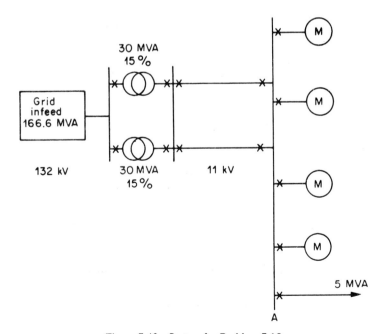

Figure 7.43   System for Problem 7.15.

frequency ω. Investigate the peculiarities arising if $\omega^2 LC < 2/3$ and $\omega^2 LC > 2/3$.

$$\left(\text{Answer:} \frac{V}{3\omega^2 LC - 2} \right)$$

7.14. In the system shown in Figure 7.42 all reactances are in per unit. Determine the p.u. fault current for a three-phase balanced fault on the terminals of the machine of reactance 0.05 p.u.
(Answer: 29 p.u.),
(*Note* This system is large enough to require programming and use of a computer.)
7.15 An industrial distribution system is shown schematically in Figure 7.43. Each line has a reactance of j0.4 p.u. calculated on a 100 MVA base; other system parameters are given in the diagram. Choose suitable short-circuit ratings for the oil circuit breakers, situated at substation A, from those commercially available, which are given in the table below.

Short circuit (MVA)	75	150	250	350
Rated current (amps)	100	300	400	600

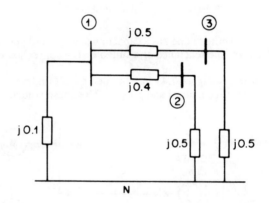

Figure 7.44   Network for Problem 7.16.

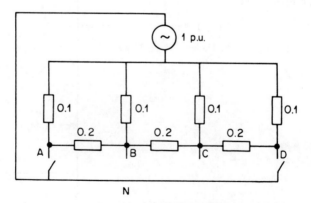

Figure 7.45   Network for Problem 7.17.

The industrial load consists of a static component of 5 MVA and four large induction motors each rated at 6 MVA. Show that only three motors can be started simultaneously given that, at starting, each motor takes five times full-load current at rated voltage.

7.16. By systematically adding one branch at a time form the Z_{bus} matrix of the network shown in Figure 7.44. All values are in per unit.

(Answer: (diagonal terms Z_{11} 0.0825 p.u.; Z_{22} 0.248 p.u.; Z_{33} 0.27 p.u.)

7.17. For the network of Figure 7.45 obtain the Z_{bus} matrix for buses A and D.

8

Stability Limits

8.1 Introduction

The stability of a system of interconnected dynamic components is its ability to return to normal or stable operation after having been subjected to some form of disturbance. The study of stability is one of the main concerns of the control engineer whose methods may be applied to electric power systems.

When the rotor of a synchronous generator advances beyond a certain critical angle, the magnetic coupling between the rotor (and hence the turbine) and the stator fails. The rotor, no longer held in synchronism with the rotating field of the stator currents, rotates relative to the field and pole slipping occurs. Each time the poles traverse the angular region where stability obtains, synchronizing forces attempt to pull the rotor into synchronism. It is general practice to disconnect the machine from the system if it commences to slip poles. However, a generator having lost synchronism may operate successfully as an induction generator for some time and then be resynchronized; the possibility of allowing controlled pole-slipping for limited periods is thus being investigated. It should be remembered, however, that an induction generator takes its excitation requirements from the network which must be capable of supplying the requisite reactive power.

There are two forms of instability in power systems: the loss of synchronism between synchronous machines, and the stalling of asynchronous loads. Synchronous stability may be divided into two regimes, steady-state and transient. Reference to the former has already been made when discussing synchronous-machine characteristics. It is basically the ability of the power system when operating under given load conditions to retain synchronism when subject to *small* disturbances such as the continual changes in load or generation and the switching out of lines. It is most likely to result from the changes in source-to-load impedance resulting from changes in the network configuration.

Transient stability is concerned with sudden and large changes in the network condition such as brought about by faults. The maximum power transmittable, the stability limit, is less than that for the corresponding steady-state condition.

The stability of an asynchronous load is controlled by the voltage across it; if this becomes lower than a critical value, induction motors may become unstable and stall. This is, in effect, the voltage instability problem already mentioned. In a power system it is possible for either synchronous or load instability to occur. The former is more probable and hence has been given much more attention.

8.2 Equation of Motion of a Rotating Machine

Before the equation is considered a revision of the definitions of certain quantities is given. The kinetic energy absorbed by a rotating mass $= \frac{1}{2}I\omega^2$ joules. The angular momentum, $M = I\omega$ joule-seconds per radian where ω is the synchronous speed of the rotor (radians/second) and I is the moment of inertia (kilogram-metre^2). The inertia constant (H) is defined as the stored energy at synchronous speed per volt-ampere of the rating of the machine. In power systems the unit of energy is taken as the kilojoule or megajoule. If G megavolt-amperes is the rating of the machine, then

GH = stored energy (megajoules)

$= \frac{1}{2}I\omega^2 = \frac{1}{2}M\omega$ and as $\omega = 180/\pi$ (pole pairs) or $360f$ electrical degrees per second,

$GH = \frac{1}{2}M\,(360f)$

$\therefore\ M = GH/180f$ megajoule-seconds/electrical degree.

The inertia constant H for steam turbo-generators decreases from 10 kW-s/kVA for machines up to 30 MVA to values in the order of 4 kW-s/kVA for large machines, the value decreasing as the capacity increases. For salient-pole water-wheel machines H depends on the number of poles; for machines in the range 200–400 rev/min the value increases from about 2 kW-s/KVA at 10 MVA rating to 3.5 at 60 MVA. A mean value for synchronous motors is 2 kW-s/kVA. The net accelerating torque on the rotor of a machine, $\Delta T =$ mechanical torque input − electrical torque output $= I\,d^2\delta/dt^2$

$$\therefore\ \frac{d^2\delta}{dt^2} = \frac{\Delta T}{I} = \frac{(\Delta T\omega)\omega}{2 \times I\omega^2/2}$$

$$= \frac{\Delta P \cdot \omega}{2 \times \text{kinetic energy}}$$

where ΔP = net power corresponding to ΔT, i.e. $P_{mech} - P_{elect}$.

$$\therefore\ \frac{d^2\delta}{dt^2} = \frac{\Delta P}{M} \tag{8.1}$$

In equation (8.1) a negative change in power output results in an increase in δ. Sometimes ΔP is considered as the change in *electrical power output* and increase in ΔP_{el}, results in increase in angle δ. As the power input is assumed constant the equation of motion now becomes

$$\frac{d^2\delta}{dt^2} = -\frac{\Delta P}{M} \quad \text{or} \quad M\frac{d^2\delta}{dt^2} + \Delta P = 0$$

8.3* Steady-state Stability—Theoretical Considerations

The power system forms a group of interconnected electromechanical elements the motion of which may be represented by the appropriate differential equations. With large disturbances in the system the equations are non-linear, but with small changes the equations may be linearized with little loss of accuracy. The differential equations having been determined, the characteristic equation of the system is formed from which information regarding stability is obtained. The solution of the differential equation of the motion is of the form

$$\delta = k_1 e^{a_1 t} + k_2 e^{a_2 t} + \ldots k_n e^{a_n t}$$

where $k_1, k_2 \ldots k_n$ are constants of integration and $a_1, a_2 \ldots a_n$ are the roots of the characteristic equation. If any of the roots have positive real terms then the quantity δ increases continuously with time and the original steady condition is not re-established. The criterion for stability is therefore that all the real parts of the roots of the characteristic equation be negative; imaginary parts indicate the presence of oscillation. Figure 8.1 shows the various types of motion. The

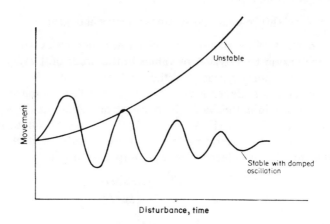

Figure 8.1 Types of response to a disturbance on a system.

determination of the roots is often difficult and tedious and indirect methods for predicting stability have been established, for example the Hurwitz–Routh criterion in which stability is predicted without the actual solution of the characteristic equation. No information regarding the degree of stability or instability is obtained, only that the system is, or is not stable. One advantage of

* This section may be omitted by readers without a basic knowledge of control theory.

this approach is that the characteristics of the control loops associated with governors and automatic voltage regulators may be incorporated in the general treatment.

For a generator connected to an infinite busbar through a network of zero resistance, with operation at P_0 and δ_0 (Figure 8.2),

$$M\frac{d^2\Delta\delta}{dt^2} = -\Delta P = -\Delta\delta\left(\frac{\partial P}{\partial\delta}\right)_0$$

where change in P causing increase in δ is positive and refers to small changes

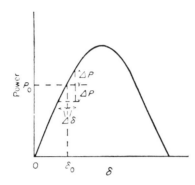

Figure 8.2 Small disturbance—initial operation on power-angle curve at P_0, δ_0. Linear movement assumed about P_0, δ_0.

in the load angle δ such that linearity may be assumed.

$$\therefore\ Mp^2\Delta\delta + \left(\frac{\partial P}{\partial\delta}\right)_0 \Delta\delta = 0 \qquad (8.2)$$

where

$$p \equiv d/dt$$

Here, $Mp^2 + (\partial P/\partial\delta)_0 = 0$ is the characteristic equation which has two roots $\pm\sqrt{\dfrac{-(\partial P/\partial\delta)_0}{M}}$. When $(\partial P/\partial\delta)_0$ is positive both roots are imaginary and the motion is oscillatory and undamped; when $(\partial P/\partial\delta)_0$ is negative both roots are real, and positive and negative respectively, and stability is lost. At $\delta = 90°$, $(\partial P/\partial\delta)_{90} = 0$ and the system is at the limit. If damping is accounted for the equation becomes

$$Mp^2\Delta\delta + K_d p\Delta\delta + \left(\frac{\partial P}{\partial\delta}\right)\Delta\delta = 0 \qquad (8.3)$$

and the characteristic equation is

$$Mp^2 + K_d p + \left(\frac{\partial P}{\partial\delta}\right) = 0 \qquad (8.4)$$

where K_d is the damping coefficient.

Again if $(\partial P/\partial \delta)$ is negative, stability is lost. The frequency of the oscillation is given by the roots of the characteristic equation.

If the excitation of the generator is controlled by a fast-acting automatic voltage regulator without appreciable dead zone, the excitation voltage E is increased as increments of load are added. Hence the actual power-angle curve pertaining is no longer that for constant E (refer to Chapter 3) and the change of power may be obtained by linearizing the P–V characteristic at the operating point (1), when

$$\Delta P = \left(\frac{\partial P}{\partial E}\right)_1 \Delta E$$

The complete equation of motion is now

$$Mp^2\Delta\delta + K_dp\Delta\delta + \left(\frac{\partial P}{\partial \delta}\right)_1 \Delta\delta + \left(\frac{\partial P}{\partial E}\right)_1 \Delta E = 0 \tag{8.5}$$

Without automatic voltage control the stability limit is reached when $\delta = 90°$; with control the criterion is obtained from the characteristic equation of (8.5).

Example 8.1 A synchronous generator of reactance 1.5 p.u. is connected to an infinite busbar system ($V = 1$ p.u.) through a line and transformers of total reactance 0.5 p.u. The no-load voltage of the generator is 1.1 p.u. and the inertia constant $H = 5$ MW-s per MVA. All per unit values are expressed on the same base; resistance and damping may be neglected. Calculate the frequency of the oscillations set up when the generator operates at a load angle of 60° and is subjected to a small disturbance. The system frequency is 50 Hz.

Solution The nature of the movement is governed by the sign of the quantity under the root sign in the equation for p_1 and p_2 (8.4). This changes when $k_d^2 = 4M(\partial P/\partial \delta)_1$; in this case $k_d = 0$. The roots of the characteristic equations give the frequency of oscillation; when $\delta_0 = 60°$,

$$\left(\frac{\partial P}{\partial \delta}\right)_{60°} = \frac{1.1 \times 1}{2}\cos 60$$

$$= 0.275$$

$$p_1 \text{ and } p_2 = \pm j\sqrt{\left(\frac{\partial P}{\partial \delta}\right)\cdot\frac{1}{M}}$$

$$= \pm j\sqrt{\frac{0.275}{5 \times (1/\pi \times 50)}}$$

$$= \pm j\sqrt{8.62}$$

Therefore frequency of oscillation

$$= 2.94 \text{ rad/s} = \frac{2.94}{2\pi} \text{ Hz}$$

$$= 0.468 \text{ Hz}$$

and the periodic time

$$= \frac{1}{0.468} = 2.14 \text{ s}$$

The stability of a two-machine system M_1 and M_2 refer to machines 1 and 2 which are connected in parallel through an impedance.
   The equations of motion are (for small changes)

$$M_1 p^2 \Delta \delta_1 + \left(\frac{\partial P_1}{\partial \delta_{12}}\right) \Delta \delta_{12} = 0$$

and

$$M_2 p^2 \Delta \delta_2 + \left(\frac{\partial P_2}{\partial \delta_{12}}\right) \Delta \delta_{12} = 0$$

As

$$\Delta \delta_1 - \Delta \delta_2 = \Delta \delta_{12}$$

$$p^2 \Delta \delta_{12} + \left[\frac{(\partial P_1 / \partial \delta_{12})}{M_1} - \frac{(\partial P_2 / \partial \delta_{12})}{M_2}\right] \Delta \delta_{12} = 0$$

the characteristic equation has two roots,

$$p_1, p_2 = \pm j \sqrt{\frac{(\partial P_1 / \partial \delta_{12})}{M_1} - \frac{(\partial P_2 / \partial \delta_{12})}{M_2}}$$

Stability is assured if the quantity under the square root is positive. Hence the stability limit for small disturbances is not the same as the maximum power limit discussed in Chapter 3 and is in fact always larger. The difference, however, is never large and when one machine is effectively an infinite busbar the two limits coincide.

Effects of governor action   In the above analysis the oscillations set up with small changes in load on a system have been considered and the effects of governor operation ignored. After a certain time has elapsed the governor control characteristics commence to influence the powers and oscillations and this will now be considered. The basic nature of governor control has been described in Chapter 3 and in Figure 8.3 the idealized speed-load characteristic is shown. The treatment below is based on the method given by Rudenburg[8].
   Consider a governor system with a time-delay constant τ_g and a speed droop of N_1 radians per second from no load to full load (P_1). δ is the change in rotor operating angle, and

$$\frac{d\delta_1}{dt} - \frac{d\delta_2}{dt} = \frac{d(\delta_1 - \delta_2)}{dt} = \frac{d\delta}{dt}$$

where $d\delta/dt$ is the speed change due to ΔP and $(d\delta/dt)(1/\omega)$ is the per unit value or slip where ω is the synchronous speed.

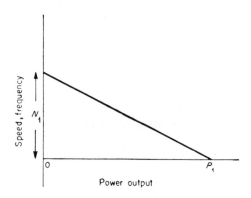

Figure 8.3   Idealized governor characteristic.

The governor causes the steam input to the turbine to change with time according to the speed deviation, i.e.

$$\frac{\mathrm{d}\Delta P}{\mathrm{d}t} = \frac{P_1}{N_1 \tau_g} \frac{\mathrm{d}\delta}{\mathrm{d}t} \frac{1}{\omega}$$

(τ_g lies between 0.5 and 10 s).

The pilot-valve movement is influenced by the supplementary return (Figure 4.1) derived from the main servo-motor position, i.e. influenced by the power.

$$\therefore \quad \frac{\mathrm{d}\Delta P}{\mathrm{d}t} = \frac{P_1}{N_1 \tau_g} \frac{\mathrm{d}\delta}{\mathrm{d}t} \frac{1}{\omega} - \frac{\Delta P}{\tau_g}$$

or

$$\tau_g \frac{\mathrm{d}(\Delta P)}{\mathrm{d}t} + \Delta P = \frac{P_1}{N_1} \cdot \frac{\mathrm{d}\delta}{\mathrm{d}t} \cdot \frac{1}{\omega} \tag{8.6}$$

For a machine connected to an infinite busbar

$$\frac{\mathrm{d}P}{\mathrm{d}\delta} = \text{synchronizing power coefficient}$$

$$= k_s$$

and

$$M\frac{\mathrm{d}^2\delta}{\mathrm{d}t^2} + k_s\delta + \Delta P = 0$$

from which

$$\frac{\mathrm{d}(\Delta P)}{\mathrm{d}t} = -M\frac{\mathrm{d}^3\delta}{\mathrm{d}t^3} - k_s\frac{\mathrm{d}\delta}{\mathrm{d}t} \tag{8.7}$$

Also

$$\Delta P = -k_s\delta - M\frac{\mathrm{d}^2\delta}{\mathrm{d}t^2} \tag{8.8}$$

From (8.8), (8.7), and (8.6)

$$\frac{d^3\delta}{dt^3}+\frac{d^2\delta}{dt^2}\cdot\frac{1}{\tau_g}+\left(\frac{k_s}{M}+\frac{P_1}{\omega N_1\tau_g M}\right)\frac{d\delta}{dt}+\frac{k_s\cdot\delta}{M\tau_g}=0 \tag{8.9}$$

The condition for no governor action is represented by $\tau_g\to\infty$ when equation (8.9) becomes,

$$\frac{d^3\delta}{dt^3}+\frac{k_s}{M}\cdot\frac{d\delta}{dt}=0$$

or

$$\frac{d^2\delta}{dt^2}+\frac{k_s}{M}=0$$

i.e. equation (8.2).

The stability of this system may be determined by the Hurwitz–Routh criterion. If the characteristic equation of a system can be expressed in the form

$$a_0 p^n + a_1 p^{n-1} + a_2 p^{n-2} + a_3 p^{n-3} + \ldots + a_{n-1}p + a_n = 0$$

then the determinant

$$\begin{vmatrix} a_1 & a_0 & 0 & 0 & \ldots & 0 & 0 & 0 \\ a_3 & a_2 & a_1 & a_0 & \ldots & \cdot & \cdot & \cdot \\ a_5 & a_4 & a_3 & a_1 & \ldots & \cdot & \cdot & \cdot \\ \cdot & \cdot & \cdot & \cdot & & \cdot & \cdot & \cdot \\ \cdot & \cdot & \cdot & \cdot & & a_n & a_{n-1} & a_{n-2} \\ 0 & 0 & 0 & 0 & & 0 & 0 & a_n \end{vmatrix} > 0$$

for stability, i.e. the roots of the characteristic equation have negative real parts. Removing the last row and column the remaining determinant must also be >0 for stability and so on, i.e.

$$\begin{vmatrix} a_1 & a_0 & 0 \\ a_3 & a_2 & a_1 \\ a_5 & a_4 & a_3 \end{vmatrix}>0 \qquad \begin{vmatrix} a_1 & a_0 \\ a_3 & a_2 \end{vmatrix}>0 \qquad a_1>0$$

In equation (8.9),

$$a_0 = 1$$

i.e. is positive,

$$a_1 = \frac{1}{\tau_g}$$

i.e. is positive,

$$a_1 a_2 = \frac{1}{\tau_g}\left(\frac{k_s N_1\tau_g + P_1}{\omega N_1\tau_g M}\right)$$

$$a_3 a_0 = \frac{k_s}{M\tau_g}\cdot 1$$

For stability,

$$a_1a_2 - a_3a_0 > 0$$

i.e.

$$P_1 + k_s\omega N_1\tau_g > k_s\tau_g\omega N_1$$

$$\therefore \ P_1 > k_s\tau_g N_1(\omega - \omega) \quad \text{i.e.} > 0$$

i.e. the system is always stable for this simple (but not always representative) governor system.

8.4 Steady-state Stability—Practical Considerations

The steady-state limit is the maximum power that can be transmitted in a network between sources and loads when the system is subject to small disturbances. The power system is of course constantly subjected to small changes as load variations occur. To obtain the limiting value of power, small increments of load are added to the system; after each increment the generator excitations are adjusted to maintain constant terminal voltages and a load flow is carried out. Eventually a condition of instability is reached.

The stability limits of synchronous machines have been discussed in Chapter 3. It was seen that provided the generator operates within the 'safe area' of the performance chart, stability is assured; usually a 20 per cent margin of safety is allowed and the limit is extended by the use of automatic voltage regulators. Often the performance charts are not used directly and the generator equivalent-circuit employing the synchronous impedance is used. The normal operating load angle for modern machines is in the order of 60 electrical degrees and for the limiting value of 90° this leaves 30° to cover the transmission network. In a complex system a reference point must be taken from which the load angles are measured; this is usually a point where the direction of power flow reverses.

The simplest criterion for steady-state synchronous stability is $(\partial P/\partial\delta) > 0$, i.e. the synchronizing coefficient must be positive. The use of this criterion involves the following assumptions:

(a) generators are represented by constant impedances in series with the no-load voltages;
(b) the input torques from the turbines are constant;
(c) changes in speed are ignored;
(d) electromagnetic damping in the generators is ignored;
(e) the changes in load angle δ are small.

The degree of complexity to which the analysis is taken has to be decided, for example the effects of machine inertia, governor action, and automatic voltage regulators can be included; these items, however, greatly increase the complexity of the calculations. The use of the criterion $(\partial P/\partial\delta) = 0$ alone gives a pessimistic or low result and hence an inbuilt factor of safety.

In a system with several generators and loads, the question as to where the increment of load is to be applied is important. A conservative method is to assume the increment applied to one machine only, determine the stability and then repeat for each of the other machines in turn. Alternatively the power outputs from all but the two generators having the largest load angles are kept constant.

For calculations made without the aid of computers it is usual to reduce the network to the simplest form which will keep intact the generator nodes. The values of load angle, power, and voltage are then calculated for the given conditions, $\partial P/\partial \delta$ determined for each machine, and if positive the loading is increased and the process repeated.

In a system consisting of a generator supplying a load through a network of lines and transformers of effective reactance X_T the value of $(dP/d\delta) = (EV/X_T) \cos \delta$, where E and V are the supply and receiving end voltages and δ the total angle between the generator rotor and the phasor of V. The power transmitted is obviously increased with higher system voltages and lower reactances, and it may be readily shown that line capacitance increases the stability limit. The determination of $dP/d\delta$ is not very difficult if the voltages at the loads can be assumed constant or if the loads can be represented by impedances. Use can be made of the P–V, Q–V characteristics of the load if the voltages change appreciably with the redistribution of the power in the network; this process, however, is extremely tedious.

Example 8.2 A system consisting of three interconnected synchronous machines is shown in the line diagram of Figure 8.4. It is required to investigate

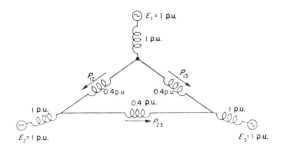

Figure 8.4 Three-machine system for Example 8.2. The output of machine (2) is constant. All values are per unit on the same voltage and MVA bases. P_{12}, P_{13}, and P_{23} are power flows in the lines.

the steady-state stability for the following operating load angles (in electrical degrees).

$$\delta_{12} = 90° \qquad \delta_{13} = 30° \qquad \delta_{23} = 60°$$

The reactance of each machine is j per unit and the generated e.m.f. or no-load voltage is 1 p.u. The resistance of the machines and lines is negligible.

310

Solution   It is necessary to determine first the transfer reactances of the network. As the network is entirely composed of reactance for clarity the j's will be omitted. The system can be represented by the equivalent circuit shown in Figure 8.5(a) when machine (1) is operative and the remaining machines are represented by their reactances only. Figure 8.5(a) can be reduced to Figure

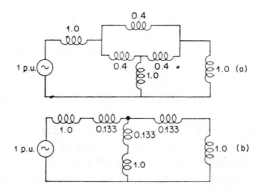

Figure 8.5(a)   Application of superposition to obtain the transfer reactances of the system in Figure 8.4. (b) Network of 8.5(a) reduced by transformation.

8.5(b) by applying the delta–star transformation on the mesh of 0.4 p.u. reactances. The reactance of each branch of the star is

$$\frac{0.4 \times 0.4}{0.4 \times 0.4 \times 0.4}$$

i.e.

$$0.133 \text{ p.u.}$$

The current form machine (1) to this circuit

$$= \frac{1}{1 + 0.133 + 0.565}$$

$$= 0.59 \text{ p.u.}$$

The current in the reactance of machine (2) due to (1),

$$I_{12} = 0.295 \text{ p.u.}$$

Similarly,

$$I_{13} = 0.295 \text{ p.u.}$$

Hence,

$$X_{12} = \text{transfer reactance from (1) to (2)}$$

$$= 3.39 \text{ per unit}$$

and similarly,

$$X_{13} = 3.39 \text{ p.u.}$$

It is evident from inspection of the circuit that,

$$X_{23} = 3.39 \text{ p.u.}$$

$$\frac{dP_{12}}{d\delta_{12}} = \frac{E_1 E_2}{X_{12}} \cos \delta_{12}$$

similarly for δ_{23} and δ_{31}.
  The output of (1)

$$P_1 = P_{12} + P_{13}$$

and

$$dP_1 = \frac{\partial P_{12}}{\partial \delta_{12}} d\delta_{12} + \frac{\partial P_{13}}{\partial \delta_{13}} d\delta_{13}$$

$$= \frac{E_1 E_2}{X_{12}} \cos \delta_{12} \cdot d\delta_{12} + \frac{E_1 E_3}{X_{13}} \cos \delta_{13} d\delta_{13}$$

Similar expressions are obtained for dP_2 and dP_3; P_2 is kept constant and hence $dP_2 = 0$.

$$\delta_{12} = 90° \qquad \delta_{13} = 30° \qquad \delta_{23} = 60°$$

$$dP_1 = \frac{1 \times 1}{3.39} \cos 90° \, d\delta_{12} + \frac{1 \times 1}{3.39} \cos 30° \, d\delta_{13}$$

$$0 = \frac{1 \times 1}{3.39} \cos 90° \, d\delta_{12} + \frac{1 \times 1}{3.39} \cos 60° \, d\delta_{23}$$

$$dP_3 = \frac{1 \times 1}{3.39} \cos 30° \, d\delta_{13} + \frac{1 \times 1}{3.39} \cos 60° \, d\delta_{32}$$

From these equations,

$$\frac{dP_3}{d\delta_{31}} = \frac{1.366}{3.39} = 0.4$$

Also,

$$\frac{dP_1}{d\delta_{13}} = 0.256.$$

Hence the system is stable.

Example 8.3   For the system shown in Figure 8.6 investigate the steady-state stability.

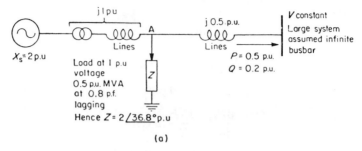

(a)

Figure 8.6(a)   Line diagram of system for Example 8.3.

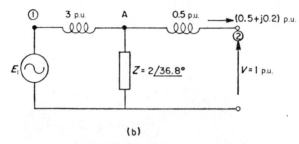

(b)

Figure 8.6(b)   Equivalent circuit for Example 8.3.

Solution   First it is necessary to determine the transfer impedance.

$$Y_{11} = \frac{1}{j3 + \dfrac{j0.5(1.6+j1.2)}{1.6+j1.2+j0.5}} = \frac{1}{3.423/88.8°} = 0.293\angle-88.8° \text{ p.u.}$$

$$Y_{12} = \frac{1}{j3 + j0.5 + \dfrac{j3 \times j0.5}{2\angle36.8°}} = \frac{1}{4\angle98.7°} = 0.25\angle-98.7° \text{ p.u.}$$

Hence,

$$\alpha_{11} = 90 - 88.8° = 1.2°$$

$$\alpha_{12} = 90 - 98.7° = -8.7°.$$

The voltage at the tapping point A,

$$V_A = \sqrt{\left[\left(V + \frac{QX}{V}\right)^2 + \left(\frac{PX}{V}\right)^2\right]}$$

$$= \sqrt{\left[\left(1 + \frac{0.2 \times 0.5}{1}\right)^2 + \left(\frac{0.5 \times 0.2}{1}\right)^2\right]}$$

$$= 1.105 \text{ p.u. at } 5.75° \text{ to } V$$

$$I_R^2 = \frac{P^2 + Q^2}{V^2}$$

and the reactive power absorbed by the line from A to the infinite bus

$$= I_R^2 X = \left(\frac{0.5^2 + 0.2^2}{1^2}\right) 0.5 \text{ p.u.}$$

$$= 0.145 \text{ p.u.}$$

If the load at A is represented by an impedance then the actual load taken at 1.105 p.u. voltage

$$= \left(\frac{1.105}{1}\right)^2 0.5 \angle 36.8° \text{ p.u.}$$

$$= 0.61 \angle 36.8° \text{ p.u.}$$

$$= 0.49 + j0.366 \text{ p.u.}$$

Total load carried by link from generator to A

$$= 0.5 + 0.49 + j(0.2 + 0.145 + 0.366)$$

$$= 0.99 + j0.71 \text{ p.u.}$$

Internal voltage of generator

$$= \sqrt{\left[\left(1.105 + \frac{0.71 \times 3}{1.105}\right)^2 + \left(\frac{0.99 \times 3}{1.105}\right)^2\right]} = \sqrt{[9.2 + 7.23]}$$

$$= 4.06 \angle 41.5°$$

This angle is with respect to A so that

$$\delta_{EV} = 41.5 + 5.75 = 47.25° = \delta_{12}.$$

Power from generator

$$= E^2 Y_{11} \sin \alpha_{11} + EVY_{12} \sin (\delta_{12} - \alpha_{12})$$

$$= 4.06 \times 0.293 \sin 1.2 + 4.06 \times 1 \times 0.25 \sin (\delta_{12} + 8.7)$$

$$= 1.09 \text{ p.u.}$$

Maximum power $= 1.26$ p.u. and the system is stable.

8.5 Transient Stability—Consideration of Rotor Angle

Transient stability is concerned with the effect of large disturbances. These are usually due to faults the most severe of which is the three-phase short circuit which governs the transient stability limits in Britain. Elsewhere limits are based on other types of fault, notably the single line to earth which is by far the most frequent in practice.

When a fault occurs at the terminals of a synchronous generator the power output of the machine is greatly reduced as it is supplying a mainly inductive circuit. However, the input power to the generator from the turbine has not time to change during the short period of the fault and the rotor endeavours to gain speed to store the excess energy. If the fault persists long enough the rotor angle will increase continuously and synchronism will be lost. Hence the time of operation of the protection and circuit breakers is all-important.

An aspect of increasing importance is the use of *auto-reclosing* circuit breakers. These open when the fault is detected and automatically reclose after a prescribed period (usually less than 1 s). If the fault persists the circuit breaker reopens and then recloses as before. This is repeated once more, when, if the fault still persists, the breaker remains open. Owing to the transitory nature of most faults often the circuit breaker successfully recloses and the rather lengthy process of investigating the fault and switching in the line is avoided. The length of the auto-reclose operation must be considered when assessing transient stability limits; in particular analysis must include the movement of the rotor over this period and not just the first swing, as is often the case. If in equation (8.1) both sides are multiplied by $2(d\delta/dt)$,

$$\left(2\frac{d\delta}{dt}\right)\left(\frac{d^2\delta}{dt^2}\right) = \frac{d}{dt}\left(\frac{d\delta}{dt}\right)^2 = \frac{2\Delta P}{M}\left(\frac{d\delta}{dt}\right)$$

$$\therefore \left(\frac{d\delta}{dt}\right)^2 = \frac{2}{M}\int_{\delta_0}^{\delta} \Delta P \, d\delta \tag{8.10}$$

The rotor will swing until its angular velocity is zero in which case the machine remains stable; if $d\delta/dt$ does not become zero the rotor will continue to move and synchronism lost. Hence the criterion for stability is that the area between the P–δ curve and the line representing the power input must be zero. This is known as the *equal-area criterion*. It should be noted that this is based on the assumption that synchronism is retained or lost on the first swing or oscillation of the rotor, which in certain cases is subject to doubt. Physically the criterion means that the rotor must be able to return to the system all the energy gained from the turbine during the acceleration period.

A simple example of the equal-area criterion may be seen by an examination of the switching out of one of two parallel lines which connect a generator to an infinite busbar (Figure 8.7). For stability to be retained the two shaded areas are equal and the rotor comes initially to rest at angle δ_2, after which it oscillates until completed damped. In this particular case the initial operating power and angle could be increased to such values that the shaded area between δ_0 and δ_1 is equal to the shaded area between δ_1 and δ_3 where $\delta_3 = 180 - \delta_1$; this would be the condition for maximum input power. Beyond δ_3 energy would again be absorbed by the rotor from the turbine.

The power angle curves pertaining to a fault on one of two parallel lines are shown in Figure 8.8. The fault is cleared in a time corresponding to δ_1 and the shaded area δ_0 to δ_1 indicates the energy stored. The rotor swings until it reaches δ_2 when the two areas are equal. In this particular case P_0 is the maximum operating power for a fault clearance time corresponding to δ_1, and conversely δ_1 is the *critical clearing angle* for P_0. If the angle δ_1 is decreased it is

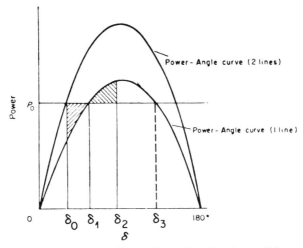

Figure 8.7 Power-angle curves for one line and two lines in parallel. Equal area criterion. Resistance neglected.

possible to increase the value of P_0 without loss of synchronism. The general case where the clearing angle δ_1 is not critical is shown in Figure 8.9. Here the rotor swings to δ_2 where the shaded area from δ_0 to δ_1 is equal to the area δ_1 to δ_2. Critical conditions are reached when $\delta_2 = 180 - \sin^{-1}(P_0/P_2)$. The time corresponding to the critical clearing angle is called the *critical clearing time* for the particular (normally full-load) value of power input. This time is of great importance to protection and switchgear engineers as it is the maximum time allowable for their equipment to operate without instability occurring. The critical clearing angle for a fault on one of two parallel lines may be determined as follows.

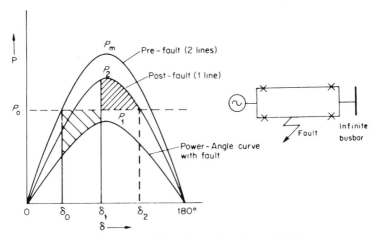

Figure 8.8 Fault on one line of two lines in parallel. Equal area criterion. Resistance neglected. δ_1 is critical clearing angle for input power P_0.

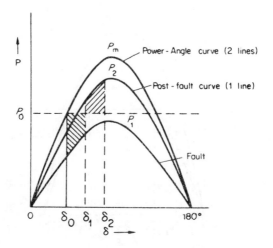

Figure 8.9   Situation as in Figure 8.8, but δ_1 not critical.

Applying the equal-area criterion to Figure 8.8.

$$\int_{\delta_0}^{\delta_1} (P_0 - P_1 \sin \delta)\, d\delta + \int_{\delta_1}^{\delta_2} (P_0 - P_2 \sin \delta)\, d\delta = 0$$

$$\therefore \ [P_0\delta + P_1 \cos \delta]_{\delta_0}^{\delta_1} + [P_0\delta + P_2 \cos \delta]_{\delta_1}^{\delta_2} = 0$$

from which as $\delta_2 = 180 - \sin^{-1}\left(\dfrac{P_0}{P_2}\right)$

$$\cos \delta_1 = \frac{P_0(\delta_0 - \delta_2) + P_1 \cos \delta_0 - P_2 \cos \delta_2}{P_1 - P_2} \qquad (8.11)$$

Hence the critical clearing angle δ_1 is determined.

Example 8.4   A generator operating at 50 Hz delivers 1 p.u. power to an infinite busbar through a network in which resistance may be neglected. A fault occurs which reduces the maximum power transferable to 0.4 p.u., whereas before the fault this power was 1.8 p.u. and after the clearance of the fault 1.3 p.u. By the use of the equal-area criterion determine the critical clearing angle.

Solution   The appropriate load angle curves are shown in Figure 8.8. $P_0 = 1$ p.u., $P_1 = 0.4$ p.u., $P_2 = 1.3$ p.u., and $P_m = 1.8$ p.u.

$$\delta_0 = \sin^{-1}\left(\frac{1}{1.8}\right) = 33.8 \text{ electrical degrees}$$

$$\delta_2 = 180 - \sin^{-1}\left(\frac{1}{1.3}\right) = 180 - 50° \, 24' = 129° \, 36'.$$

Applying equation (8.11),

$$\cos \delta_1 = \frac{1(-1.6718) + 0.4 \times 0.831 + 1.3 \times 0.77}{0.4 - 1.3}$$

$$= 0.377$$

$$\therefore \quad \delta_1 = 67° 51' \text{ (electrical degrees)}.$$

In a large system it is usual to divide the generators and loads into a single equivalent generator connected to an equivalent motor. The main criterion is that the machines should be electrically close when forming an equivalent generator or motor. If stability with faults in various places is investigated the position of the fault will decide the division of machines between the equivalent generator and motor. A power system (including generation) at the receiving end of a long line would constitute an equivalent motor if not large enough to be an infinite busbar.

Reduction to simple system With a number of generators connected to the same busbar the inertia constant (H) of the equivalent machine,

$$H_e = H_1 \frac{S_1}{S_b} + H_2 \frac{S_2}{S_b} \dots H_n \frac{S_n}{S_b},$$

where $S_1 \dots S_n = $ MVA of the machines, $S_b = $ base MVA. This is obtained by equating the stored energy of the equivalent machine to the total of the individual machines. For example, consider six identical machines connected to the same busbar each having an H of 5 MW-s/MVA and rated at 60 MVA. Making the base MVA equal to the combined rating of the machines the effective value of

$$H = 5 \times \frac{60}{6 \times 60} \times 6 = 5 \frac{\text{MW-s}}{\text{MVA}}$$

It is important to remember that the inertia of the loads must be included; normally this will be the sum of the inertias of the induction motors connected.

Two synchronous machines connected by a reactance may be reduced to one equivalent machine feeding through the reactance to an infinite busbar system. The properties of the equivalent machine are found as follows.

The equation of motion for the two-machine system is

$$\frac{d^2\delta}{dt} = \frac{\Delta P_1}{M_1} - \frac{\Delta P_2}{M_2} = \left(\frac{1}{M_1} + \frac{1}{M_2}\right)(P_0 - P_m \sin \delta)$$

where δ is the relative angle between the machines. Note that

$$\Delta P_1 = -\Delta P_2 = (P_0 - P_m \sin \delta)$$

where P_0 is the input power and P_m the maximum transmittable power.

For a single generator of M_e and the same input power connected to the infinite busbar system,

$$M_e \frac{d^2\delta}{dt^2} = P_0 - P_m \sin \delta$$

therefore

$$M_e = \frac{M_1 M_2}{M_1 + M_2} \qquad (8.12)$$

This equivalent generator has the same mechanical input as the actual machines and the load angle δ it has with respect to the busbar is the angle between the rotors of the two machines.

So far the maximum powers transferable before, during, and after the fault have been assumed. These values can be obtained as described in Chapter 3. For the common case of a three-phase fault in the middle of one of two parallel lines as shown in Figure 8.8, the mesh formed by X_1 and $X_2/2$ (Figure 8.10(a)) is replaced by the equivalent star as shown in Figure 8.10(b). The power transmitted during the fault is $E_g V / X_e \sin \delta$ as X_A and X_B are purely inductive loads. The transfer reactance before and after the faults is obtained readily

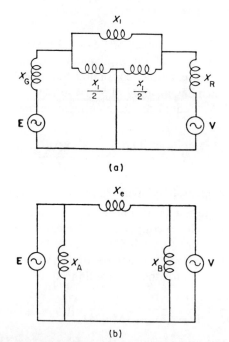

(a)

(b)

Figure 8.10(a) Equivalent circuit of a generator connected to an infinite busbar through two lines, one of which has a three-phase fault midway along its length. (b) Figure 8.10(a) after reduction. $X_1 =$ reactance of line; $X_G =$ reactance of generator plus transformer; $X_R =$ reactance of receiving end transformer.

from the equivalent circuit. With unbalanced faults more power is transmitted during the fault period than with three-phase short circuits and the stability limits are higher.

Effect of automatic voltage regulators and governors   These may be represented in the equation of motion as follows,

$$M\frac{d^2\delta}{dt^2} + K_d\frac{d\delta}{dt} = (P_0 - \Delta P_0) - P_e \qquad (8.13)$$

in which K_d = damping coefficient,

   P_0 = power input,

   ΔP_0 = change in input power due to governor action,

   P_e = electrical power output modified by the voltage regulator.

Equation (8.13) is best solved by means of an analogue or digital computer. In most studies the governor and voltage regulator effects are ignored, but their influence on stability limits have been analysed in recent investigations.

8.6   Transient Stability—Consideration of Time

The swing curve   In the previous section attention has been mainly directed towards the determination of rotor angular position; in practice the corresponding times are more important. The protection engineer requires allowable times rather than angles when specifying relay settings. The solution of equation (8.1) with respect to time is performed by means of numerical methods and the resulting time–angle curve is known as the swing curve. A simple step-by-step method by Dahl[6] will be given in detail and references made to more sophisticated methods suitable for digital computation.

   In this method the change in the angular position of the rotor over a short time interval is determined. In performing the calculations the following assumptions are made:

(a) the accelerating power ΔP at the commencement of a time interval is considered constant from the middle of the previous interval to the middle of the interval considered;
(b) the angular velocity is constant over a complete interval and is computed for the middle of this interval. These assumptions are probably better understood by reference to Figure 8.11.

From Figure 8.11,

$$\omega_{n-\frac{1}{2}} - \omega_{n-\frac{3}{2}} = \frac{d^2\delta}{dt^2}\Delta t = \frac{\Delta P_{n-1}}{M}\Delta T$$

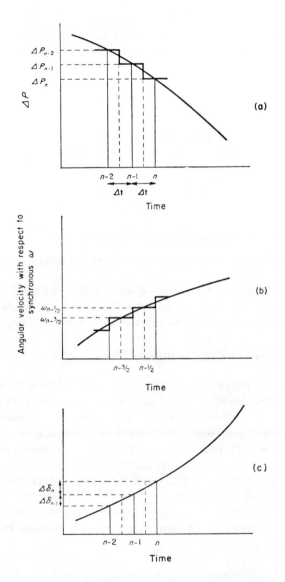

Figure 8.11(a), (b) and (c)   Variation of ΔP, ω, and $\Delta\delta$ with time.
Illustration of step-by-step method to obtain δ-time curve.

The change in δ over the $(n-1)$th interval, i.e. from times $(n-1)$ to $(n-2)$

$$= \delta_{n-1} - \delta_{n-2} = \Delta\delta_{n-1} = \omega_{n-\frac{3}{2}}\Delta t$$

Over the nth interval,

$$\Delta\delta_n = \delta_n - \delta_{n-1} = \omega_{n-\frac{1}{2}}\Delta t$$

From the above,

$$\Delta\delta_n - \Delta\delta_{n-1} = \Delta t(\omega_{n-\frac{1}{2}} - \omega_{n-\frac{3}{2}})$$

$$= \Delta t \cdot \Delta t \cdot \frac{\Delta P_{n-1}}{M}$$

$$\therefore \quad \Delta\delta_n = \Delta\delta_{n-1} + \frac{\Delta P_{n-1}}{M}(\Delta t)^2 \qquad (8.13)$$

It should be noted that $\Delta\delta_n$ and $\Delta\delta_{n-1}$ are the *changes* in angle.

Equation (8.13) is the basis of the numerical method. The time interval Δt used should be as small as possible (the smaller Δt, however, the larger the amount of labour involved) and a value of 0.05 s is frequently used. Any change in the operational condition causes an abrupt change in the value of ΔP. At the commencement of a fault for example ($t = 0$), the value of ΔP is initially zero and then immediately after the occurrence takes a definite value. When two values of ΔP apply the mean is used. The procedure is best illustrated by an example.

Example 8.5 In the system described in Example 8.4 the inertia constant of the generator plus turbine is 2.7 p.u. Obtain the swing curve for a fault clearance time of 125 ms.

Solution $H = 2.7$ p.u., $f = 50$ Hz, $G = 1$ p.u.

$$\therefore \quad M = \frac{HG}{180f} = 3 \times 10^{-4} \text{ p.u.}$$

A time interval $\Delta t = 0.05$ s will be used. Hence

$$\frac{(\Delta t)^2}{M} = 8.33.$$

The initial operating angle

$$\delta_0 = \sin^{-1}\left(\frac{1}{1.8}\right) = 33.8°.$$

Just before the fault the accelerating power $\Delta P = 0$. Immediately after the fault,

$$\Delta P = 1 - 0.4 \sin \delta_0$$

$$= 0.78 \text{ p.u.}$$

The first value is that for the middle of the preceding period and the second for the middle of the period under consideration. The value to be taken for ΔP at the commencement of this period is $(0.780/2)$, i.e. 0.39 p.u. At $t = 0$, $\delta = 33.8°$.

$$\Delta t_1 = 0.05 \text{ s} \qquad \Delta P = 0.39 \text{ p.u.}$$

$$\therefore \ \Delta \delta_n = \Delta \delta_{n-1} + \frac{(\Delta t)^2}{M} \Delta P_{n-1}$$

$$\therefore \ \Delta \delta_n = 8.33 \times 0.39 = 3.23°$$

$$\therefore \ \delta_{0.05} = 33.8 + 3.23 = 37.03°$$

$$\underline{\Delta t_2} \colon \Delta P = 1 - 0.4 \sin 37.03° = 0.76 \text{ p.u.}$$

$$\therefore \ \Delta \delta_n = 3.23 + 8.33 \times 0.76 = 9.56°$$

$$\therefore \ \delta_{0.1} = 37.03 + 9.56 = 46.59°$$

$$\underline{\Delta t_3} \colon \Delta P = 1 - 0.4 \sin 46.59° = 0.71 \text{ p.u.}$$

$$\Delta \delta_n = 9.56 + 5.9 = 15.46°$$

$$\delta_{0.15} = 46.59 + 15.46 = 62.05°$$

The fault is cleared after a period of 0.125 s. As this discontinuity occurs in the middle of a period (0.1 to 0.15 s) no special averaging is required (Figure 8.12).

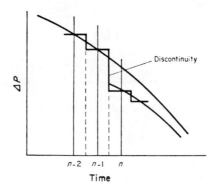

Figure 8.12   Discontinuity of ΔP in middle of a period of time.

If on the other hand the fault is cleared in 0.15 s an averaging of two values would be required.

From $t = 0.15$ s onwards,

$$P = 1 - 1.3 \sin \delta$$

$$\Delta t_4 \colon \qquad \Delta P = 1 - 1.3 \sin 62.05° = -0.145 \text{ p.u.}$$

$$\Delta \delta_n = 15.46 + 8.33(-0.145) = 14.25°$$

$$\therefore \ \delta_{0.2} = 14.25 + 62.05 = 76.3°$$

$\underline{\Delta t_5}$: $\qquad \Delta P = 1 - 1.3 \sin 76.3° = -0.26$ p.u.

$\qquad\qquad \Delta \delta_n = 14.25 - (8.33 \times 0.26) = 12.09°$

and $\qquad\qquad \delta_{0.25} = 88.39°$

$\underline{\Delta t_6}$: $\qquad \Delta P = 1 - 1.3 \sin 88.39 = -0.3$ p.u.

$\qquad\qquad \Delta \delta_n = 12.09 - 2.5 = 9.59°$

and $\qquad\qquad \delta_{0.3} = 97.98°$

If this process is continued δ commences to decrease and the generator remains

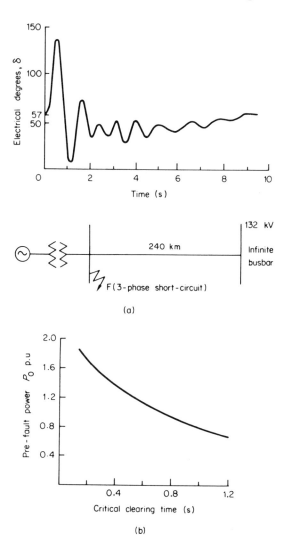

Figure 8.13(a) Typical swing curve. (b) Typical stability boundary.

stable. Different curves will be obtained for other values of clearing time. It is evident from the way the calculation proceeds that for a sustained fault δ will continuously increase and stability will be lost. The critical clearing time should be calculated for conditions which allow the least transfer of power from the generator. Circuit breakers and the associated protection operate in times dependent upon their sophistication; these times can be in the order of a few cycles of alternating voltage. A typical swing curve for a large machine over several seconds is shown in Figure 8.13(a). For a given fault position a faster clearing time implies a greater permissible value of input power P_0. A typical relationship between the critical clearing time and input power is shown in Figure 8.13(b); this is often referred to as the stability boundary. The critical clearing time increases with increase in the inertia constant (H) of turbine-generators. Often the first swing of the machine is sufficient to indicate stability. If, however, the result is marginal the step-by-step process should be continued for several swings.

8.7 The use of Computers in Transient Stability Studies

Alternating current analysers The analyser is used to obtain the new value of ΔP for each value of δ. In effect, therefore, a load flow is carried out for each new set of rotor angles, $\Delta\delta$ having been obtained in the same way as in Example 8.5. In a large multi-machine system this process is extremely tedious and the simplest possible representation of machines and loads is used, e.g. saliency in the generators is ignored. On the first balance of the analyser the pre-fault conditions will be obtained and the power outputs, voltages, and load angles of the various machines recorded. The fault is then applied and the values of ΔP and δ obtained. From these the new values of the load angles are obtained and set up on the analyser and the process repeated. The faulty line is finally removed from the network and steady-state stability assessed.

Analogue computers The network is reduced to the simplest form compatible with the retention of the machine nodes. The differential equations are then set up on the computer and solutions obtained. A system with several machines will strain the resources of the largest computer.

Micro-machines Micro-réseaux studies are based on the use of electrically scaled models of the actual machines. Difficulties are experienced in reproducing in a small laboratory-size generator the time constants of the actual machine, nevertheless much useful work has been carried out on such models.

Digital computers The methods used are basically those used for alternating-current analysers. The computer can either reduce the network to the machine nodes first and then solve the equations, or perform a complex load flow for each new load angle on the complete system. The former method is quicker, but information regarding power flows during rotor swinging is lost. Most of the generator and load assumptions used for analysers are also used for digital computers. The Runge–Kutta method is widely used in the digital programs for the integration process and predictor–corrector routines are also used.

8.8 Stability of Loads

In Chapter 5 the power–voltage characteristics of a line supplying a load were considered. It was seen that for a given load power-factor a value of transmitted power was reached, beyond which further decreases of the load impedance produce greatly reduced voltages, i.e. voltage instability. If the load is purely static, e.g. represented by an impedance, the system will operate stably at these lower voltages. Sometimes, when using alternating current network analysers, this lower-voltage condition is unknowingly obtained and unexpected load flows result. If the load contains non-static elements such as induction motors, the nature of the load characteristics is such that beyond the critical point the motors will run down to a standstill or stall. It is therefore of importance to consider the stability of composite loads which will normally include a large proportion of induction motors.

The process of voltage collapse may be seen from a study of the V–Q characteristics of an induction motor (Figure 3.48) from which it is seen that below a certain voltage the reactive power consumed increases with decrease in voltage until $(\mathrm{d}Q/\mathrm{d}V) = \infty$ when the voltage collapses. In the power system the problem arises owing to the impedance of the connexion between the load and infinite busbar and is obviously aggravated when this impedance is high (connexion electrically weak). The usual cause of an abnormally high impedance is the loss of one line of two or more forming the connexion. It is profitable, therefore, to study the process in its basic form—that of a load supplied through a reactance from a constant voltage source (Figure 8.14).

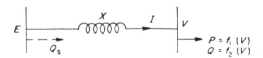

Figure 8.14   System with a load dependent on voltage as follows:
$P = f_1(V)$ and $Q = f_2(V)$; $Q_s = $ supply vars $= Q + I^2X$; $E = $ supply voltage.

Already two criteria for load instability have been given, i.e. $(\mathrm{d}P/\mathrm{d}s) = 0$ and $(\mathrm{d}Q/\mathrm{d}V) = \infty$; from the system viewpoint, voltage collapse takes place when $(\mathrm{d}E/\mathrm{d}V) = 0$ or $(\mathrm{d}V/\mathrm{d}E) = \infty$. Each value of E yields a corresponding value for V and the plot of E–V is shown in Figure 8.15, also the plot of V–X for various values of E is shown in Figure 8.16. In these graphs the critical operating condition is clearly shown and the improvement produced by higher values of E apparent, indicating the importance of the system operating voltage from the load viewpoint.

326

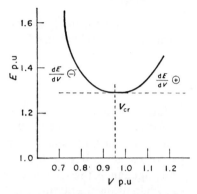

Figure 8.15 E–V relationship per system in Figure 8.14, V_{cr} = critical voltage after which instability occurs.

In the circuit shown in Figure 8.14

$$E = V + \frac{QX}{V} \quad \text{if} \quad \frac{PX}{V} \ll \frac{V^2 + QX}{V} \tag{2.6}$$

$$\therefore \quad \frac{dE}{dV} = 1 + \left(\frac{dQ}{dV}XV - QX\right)\frac{1}{V^2}$$

which is zero at the stability limit and negative in the unstable region.

At the limit,

$$\frac{dE}{dV} = 0 \quad \text{and} \quad \frac{dQ}{dV} = \left(\frac{QX}{V^2} - 1\right)\frac{V}{X} = \frac{Q}{V} - \frac{V}{X}$$

From (2.6)

$$\frac{Q}{V} = \frac{E}{X} - \frac{V}{X}$$

$$\therefore \quad \frac{dQ}{dV} = \frac{E}{X} - \frac{2V}{X} \tag{8.15}$$

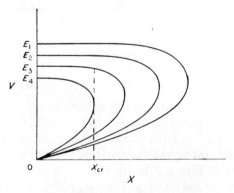

Figure 8.16 V–X relationship. Effect of change in supply voltage E. X_{cr} = critical reactance of transmission link.

Example 8.6   A load is supplied from a 275 kV infinite busbar through a line of reactance 70 Ω phase-to-neutral. The load consists of a constant power demand of 200 MW and a reactive power demand Q which is related to the load voltage V by the equation:

$$(V-0.8)^2 = 0.2(Q-0.8)$$

This is shown in Figure 8.17 where the base quantities for V and Q are 275 kV and 200 MVAr respectively.

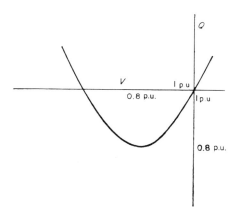

Figure 8.17   Example 8.6. Reactive power-voltage characteristic.

Examine the voltage stability of this system, indicating clearly any assumptions made in the analysis.

Solution   It has been shown that

$$E = \sqrt{\left[\left(V+\frac{QX}{V}\right)^2 + \left(\frac{PX}{V}\right)^2\right]}$$

If

$$\frac{PX}{V} < \frac{V^2+QX}{V} \quad \text{then} \quad E = \frac{V^2+QX}{V}$$

and

$$\frac{dE}{dV} = 1 + \left(\frac{dQ}{dV}XV - QX\right)\frac{1}{V_2} = 0$$

In this problem. $P = 200$ MW $\left.\begin{array}{l}\\ \\ \end{array}\right\}$three-phase values
and   1 pu $Q = 200$ MVAr

$$X = \frac{70 \times 200}{275^2} \text{ p.u.}$$

$$= 0.185 \text{ p.u.}$$

$$E = V + \frac{QX}{V}$$

$$\therefore \quad 1 = V + \frac{0.185}{V}\left(\frac{(V-0.8)^2}{0.2} + 0.8\right)$$

$$\therefore \quad V^2 = V - 0.925\,V^2 - 0.59 + 1.48\,V - 0.148$$

$$1.925\,V^2 - 2.48\,V + 0.74 = 0$$

$$\therefore \quad V = \frac{+2.48 \pm \sqrt{(2.48^2 - 4 \times 1.925 \times 0.74)}}{2 \times 1.925}$$

$$= \frac{2.48 \pm 0.67}{3.85}$$

Taking the upper value, $V = 0.818$ p.u.

$$Q \text{ at this } V = \frac{(0.818 - 0.8)^2}{0.2} + 0.8$$

$$= 0.8016 \text{ p.u.}$$

$$\frac{dE}{dV} = 1 + [10(0.818 - 0.8) \times 0.185 \times 0.818 - 0.8016 \times 0.185/V^2]$$

$$= 1 + \left(\frac{0.027 - 0.148}{0.818^2}\right)$$

which is positive, i.e. the system is stable.

Note $PX/V \doteqdot 0.16$ and $(V^2 + QX)/V \doteqdot 1$, therefore approximation is reasonable.

When the reactance between the source and load is very high the use of tap-changing transformers is of no assistance. Large voltage drops exist in the supply lines and the 'tapping-up' of transformers increases these due to the increased supply currents. Hence the peculiar effects from tap changing noticed when conditions close to a voltage collapse have occurred in practice, i.e. tapping-up reduces the secondary voltage and vice versa. One symptom of the approach of critical conditions is sluggishness in the response of tap-changing transformers.

The above form of instability is essentially a form of steady-state instability, and a full assessment of the latter should not only take into account the synchronous criterion $(dP/d\delta) \geqslant 0$, although this is the more likely to occur, but also the criterion for load instability $(dV/dE) = \infty$.

The use of computers It is difficult to detect the power transmitted in the region near to the condition of voltage collapse. A computer will detect this condition by a divergence and it is difficult to assess if this is due to the network or to other conditions. One possible method is to program the power–voltage characteristics of the line and of the load; the intersection of these two characteristics will give the operating conditions of the system.

In steady-state stability studies the increments of load added at each step are small and this makes the results of alternating-current network analysers of doubtful accuracy. Using a digital computer, however, successive load flows can be performed with high accuracy in computation.

8.9 Further Aspects

Faults on the feeders to induction motors A more common cause of the stalling of induction motors (or the low-voltage releases operating and removing them from the supply) occurs when the supply voltage is either zero or very low for a brief period due to a fault on the supply system. When the supply voltage is restored the induction motors accelerate and endeavour to attain their previous operating condition. In accelerating, however, a large current is taken and this, plus the fact that the system impedance has increased due to the loss of a line, results in a greatly depressed voltage at the motor terminals. If this voltage is too low the machines will stall or cut out of circuit.

Steady-state instability due to voltage regulators Consider a generator supplying an infinite busbar through two lines, one of which is suddenly removed. The load angle of the generator is instantaneously unchanged and therefore the power output decreases due to the increased system reactance, thus causing the generator voltage to rise. The automatic voltage regulator of the generator then weaken its field to maintain constant voltage, i.e. decreases the internal e.m.f., and synchronous instability may result.

Dynamic stability The control circuits associated with generator automatic voltage regulators, although improving steady-state stability, can introduce problems of poorly damped response and even instability. For this reason dynamic stability studies are performed, i.e. steady-state stability analysis including the automatic control features of the machines. The stability is assessed by determining the response to small step changes of rotor angle and hence the machine and control system equations are often linearized, i.e. constant machine parameters and linear AVR characteristic. The study usually extends over several seconds of real system time.

Improvement of system stability Apart from the use of fast acting AVRs the following techniques are in use or being considered.

(a) Reduction of fault clearance times, 80 ms is being approached in some new schemes.

(b) Turbine fast-valving or bypass valving—this controls the accelerating power by closing steam valves. Valves which can close or open in 0.2 s are available.

(c) Dynamic breaking by the use of shunt resistors across the generator terminals; this would limit rotor swings. In future the switching in such resistors could be achieved by thyristors.

(d) High-speed reclosure (independent pole tripping) in long (point-to-point) lines. In highly interconnected systems the increase in overall clearance times on unsuccessful reclosure makes this technique of dubious value.

(e) Increased use of HV direct-current links using thyristors would alleviate stability problems.

References

BOOKS

1. Crary, S. B., *Power System Stability*, Vol. I (*Steady State Stability*), Vol. II (*Transient Stability*), Wiley, New York, 1945.
2. Kimbark, E. W., *Power System Stability*, Vols. 1, 2, and 3, Wiley, New York, 1948.
3. Stagg, C. W., and A. H. El-Abiad, *Computer Methods in Power System Analysis*, McGraw-Hill, New York, 1968.
4. *Electrical Transmission and Distribution Reference Book*, Westinghouse Electric Corp., East Pittsburgh, Pennsylvania, 1964.
5. Mortlock, J. R., and M. W. Humphrey Davies, *Power System Analysis*, Chapman and Hall, London, 1952.
6. Dahl, O. G. C., *Electric Power Circuits*, Vol. II. McGraw-Hill, New York, 1938.
7. Venikov, V. A., *Transient Phenomena in Electrical Power Systems* (translated from the Russian), Pergamon, London, 1964.
8. Rudenberg, R., *Transient Performance of Electric Power Systems*, M.I.T. Press, 1969.
9. Adkins, B., *The General Theory of Electrical Machines*, Chapman and Hall, London, 1957.

PAPERS

10. Miles, J. G., 'Analysis of overall stability of multimachine power systems', *Proc. I.E.E.*, **109A** (1961), 203.
11. Lane, C. M., R. W. Long, and J. N. Powers, 'Transient-stability studies, II', *Trans. A.I.E.E.*, **78,** Pt. III (1959), 1291.
12. Taylor, D. G., 'Analysis of synchronous machines connected to power-system networks', *Proc. I.E.E.*, **110C** (1963), 606.
13. Kapoor, S. C., 'Dynamic stability of long transmission systems with static compensators and synchronous machines', *I.E.E.E. Trans*, **PAS-98** (1979), 124.
14. Humpage, W. D., and B. Stott, 'Predictor-correct methods of numerical integration in digital-computer analyses of power-system transient stability,' *Proc. I.E.E.*, **112** (1965), 1557.
15. Scott, E. C., W. Canon, A. Chorlton, and J. H. Banks, 'Multigenerator transient-stability performance under fault conditions', *Proc. I.E.E.*, **110** (1963), 1051.
16. Sicy, J., 'The influence of regulating transformers for the excitation of alternators on system voltages and stability', *C.I.G.R.E.*, **1962,** No. 323.
17. Stagg, G. W., A. F. Gabrielle, D. R. Moore, and J. F. Hohenstein, 'Calculation of transient stability problems using a high speed digital computer', *Trans. A.I.E.E.*, **78** (1959).

18. Concordia, C., and Ihara, S., 'Load representation in power system stability studies', *I.E.E.E. Trans.*, **PAS-101** (1982), 969.
19. Abe, S., *et al.*, 'Power system voltage stability', *I.E.E.E. Trans.*, **PAS-101** (1982), 3830.
20. Brereton, D. S., D. G. Lewis, and C. C. Young, 'Representation of induction-motor loads during power-system stability studies', *Trans. A.I.E.E.*, **76**, Pt. III (1957), 451.
21. Gross, G., *et al.*, 'A tool for the comprehensive analysis of power system dynamic stability', *I.E.E.E. Trans.*, **PAS-101** (1982), 226.
22. Brown, H. E., H. H. Happ, C. E. Person, and C. C. Young, 'Transient stability solution by an impedance matrix method', *Trans. I.E.E.E., P.A.S.*, **84** (1965), 1204.
23. Frowd, R. J., *et al.*, 'Transient stability and long term dynamics unified', *I.E.E.E. Trans.*, **PAS-101** (1982), 3841.
24. Schackshaft, C., 'General purpose turbo-alternator model', *Proc. I.E.E.*, **110** (1963), 703–713.
25. Humpage, W. D. *et al.*, 'Multi power-system dynamic analysis', *Proc. I.E.E.*, **119** (1972) 1167–1175.
26. I.E.E.E. Committee Report, 'Dynamic models for steam and hydroturbine in power system studies', *Trans. I.E.E.E.* **PAS–92** (1973), 1904–1915.
27. Dharma Rao, N., 'Routh–Hurwitz condition and Lyapunov methods for the transient-stability problem', *Proc. I.E.E.*, **116**, (1969), 533.
28. I.E.E.E. Committee Report, 'Computer representation of excitation systems', *Trans. I.E.E.E.* **PAS–87** (1968) 1460–1464.
29. Smith, J. R. *et al.*, 'Dynamic simulation of synchronous machines interconnected by long, compensated transmission circuits', *Proc. I.E.E.*, 123 (1976), 223–228.

Problems

8.1. A round-rotor generator of synchronous reactance 1 p.u. is connected to a transformer of 0.1 p.u. reactance. The transformer feeds a line of reactance 0.2 p.u. which terminates in a transformer (0.1 p.u. reactance) to the L/V side of which a synchronous motor is connected. The motor is of the round-rotor type and of 1 p.u. reactance. On the line-side of the generator transformer a three-phase static reactor of 1 p.u. reactance per phase is connected via a switch. Calculate the steady-state power limit with and without the reactor connected. All per unit reactances are expressed on a 10 MVA base and resistance may be neglected. The internal voltage of the generator is 1.2 p.u. and of the motor 1 p.u.
(Answer: 5 MW and 3.14 MW for shunt reactor)

8.2. In the system shown in Figure 8.18 two equivalent round-rotor generators feed into a load which may be represented by the constant impedance shown. Determine Z_{11},

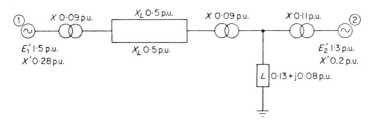

Figure 8.18 Line diagram of system in Problem 8.2.

Z_{12}, and Z_{22} for the system and calculate δ_{12} if P_1 is 1.2 p.u. All per unit values are expressed on the same base and the resistance of the system (apart from the load) may be neglected.

(Answer: $Z_{11} = 0.78\angle87.8°$; $Z_{12} = 1.47\angle112°$; $Z_{22} = 0.22\angle62.5°$)

8.3. For the system of three synchronous machines shown in Figure 8.4 investigate the steady-state stability for the following operating angles: (a) δ_{12} 75°. δ_{13} 45°, δ_{23} 30°; (b) δ_{12} 85°, δ_{13} 90°, δ_{23} − 10°.

(Answer: (a) stable; (b) unstable)

8.4. A hydroelectric generator feeds a load through two 132 kV lines in parallel, each having a total reactance of 70 Ω phase-to-neutral. The load consists of induction motors operating at three-quarters of full load and taking 30 MVA. A local generating station feeds directly into the load busbars. Determine the parameters of an equivalent circuit, consisting of a single machine connected to an infinite busbar through a reactance, which represents the above system when a three-phase symmetrical fault occurs halfway along one line.

The machine data are as follows:

Hydro-electric generator:	rating 60 MW at power factor 0.9 lagging; transient reactance 0.3 p.u.; inertia constant 3 kW-s/kVA.
Induction motors:	composite transient reactance 3 p.u.; inertia constant 1 kW-s/kVA.
Local generator:	rating 30 MVA; transient reactance 0.15 p.u. inertia constant 10 kW-s/kVA.

(Answer: $X = 9.9$ p.u., $M = 0.000121$ p.u., on a 100 MVA base).

8.5. For the system described in Example 8.4 determine the swing curves for fault clearance times of 250 and 500 ms.

8.6. The $P-V$, $Q-V$ characteristics of a substation load are as follows:

V	1.05	1.025	1	0.95	0.9	0.85	0.8	0.75
P	1.03	1.017	1	0.97	0.94	0.92	0.98	0.87
Q	1.09	1.045	1	0.93	0.885	0.86	0.84	0.85

The substation is supplied through a link of total reactance 0.8 p.u. and negligible resistance. With nominal load voltage $P = 1$ and $Q = 1$ p.u. By determining the supply voltage–received voltage characteristic, examine the stability of the system by the use of dE/dV. All quantities are per unit.

8.7. An induction motor load is supplied through a transformer of 0.1 p.u. reactance from a receiving-end substation which is supplied from a generator through a transmission link of total reactance 0.3 p.u. Plant data are as follows:

Generator $X_s = 1.1$, $X' = 0.3$. Induction motor load $X_s = 0.2$, $R_2 = 0.03$, mechanical power output independent of speed. All per unit values are on a 60 MVA base. Examine the stability of the load when the motor takes 50 MW when the generator has, (a) no AVR; (b) a continuously acting AVR.

8.8. A large synchronous generator, of synchronous reactance 1.2 p.u., supplies a load through a link comprising a transformer of 0.1 p.u. reactance and an overhead line of initially 0.5 p.u. reactance; resistance is negligible. Initially the voltage at the load busbar is 1 p.u. and the load $P+jQ$ is $(0.8+j0.6)$ p.u. regardless of the voltage. Assuming the internal voltage of the generator to remain unchanged, determine the value of line reactance at which voltage instability occurs.

(Answer: Unstable when $X = 2.0$ p.u.)

8.9. A load is supplied from an infinite busbar of voltage 1 p.u. through a link of series reactance X p.u. and of negligible resistance and shunt admittance. The load consists of a constant power component of 1 p.u. at 1 p.u. voltage and a per unit reactive power component (Q) which varies with the received voltage (V) according to the law

$$(V-0.8)^2 = 0.2(Q-0.8)$$

All per unit values are to common voltage and MVA bases.

Determine the value of X at which the received voltage has a unique value and the corresponding magnitude of the received voltage.

Explain the significance of this result in the system described. Use approximate voltage drop equations.

(Answer: $X = 0.25$ p.u.; $V = 0.67$ p.u.)

9

Direct Current Transmission

9.1 Introduction

The drawbacks to the use of alternating current for transmission of large powers over long distances will now be apparent to the reader. However, large series inductive reactances can be substantially reduced by series capacitors along the line and the stability limit greatly increased. For this form of transmission the decision to use either alternating current or direct current is entirely economic in nature. Critical lengths of lines have been quoted above which the use of d.c. is more economical (e.g. for 750 MW a critical distance of 550–750 km), but these are largely dependent on the cost of the valves and associated devices which, with the increased use of direct current, will probably become cheaper.

In two cases there are strong technical reasons for the use of direct-current transmission. These are as follows:

(a) For the connexion of large systems through links of small capacity. An example is the Britain–France cross-Channel link where slightly different frequencies in the two large systems would produce serious problems of power transfer control in the small capacity link. A d.c. line is an asynchronous or flexible link between two rigid systems.

(b) Where high-voltage underground cables are needed for reasonable transmission distances. The limitations of cables due to charging current with a.c. have been discussed and to increase lengths either artificial reactors or d.c. must be used. D.C. transmission by cables inland may be expected to take place in areas where amenity considerations restrict the use of overhead lines. The use of cables for cross-Channel crossings is well established. A comparison between the use of a.c. and d.c. with underground cables is interesting. Six 275 kV, 3 in^2,* cables in two groups of three in horizontal formation (total width of trench 5.2 m) in soil of resistivity 120°C cm/W have an a.c. capacity of 1520 MVA. Two cables at ±500 kV d.c. have a capacity of 1600 MW with a trench width of only 0.68 m.

* 3 in^2 = 1935 mm^2 = 3.8×10^6 cmil.

Further advantages in the use of d.c. are as follows: the corona loss in a d.c. line operating at a voltage corresponding to the peak value of the equivalent alternating voltage is substantially less than for the a.c. line. This is important, not so much because of the loss of power, but due to the resulting interference with radio and television transmissions. Generally the line loss will be smaller than for the equivalent a.c. line. Investigations have shown that the fault levels in an a.c. system with a d.c. link operating are less than with the link replaced by an a.c. equivalent. This is of great importance when the use of higher voltages and many interconnexions has greatly increased the fault MVA to be withstood by circuit breakers.

Disadvantages are as follows:

(a) The much more onerous conditions for circuit breaking when the current does not reduce to zero twice a cycle. Because of this, switching is not carried out on the d.c. link but effected by means of the terminal rectifiers and inverters. This severely hampers the creation of an interconnected d.c. system with tee-junctions. However, the use of multiple links with converter stations in series or parallel with each other and with the valves used as switches is likely.

(b) Voltage transformation has to be provided on the a.c. sides of the system.

(c) Rectifiers and inverters absorb reactive power and this must be supplied locally.

(d) D.C. converting stations are much more expensive than conventional a.c. substations.

In the past the *mercury-arc valve* has been used for high-voltage conversion. The main requirements are as follows:

(a) It must withstand the high inverse peak voltages between cathode and anode when not conducting.

(b) When a small negative voltage is applied to the grid the valve must not conduct with peak positive voltage on the anode. The instant of firing must be accurately controllable by means of the grid.

(c) The arc between anode to cathode must be stable and not subject to self-quenching.

Extensive research has gone on for many years to develop valves for the very high voltages employed in transmission links. A diagram of the basic form of the valve is shown in Figure 9.1. In new schemes solid-state thyristors are used and these will be discussed later.

A converter is required at each end of a d.c. line and operates as a rectifier (a.c. to d.c.) or an inverter (power transfer from d.c. to a.c.). The valves at the sending end of the link rectify the alternating current providing direct current which is transmitted to the inverter. Here it is converted back into alternating current which is fed into the connected a.c. system (Figure 9.2). If a reversal of power flow is required the inverter and rectifier exchange roles and the direct voltages at each end are reversed (Figure 9.2(b)). This is necessary because the

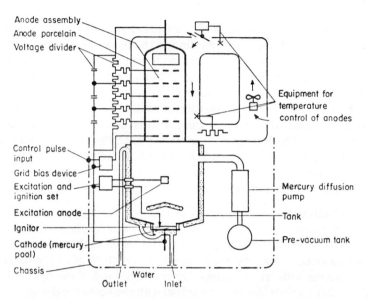

Anode assembly
Anode porcelain
Voltage divider
Control pulse input
Grid bias device
Excitation and ignition set
Excitation anode
Ignitor
Cathode (mercury pool)
Chassis
Water
Outlet
Inlet
Equipment for temperature control of anodes
Mercury diffusion pump
Tank
Pre-vacuum tank

Figure 9.1   Cross-section of a mercury arc valve. (*Permission of Institution of Electrical Engineers.*)

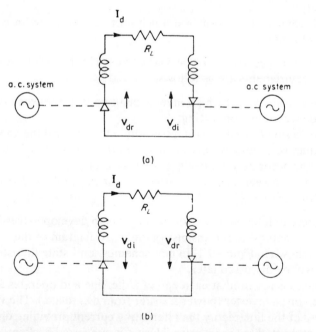

(a)

(b)

Figure 9.2(a)   Symbolic representation of two alternating current systems connected by a direct-current link; V_{dr} = direct voltage across rectifier, V_{di} = direct voltage across inverter. (b) System as in (a) but power flow reversed.

direct current can flow only in one direction (anode to cathode in the valves), so to reverse the direction (or sign) of power the voltage direction must be reversed.

The alternating current waveform injected by the inverter into the receiving-end a.c. system and taken by the rectifier is roughly trapezoidal in shape, and thus produces not only a fundamental sinusoidal wave but harmonics of an order dependent on the number of valves. For a six-valve bridge the order is $6n \pm 1$, i.e. 5, 7, 11, 13, etc. Filters are incorporated to tune out harmonics up to the twenty-fifth. With the continual increase in the ratings of thyristors they are supplanting mercury valves. In this text the term 'valve' will be assumed to embrace solid-state devices.

In the following sections the processes of rectification and inversion will be analysed and the control of the complete link discussed.

9.2 Rectification

Transformer secondary connexions may be arranged to give several phases for supplying the valves. Common arrangements are three, six, and twelve phases and high numbers are possible. Six-phase is popular because of the better output-voltage characteristics. To begin with a three-phase arrangement will be described, but most of the analysis will be for n phases so that results are readily adaptable for any system. In Figure 9.3(a) a three-phase rectifier is shown and in Figure 9.3(b) the current and voltage variation with time in the three phases of the supply transformer. With no grid control, conduction will take place between the cathode and the anode of highest potential. Hence the output voltage wave is the thick line and the current output continuous. In an n-phase system the anode changeover occurs at $(\pi/2 - \pi/n)$ degrees at a voltage $\hat{V} \sin \pi/n$ and the mean value of the direct output voltage is

$$V_O = \frac{1}{2\pi/n} \int_{\pi/2-\pi/n}^{\pi/2+\pi/n} \hat{V} \sin \omega t \, d(\omega t)$$

$$= \frac{\hat{V} \sin (\pi/n)}{\pi/n} \tag{9.1}$$

For three phases,

$$V_O = \frac{\hat{V} \sin 60°}{\pi/3}$$

$$= \hat{V} \frac{3\sqrt{3}}{2\pi} = 0.83 \hat{V}$$

and for six phases,

$$V_O = \frac{3\sqrt{2}}{\pi} V = \frac{3}{\pi} \hat{V} = 0.955 \hat{V}$$

where $V = $ r.m.s. voltage.

338

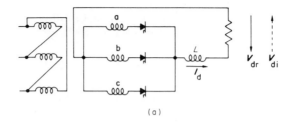

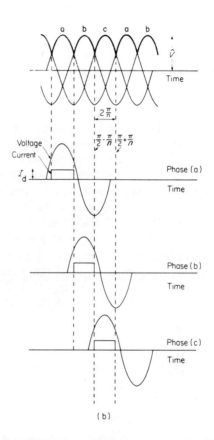

(a)

(b)

Figure 9.3(a) Three-phase rectifier; V_{di} = voltage of inverter. (b) Waveforms of anode voltage and rectified current in each phase.

Owing to the inductance present in the circuit the current cannot change instantaneously from $+I_d$ to 0 in one anode and from 0 to I_d in the next. Hence two anodes conduct simultaneously over a period known as the commutation time or overlap angle (γ). When valve (b) commences to conduct it short-circuits the a and b phases, the short-circuit current eventually becoming zero in valve (a) and I_d in (b). This is shown in Figure 9.4.

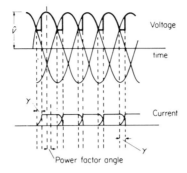

Figure 9.4   Waveforms of voltage and current showing effect of the
commutation angle γ. A lagging power factor is produced.

Grid control

A positive pulse applied to a grid situated between anode and cathode controls
the instant at which conduction commences, and once conduction has occurred
the grid exercises no further control. In the voltage waveforms shown in Figure
9.5 the conduction in the valves has been delayed by an angle α by suitably

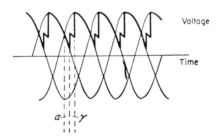

Figure 9.5   Waveforms of rectifier with instant of firing delayed by an
angle α by means of grid control.

delaying the application of positive voltage to the grids. Considering n phases
and ignoring the commutation angle γ, the new direct output voltage with a
delay angle of α is,

$$V'_{\mathrm{O}} = \frac{1}{2\pi/n} \int_{\pi/2-\pi/n+\alpha}^{\pi/2+\pi/n+\alpha} \hat{V} \sin \omega t \, \mathrm{d}(\omega t)$$

$$= \frac{n\hat{V}}{2\pi} \int_{-\pi/n+\alpha}^{\pi/n+\alpha} \cos \omega t \, \mathrm{d}(\omega t)$$

$$= \frac{n\hat{V}}{2\pi} \cdot 2 \cdot \sin \left(\frac{\pi}{n}\right) \cos \alpha$$

$$\therefore \ V'_{\mathrm{O}} = V_{\mathrm{O}} \cos \alpha \qquad\qquad (9.2)$$

340

where V_O is the maximum value of direct output voltage as defined by equation (9.1).

Bridge connexion

To avoid undue complexity in describing the basic operations the converter arrangement used so far in this chapter is simple and has disadvantages in practice. Mainly because the d.c. output voltage is doubled the bridge arrangement shown in Figure 9.6 is favoured in which there are always two

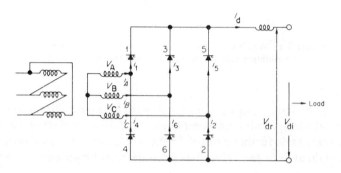

Figure 9.6   Bridge arrangement of valves.

valves conducting in series. The corresponding voltage waveforms are shown in Figure 9.7 along with the currents (assuming ideal rectifier operation).

The sequence of events in the bridge connexion is as follows (see Figures 9.6 and 9.7). Assume that the transformer voltage V_A is most positive at the beginning of the sequence, then valve (1) conducts and the current flows through (1) and the load then returns through valve (6) as V_B is most negative. After this period V_C becomes the most negative and current flows through (1)

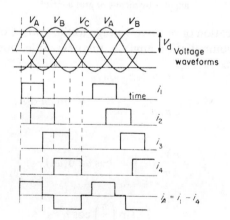

Figure 9.7   Idealized voltage and current waveforms for bridge arrangement.

and (2). Next (3) takes over from (1), the current still returning through (2). The complete sequence of valves conducting is therefore: 1 and 6, 1 and 2, 3 and 2, 3 and 4, 5 and 4, 5 and 6, 1 and 6. Grid control may be obtained in exactly the same manner as previously described and the voltage waveforms with grid delay and commutation time accounted for are shown in Figure 9.8.

The direct-voltage output with the bridge may be calculated either by using the line voltage (phase-to-phase) in the formula for six-valve, six-phase

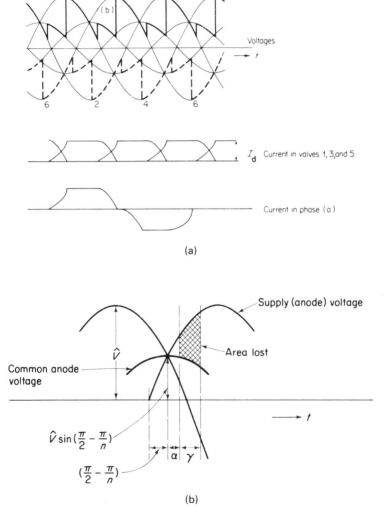

Figure 9.8(a) Voltage and current waveforms in the bridge connexion, including commutation (γ) and delay (α). Rectifier action. (b) Expanded waveforms showing voltage drop due to commutation.

rectification or determining the magnitude for three-valve operation and doubling it as both sides of the bridge contribute to the direct voltage. Hence,

$$V_0 = \frac{\hat{V}3\sqrt{(3)}}{2\pi} \times 2 = \sqrt{(2)} V \times 3\sqrt{(3)}/\pi$$

$$= \sqrt{(2)} V_L \times 3/\pi \quad (\text{as } V_L = \sqrt{(3)} V)$$

If the analysis used to obtain equation (9.2) is repeated for the bridge it will be shown that $V_d = V_0 \cos \alpha$, where V_0 is the maximum direct-voltage output for the bridge connexion.

Current relationships in the bridge circuit

During the commutation process when two valves are conducting simultaneously the two corresponding secondary phases of the supply transformer are short-circuited and if the voltage drop across the valves is neglected the following analysis applies.

When two phases of the transformer each of inductance L henries are effectively short-circuited, the short-circuit current (i_s) is governed by the equation,

$$2L \frac{di_s}{dt} = \hat{V}_L \sin \omega t = \text{resultant voltage between the two phases.}$$

$$\therefore \; i_s = -\frac{\hat{V}_L}{2L} \frac{\cos \omega t}{\omega} + A$$

where A is a constant of integration and $\hat{V}_L$ the crest value of the line-to-line voltage.

$$\omega t = \alpha \quad \text{when} \quad i_s = 0$$

$$\therefore \; A = \frac{\hat{V}_L}{2\omega L} \cos \alpha$$

Also, when

$$\omega t = \alpha + \gamma \qquad i_s = I_d$$

$$\therefore \; I_d = \frac{\hat{V}_L}{2\omega L} [\cos \alpha - \cos (\alpha + \gamma)]$$

$$= \frac{V_L}{\sqrt{2}\omega L} [\cos \alpha - \cos (\alpha + \gamma)],$$

where $V_L = $ r.m.s. line-to-line voltage and, as for the bridge circuit,

$$V_0 = \frac{3\sqrt{2}}{\pi} V_L$$

$$I_d = \frac{\pi V_0}{3\sqrt{2}} \cdot \frac{1}{\sqrt{2}X} [\cos \alpha - \cos (\alpha + \gamma)] = \frac{\pi V_0}{6X} [\cos \alpha - \cos (\alpha + \gamma)] \quad (9.3)$$

The mean direct output voltage, with grid delay angle α only considered, has been shown to be $V_0 \cos \alpha$. With both α and the commutation angle γ, the voltage with α only, will be modified by the subtraction of a voltage equal to the mean of the area under the anode voltage curve lost due to commutation (see Figures 9.4 and 9.8).

Referring to Figure 9.4 ($\alpha = 0$), the voltage drop due to commutation,

$$\Delta V_0 = \frac{\text{area lost}}{2\pi/n}$$

$$\Delta V_0 = \frac{n}{2\pi} \int_0^\gamma \hat{V} \sin \frac{\pi}{n} \sin \omega t \, d(\omega t)$$

$$= \frac{n}{2\pi} \hat{V} \sin \frac{\pi}{n} (1 - \cos \gamma) = \frac{V_0}{2}(1 - \cos \gamma)$$

When $\alpha > 0$ (see Figure 9.8(b)) the voltage drop is obtained as follows: area between input voltage wave and common anode voltage

$$= \int_{\pi/2 - \pi/2 + \alpha}^{\pi/2 - \pi/n + \gamma + \alpha} \hat{V} \sin \omega t \, d(\omega t) - \int_\alpha^{\alpha + \gamma} \hat{V} \sin \left(\frac{\pi}{2} - \frac{\pi}{n}\right) \cos \omega t \, d(\omega t)$$

$$= \hat{V}\left[-\cos \left(\frac{\pi}{2} - \frac{\pi}{n} + \alpha + \gamma\right) + \cos \left(\frac{\pi}{2} - \frac{\pi}{n} + \alpha\right)\right.$$

$$\left. -\sin \left(\frac{\pi}{2} - \frac{\pi}{n}\right) \sin (\alpha + \gamma) + \sin \left(\frac{\pi}{2} - \frac{\pi}{n}\right) \sin (\alpha)\right]$$

$$= \hat{V}\left[-\sin \left(\frac{\pi}{n} - \alpha - \gamma\right) + \sin \left(\frac{\pi}{n} - \alpha\right) - \cos \frac{\pi}{n} \sin (\alpha + \gamma)\right.$$

$$\left. + \cos \frac{\pi}{n} \sin \alpha\right]$$

$$= \hat{V}\left[\sin \frac{\pi}{n}\right][\cos \alpha - \cos (\alpha + \gamma)]$$

Voltage drop (mean value of lost area)

$$= \frac{\hat{V} \sin (\pi/n)}{2(\pi/n)} [\cos \alpha - \cos (\alpha + \gamma)]$$

$$= \frac{V_0}{2} [\cos \alpha - \cos (\alpha + \gamma)]$$

The direct-voltage output,

$$V_d = V_o \cos \alpha - \frac{V_o}{2}[\cos \alpha - \cos (\alpha + \gamma)]$$

$$= \frac{V_o}{2}[\cos \alpha + \cos (\alpha + \gamma)] \tag{9.4}$$

Adding equations (9.3) and (9.4),

$$\frac{I_d 3X}{\pi} + V_d = V_0 \cos \alpha$$

$$\therefore \quad V_d = V_o \cos \alpha - \frac{3XI_d}{\pi} \tag{9.5}$$

The power factor is given approximately by

$$\cos \phi = \tfrac{1}{2}[\cos \alpha + \cos (\alpha + \gamma)] \tag{9.6}$$

Equation (9.5) may be represented by the equivalent circuit shown in Figure 9.9, the term $(3X/\pi)I_d$ represents the voltage drop due to commutation and not a physical resistance drop. It should be remembered that V_O is the theoretical maximum value of direct output voltage and it is evident that V_d can be varied by changing V_0 (control of transformer secondary voltage by tap changing) and by changing α.

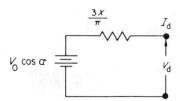

Figure 9.9 Equivalent circuit representing operation of a bridge rectifier. Reactance per phase X ohms.

9.3 Inversion

With rectifier operation the output current I_d and output voltage V_d are such that power is absorbed by a load. For inverter operation it is required to transfer power from the direct current to the alternating-current systems and as current can only flow from anode to cathode (i.e. in the same direction as with rectification) the direction of the associated voltage must be reversed. An alternating-voltage system must exist on the primary side of the transformer, and grid control of the converters is essential.

As the bridge connexion is in common use it will be used to explain the inversion process. If the bridge rectifier is given progressively greater delay the output voltage decreases becoming zero when α is 90° as shown in Fig. 9.10.

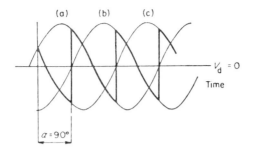

Figure 9.10 Waveforms with operation with $\alpha = 90°$, direct voltage
zero. Transition from rectifier to inverter action.

With further delay the average direct voltage becomes negative and the applied direct voltage (from the rectifier) forces current through the valves against this negative or back voltage. The converter thus receives power and inverts. The inverter bridge is shown in Figure 9.11(a) and the voltage and current waveforms in Figure 9.11(b). Commutation from valve (5) to valve (1) is possible only when phase (a) is positive with respect to (b) and the current changeover must be complete before (F) by a time (δ_0) equal to the recovery time of the valves. From the current waveforms it is seen that the current supplied by the inverter to the a.c. system *leads* the voltage and hence the inverter may be considered as a generator of leading vars or an absorber of lagging vars.

The power factor $\cos \phi \simeq (\cos \delta + \cos \beta)/2$, where δ and β are defined in Figure 9.11(a). Valve 5 is triggered at time A and as the cathode is held negative to the anode by the applied direct voltage (V_d) current flows, limited only by the circuit impedance. If the voltage drop across the arc is neglected the cathode and anode are at the same potential in valve (5). When time B is reached the anode to cathode open-circuit voltage is zero and the valve endeavours to cease conduction. The large transformer inductance (L), however, which has previously stored energy now maintains the current constant ($e = -L(di/dt)$ and if $L \to \infty$, $di/dt \to 0$). Conduction in (5) continues until time C when valve (1) is triggered. As the anode to cathode voltage for (1) is greater than for (5), (1) will conduct, but for a time (5) and (1) conduct together (commutation time), the current gradually being transferred from (5) to (1) until (5) is non-conducting (D). If triggering is delayed to F, valve (5) which is still conducting would be subject to a positively rising voltage and would continue to conduct into the positive half-cycle with breakdown of the inversion process. Hence, triggering must allow cessation of current flow before time F.

The angle (δ) between the extinction of valve (1) and the point (F) where the anode voltages are equal is called the extinction angle, i.e. sufficient time must be allowed for the grid to regain control. It is usual to replace the delay angle α by $\beta = 180 - \alpha$, hence β is equal also to $(\gamma + \delta)$. The minimum value of δ is δ_0.

The action of the inverter is essentially that of the rectifier, but with the delay angle α greater than 90° the direct voltage output (V_d) is in a certain direction,

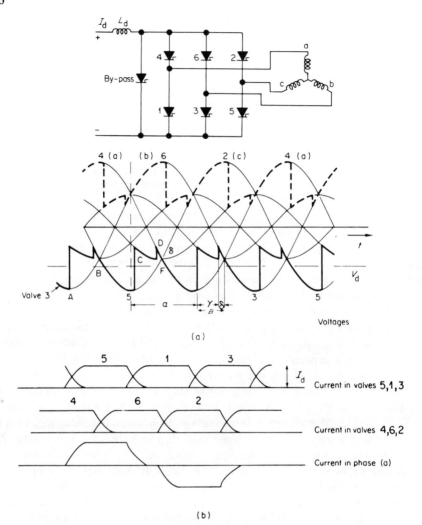

Figure 9.11(a) Bridge connexion—inverter operation. (b) Bridge connexion—inverter voltage and current waveforms.

as α increases V_d decreases and when $\alpha = 90°$, $V_d = 0$ V; with further increase in α, V_d reverses and inverter action is obtained. Hence the change from rectifier to inverter action and vice versa is smoothly obtained by control of α. This may be seen by consulting Figures 9.4 ($\alpha = 0$), 9.10 ($\alpha = 90°$) and 9.11(b) ($\alpha > 90°$).

Equations (9.3) and (9.4) may be used to describe inverter action. Replacing α by $(180 - \beta)$ and γ by $(\beta - \delta)$ the following are obtained:

$$V_d = -[V_O \cos \beta + I_d R_c] \tag{9.7}$$

$$V_d = -[V_O \cos \delta - I_d R_c] \tag{9.8}$$

where

$$R_c = \frac{3X}{\pi}$$

Therefore two equivalent circuits are obtained for the bridge circuit as shown in Figure 9.12(a) for constant β and Figure 9.12(b) for constant δ.

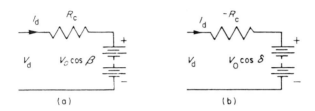

(a)   (b)

Figure 9.12(a) Equivalent circuit of inverter in terms of angle β. (b) Equivalent circuit of inverter in terms of angle δ.

9.4 Complete Direct Current Link

The complete equivalent circuit for d.c. transmission link under steady-state operation is shown in Figure 9.13. If both inverter and rectifier operate at constant delay angles the current transmitted

$$= I_d = \frac{V_{dr} - V_{di}}{R_L}$$

or

$$\frac{V_{Or} \cos \alpha - V_{Oi} \cos \beta}{R_L + R_{ci} + R_{ci}}$$

where R_L is the loop resistance of the line or cable, R_{cr} and R_{ci} are the effective commutation resistances of the rectifier and inverter, respectively.

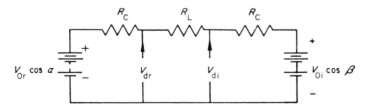

Figure 9.13   Equivalent circuit of complete link with operation with given delay angles $V_{Or} = V_O$ for rectifier, $V_{Oi} = V_O$ for inverter.

The magnitude of direct current can be controlled by variation of α, β, V_{Or} and V_{Oi} (the last two by tap changing of the supply transformers). Inverter control using constant delay angle has the disadvantage that if δ and hence β are too large excessively high reactive-power demand results (it will be seen from Figure 9.11(b) that the inverter currents are considerably out of phase with the anode voltages and hence a large requirement for reactive power established). Also, a reduction in the direct voltage to the inverter results in an increase in the commutation angle γ and if β is made large to cover this the reactive-power demand will be excessive again. In view of this it is more usual to operate the inverter with a constant δ which is achieved by the use of suitable control systems (called *compounding*).

The equations governing the operation of the inverter may be summarized as follows:

$$V_d = \frac{3\sqrt{(2)}\,V_L}{\pi}\cos\beta + \frac{3\omega L}{\pi}I_d$$

$$= \frac{3\sqrt{(2)}\,V_L}{\pi}\cos\delta - \frac{3\omega L}{\pi}I_d$$

$$= \frac{3\sqrt{(2)}\,V_L}{2\pi}(\cos\beta + \cos\delta)$$

and the power factor, $\cos\phi = (\cos\beta + \cos\delta)/2$ leading.

The advantages of operating with δ fixed and as small as possible have been discussed; it is also advisable to incorporate a facility for constant-current operation. The rectifier and inverter change roles as the required direction of power flow dictates and it is necessary for each device to have dual-control systems. A schematic diagram of the control systems is shown in Figure 9.14. In

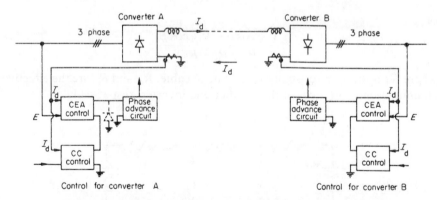

Figure 9.14 Schematic diagram of control of an HVDC system.
CEA = constant extinction angle; CC = constant current.

Figure 9.15 the full characteristics of the two converters of a link are shown with each converter operating as rectifier and inverter in turn. In the top half of the diagram converter (A) acts as a rectifier and the optimum characteristic with an α of zero is shown. With constant-current control, α is increased and the output voltage–current characteristic crosses the $V_d = 0$ axis, below which

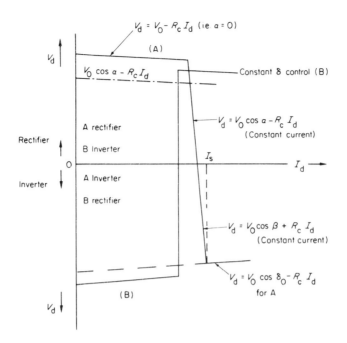

$$V_d = V_0 - R_c I_d \quad (\text{ie. } a = 0)$$

(A)

$V_0 \cos a - R_c I_d$ — Constant δ control (B)

A rectifier

Rectifier

B Inverter

$V_d = V_0 \cos a - R_c I_d$
(Constant current)

I_s

O · · · · · · · · · · · · · · · · · · · I_d →

Inverter

A Inverter

B rectifier

$V_d = V_0 \cos \beta + R_c I_d$
(Constant current)

$V_d = V_0 \cos \delta_0 - R_c I_d$
for A

(B)

V_d

Figure 9.15 Voltage–current characteristics of converters with compounding.

(A) acts as an inverter and can be operated on constant-δ control. A similar characteristic is shown for converter (B) which commences as an inverter and with constant-current control (β increasing) eventually changes to inverter operation. It is seen that the current setting for A is larger than that for B.

In Figure 9.16 the inverter characteristic for a power flow from A to B is shown drawn in the upper half of the graph and this will facilitate the discussion of the operational procedure.

Methods of control

To operate an inverter at a constant δ the instant of valve firing is controlled by a computer which takes into account variations in the instantaneous values of voltage and current. The computer then controls the firing times such that the extinction angle δ is slightly larger than the deionization angle of the valve. As the current rating of the valves should not be exceeded some measure of current control is desirable, the ideal being constant-current operation. This may be achieved in the inverter by increasing β beyond the constant-δ value, thus decreasing the back-voltage developed. Similarly constant-current compounding can be incorporated in the rectifier. In Figure 9.16 the voltage–current characteristics with both inverter and rectifier current-compounding are shown, and it is evident that the transmitted current cannot increase beyond the prescribed value.

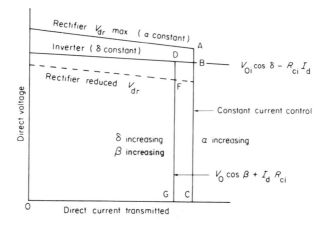

Figure 9.16 Inverter and rectifier operation characteristics with constant-current compounding. Point of operation is where the two characteristics intersect.

The methods of control may be summarized as follows, making reference to Figure 9.16.

(a) In the rectifier the transmitted current is regulated by varying the delay angle and hence V_{dr}. This is described by ABC. The rectifier tap-changing transformer is used to keep the delay angle within reasonable limits and also to give an excess in voltage over that of the inverter to cover sudden falls in rectifier output voltage which can be recovered in time by the tap changer. It should be remembered that the greater the delay angle the greater the reactive power consumed. Control is normally performed by the rectifier with inverter control used when necessary as follows.

(b) The inverter transformer gives the required back direct-voltage by tap-changing. Again this is slow acting and constant-current control is required giving a characteristic indicated by DG. The reference value of current is lower than that for the rectifier (a margin of 10–20 per cent of the rated current).

(c) Point B shows the operating condition with normal rectifier control. Should the rectifier voltage fall below the allowed margin to avoid I_d becoming zero the inverter back direct-voltage (i.e. the mean voltage of the negative anode voltages) is decreased and operation takes place along DFG and the current maintained at value OG. The new operating point is F and the power transmitted is smaller than before. Eventually the rectifier tap changer restores the original conditions. The value of the voltage margin is chosen to avoid frequent operations in the inverter control region.

Summarizing: with normal operation the rectifier operates at constant current and the inverter at constant δ; under emergency conditions the rectifier at zero firing-delay and the inverter at constant current.

9.5   Solid-state Converters

Experience is being gained on prototype installations where the valves comprise stacks *thyristors* in series instead of the mercury-arc devices. Two major problems, which have been overcome are (a) the difficulty in obtaining simultaneous firing of all thyristors, and (b) ensuring correct voltage division across them (due to capacitance effects).

A typical prototype employs a spiral assembly of 144 thyristors in series, employing oil for insulation and cooling. Each thyristor is rated at 2 kV peak-inverse voltage and 250 A. The thyristors are triggered by a hybrid optical/magnetic signal isolating system in which light pulses generated by high-power photodiodes are carried by fibre-optic guides to 12 group-potential levels in the stack. These pulses are then distributed by pulse transformers to the gate circuit of the 12-thyristor modules in each group. A typical design is for a stack rated at 150 kV, 1250 A, d.c. using thyristors rated at 4 kV.

The problem of obtaining a uniform voltage distribution with many thyristors in series is well known in low-voltage techniques. Because of the widely differing capacitances, the low-voltage technique of connecting R–C circuits in parallel with each thyristor does not suffice and additional capacitance must be incorporated. Also, it is necessary to retain uniform transient voltage distributions with time. To achieve this inductors with a non-linear characteristic may be used. On a sudden rise in voltage the inductor first absorbs the voltage, the capacitors are charged in the opposite direction with the inductor magnetizing current only, and the voltage rise across the thyristor is delayed. Similarly, the rate of change of current may be controlled on switching off.

Obviously it is advantageous to have as high a voltage per thyristor as possible. This cannot be achieved by increasing the silicon wafer thickness because of the resulting increased temperature rise. The device can only block the voltage if the temperature is below 125°C. The voltage collapse across the wafer when a positive impulse is applied to the gate must not be accompanied

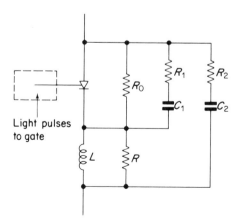

Figure 9.17   Circuitry associated with each thyristor.

by relatively high currents until conduction exists over the entire wafer. Only about a third of the possible reverse voltage is used for normal working due to overvoltages, etc.

The electrical circuitry associated with each thyristor is shown in Figure 9.17. The inductance limits the current rise on firing and on the discharge of C_2. The chain R_1, C_1 bypasses current from the thyristor which is not yet or is no longer conducting, thus avoiding unacceptable voltages. At the end of the working interval the current is allowed to flow in the reverse direction until a barrier layer is established in the wafer which enables voltage to be built up and current to cease (see Figure 9.18). After this, the element can withstand negative

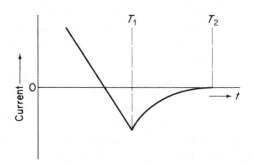

Figure 9.18   End of conduction period in thyristor–recovery of blocking property.

voltage, but the current must fall to a very low value (T_2 in the figure) if the thyristor is to block positive voltage whilst waiting for a new gate firing impulse. This time interval T_2 is analogous to the deionization time in mercury-arc valves and, if anything, is longer than the latter. It also implies that the theory presented for converter operation is applicable to both devices. Extreme care must be taken to avoid thermal overload of the thyristors because of the resulting damage.

The merits of the thyristor are summarized as follows:

(a) The serious disadvantage in mercury-arc valves is the possibility of backfire. This does not occur in thyristors.
(b) High-vacuum equipment is needed for mercury-arc values and it is also necessary to bring them up to working temperature before operation can commence. Thyristors require neither.
(c) The failure of an individual thyristor does not impair the working of the overall thyristor converter as allowance can be made for this in the design and rating of the converter.
(d) The cooling requirements and temperature control for mercury valves are very refined. With thyristors immersion in ordinary insulating oil is adequate.

(e) Because of their temperature requirements mercury-arc valves are housed inside buildings, whereas thyristor stacks may be designed for outside installation.

9.6   Faults and Harmonics

Transmission systems

The cheapest arrangement is a single conductor with a ground return. This, however, has various disadvantages. The ground-return current results in the corrosion of buried pipes, cable sheaths, etc. due to electrolysis. With submarine cables the magnetic field set up may cause significant errors in ships' compass readings, especially when the cable runs north–south. This system is

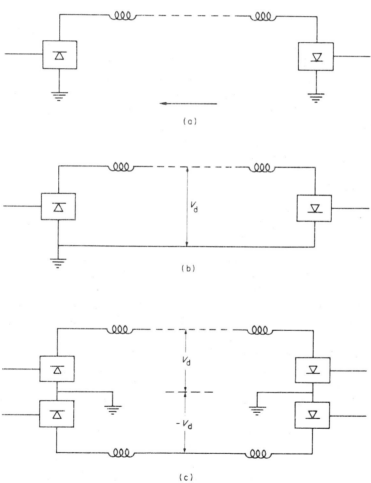

Figure 9.19   Possible conductor arrangements for d.c. transmissions: (a) ground return; (b) two conductors, return earthed at one end; (c) double-bridge arrangement.

shown in Figure 9.19(a). Two variations on two-conductor schemes are shown in Figure 9.19(b) and (c). The latter has the advantage that if the ground is used in emergencies, a double-circuit system is formed and it is the most popular system.

Converter faults

The main types of faults are as follows:

(a) Uncontrolled firing of a valve (valve made to conduct), usually caused by the failure of the grid-bias circuit, inadequate deionization time, and overvoltages.
(b) Failure of a valve to fire, often caused by arc quenching in the valve caused by high rate of rise of current or failure of the excitation circuits.
(c) Backfire, i.e. conduction through a valve in the reverse direction. This is caused by too high a reverse voltage or too rapid a rate of rise of voltage with time. It results in a short circuit between phases and is the only fault which may result in serious damage. In a bridge inverter, backfire results in a short circuit of two of the transformer phases. The reverse current in the faulty valve reaches its peak when the phase-to-phase voltage is zero (because of the highly inductive current path) and this is normally before the valve would have deionized in normal operation. The result is commutation failure and rapid corrective action is required.

Bypass valve

Due to the various faults which may occur in the d.c. system, it may be necessary to stop the bridge converter from operating. This may be rapidly achieved by the cessation of the pulses controlling the grids, accompanied by the firing of a valve connected across the bridge d.c. terminals known as the bypass valve (see Figure 9.11(a)). For a rectifier unit, although blocking of the gates would cease the operation, normally both converters in the system incorporate constant-current control. This results in the two halves which will remain conducting when the bridge is blocked trying to provide this specified current and continuing to conduct indefinitely. With an inverter the two valves which conduct at the instant of grid blocking will continue to conduct because of the applied voltage from the rectifier. In both cases, if the bypass valve is made to conduct at the instant of the blocking of the bridges, the d.c. terminals are short-circuited and the bypass valve takes over the current from the main bridge valves.

When recommencing normal operation it is necessary to unblock the valve grids and open-circuit the bypass valve. The latter, however, will not cease conduction until the anode voltage is negative with respect to the cathode when the grid will regain control. With the rectifier, the bridge valves establish a voltage across the bypass valve making the cathode positive with respect to the anode, and the bridge valves take over the complete current. On the other hand, the required polarity can only be obtained by a reversal of the polarity of the inverter-bridge voltage by advancing the angle β to greater than 60°. This angle is required only for about one cycle by which time the bypass valve will have ceased conduction.

It should be noted that the inverter analysis given in section 9.3 holds only for *angle β less than 60°*. As indicated above, if β becomes greater than 60° the mode of operation becomes very different and a new analysis will be required.

Harmonics

A knowledge of the harmonic components of voltage and current in a power system is necessary because of the possibility of resonance and also the enhanced interference with communication circuits. The direct-voltage output of a converter has a waveform containing a harmonic content, which results in current and voltage harmonics along the line. These are normally reduced by a smoothing choke.

The currents produced by the converter currents on the a.c. side contain harmonics. The current waveform in the a.c. system produced by a delta–star transformer bridge converter is shown in Figures 9.8 and 9.11. The order of the harmonics produced is $6n \pm 1$, where n is the number of valves. By the use of the Fourier series the equation for the current (i) is given by

$$i = \frac{2\sqrt{3}}{\pi} I_d \left(\cos \omega t + \frac{1}{5} \cos 5\omega t - \frac{1}{7} \cos 7\omega t - \frac{1}{11} \cos 11\omega t + \ldots \right)$$

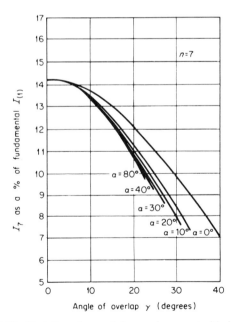

Figure 9.20   Variation of seventh harmonic current with delay angle α and commutation angle. (*Permission of International Journal of Electrical Engineering Education.*)

356

In Figure 9.20 is shown the variation of the seventh harmonic component with both commutation (overlap) angle (γ) and delay angle (α). Generally the harmonics decrease with decrease in γ this being more pronounced at higher harmonics. Changes in α for a given γ do not cause large decreases in the harmonic components, the largest change being for α's between 0 and 10°. For normal operation α is less than 10° and γ is perhaps of the order of 20°, hence the harmonics are small. During faults, however, α may reach nearly 90°, γ is small and the harmonics produced are large.

The harmonic voltages and currents produced in the a.c. system by the converter current waveform may be determined by representing the system components by their reactances at the particular harmonic frequency. Most of the system components have resonance frequencies between the fifth and eleventh harmonics.

It is usual to provide filters (L–C shunt resonance circuits) tuned to the harmonic frequencies. A typical installation is that for the England–France link (see Figure 9.22) and is shown in detail in Figure 9.21. At the fundamental frequency the filters are capacitive and help to meet the reactive power requirements of the converters.

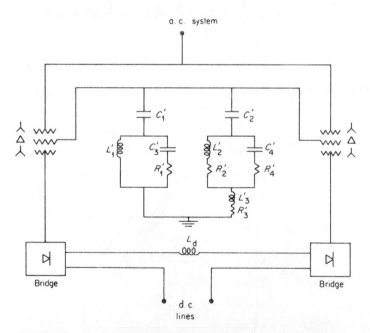

Figure 9.21 Schematic diagram of harmonic filters at Lydd station of U.K.–France scheme. Component values per phase, based on 33 kV: $C_1' = 36.1\ \mu\text{F}$; $C_2' = 24.23\ \mu\text{F}$; $C_3' = 36.1\ \mu\text{F}$, $C_4' = 192.53\ \mu\text{F}$; $L_1' = 1.44\ \text{mH}$; $L_3' = 13.05\ \text{mH}$; $R_1' = 6.3\ \Omega$; $R_2' = 0.1041\ \Omega$; $R_3' = 0.2646\ \Omega$; $L_2' = 1.5\ \text{mH}$. (*Permission of International Journal of Electrical Engineering Education.*)

9.7   Practical Schemes

The basic six-valve bridge connexion is in general use throughout the world. In order to cater for the high voltages in use several valves are often used in series to form one element of the bridge. The bridges may be interconnected in a number of ways, a common arrangement being shown in Figure 9.22 in which the centre connexion at each end of the line is earthed. In the event of a fault on one side of the system the other can continue operating using the earth as the second line.

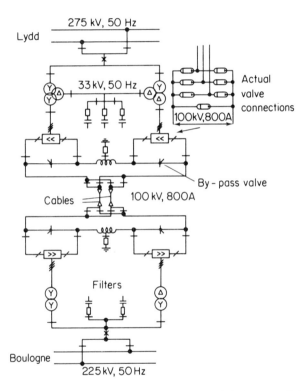

Figure   9.22   Schematic   arrangement   of   Britain–France   link.
(*Permission of A.S.E.A.*)

Several schemes are either in operation or under construction. Brief details of some of these are as follows:

(a) Donbass–Volgograd (U.S.S.R.): 750 MW, eight valve groups per station, 800 kV d.c., 470 km (overhead line).
(b) Sardinia: 200 MW, two valve groups per station, 200 kV d.c., 104 km of cable, 309 km (overhead line).
(c) Japan: 300 MW, two valve groups per station 250 kV d.c., connexion between 50 Hz and 60 Hz systems.

358

(d) New Zealand: 600 MW, four valve groups per station, 500 kV (d.c.), 40 km of cable, 576 km (overhead line).

(e) Great Britain–France: the main circuit diagram for this link across the English Channel is shown in Figure 9.22.

(f) Britain: 81 km underground cable link between Kingsnorth Power Station and two substations in the London area delivers 640 MW at ±266 kV and uses 56 valves.

(g) U.S.A.: Pacific coast interties shown in Figure 9.23. Six main valves plus bypass valve in each station, each valve rated at 133 kV, 1800 A (six anodes, 300 A each). Reactive power requirement is 840 MVAr during rectification; of this filters provide 180 MVAr, a.c. system, 200 MVAr and shunt capacitor banks, 460 MVAr.

Figure 9.23   Map of the Pacific Coast interties indicating routes of the 800 kV, 1440 MW d.c. links from The Dalles to Los Angeles and Mead.

A new and important aspect is that several manufacturers are now actively engaged in the development and provision of solid-state schemes. This international competition must further improve the competitiveness of d.c. compared with the mercury-valve situation of very concentrated expertise. A further advantage is the compactness of the solid-state conversion station, giving less environmental impact. One of the first solid-state schemes was the Eel River system which receives power at 230 kV from Hydro-Quebec, rectifies and transmits d.c. at 80 kV, 2000 A, and inverts to 230 kV a.c at the correct angle to match the U.S. and New Brunswick systems. The station used 10,000 thyristors. The Inga–Shaba scheme in Zaire is rated at 560 MW, ±500 kV and uses fibre-optic connexions to the thyristor gates. It comprises 1700 km of overhead line.

Multi-terminal systems These are feasible, but very elaborate coordination of the control systems is necessary. Obviously the great requirement of systems with more than two terminals is a d.c. circuit breaker. Intensive development on such a breaker has been in progress for some time based on a vacuum interrupter in which the discharge of a capacitor through a triggered vacuum gap produces an oscillation in the interrupter gap, thereby creating a current zero.

References

1. Adamson, C., and N. G. Hingorani, *High Voltage Direct Current Power Transmission*, Garraway Ltd., London, 1960. (*This gives a comprehensive list of references.*)
2. Cory, B. J. (Editor), *High Voltage Direct Current Converters and Systems*, Macdonald, London, 1965.
3. Uhlmann, E., *Power Transmission by Direct Current*, Springer-Verlag, Berlin, 1975.
4. Kimbark, E. W., *Direct Current Transmission*, Vol. 1, Wiley New York, 1971.
5. Arrillaga, J., *et al.*, *Computer Modelling of Electric Power Systems*, Wiley, Chichester, 1983.
6. Breuer, G. D., *et al.*, 'HVDC–AC harmonic interaction', *I.E.E.E. Trans.*, **PAS-101** (1982), 701.
7. Sebo, S. A., *et al.*, 'Model study of hvdc electric field effects', *I.E.E.E. Trans.*, **PAS-101** (1982), 1743.
8. Hingorani, N. G., and J. L. Hay, 'Representation of faults in the dynamic simulation of h.v. d.c. systems by digital computer', *Proc. I.E.E.*, **114** (1967), 629.
9. Adamson, C., and J. Arrillaga, 'Behaviour of multi-terminal a.c.–d.c. interconnexions with series-connected stations', *Proc. I.E.E.*, **115** (1968), 1685.
10. Persson, E. V., 'Calculation of transfer functions in grid-controlled converter systems. With special reference to h.v. d.c. transmission', *Proc. I.E.E.*, **117** (1970), 989.
11. Hingorani, N. G., Series of five articles in the *International Journal of Electrical Engineering Education*, Vols. 1 and 2 during 1963, 1964 and 1965, **1**, 273, 393; **2**, 37, 241, 355.
12. Reeve, J., 'Load flow method for HVDC multiterminal system control', *I.E.E.E. Trans.*, **PAS 96** (1977), 619–627.

Problems

9.1. A bridge-connected rectifier is fed from a 230 kV/120 kV transformer from the 230 kV supply. Calculate the direct voltage output when the commutation angle is 15° and the delay angle (a) 0°, (b) 30°, (c) 60°.
(Answer: (a) 160 kV; (b) 127 kV; (c) 61.5 kV.)
9.2. It is required to obtain a direct voltage of 100 kV from a bridge connected rectifier operating with $\alpha = 30°$ and $\gamma = 15°$. Calculate the necessary line secondary voltage of

the rectifier transformer which is nominally rated at 345 kV/150 kV; calculate the tap ratio required.

(Answer: 94 kV line and 1.6.)

9.3. If the rectifier in problem 9.2 delivers 800 A d.c., calculate the effective reactance X (Ω) per phase.

(Answer: $X = 13.0 \, \Omega$.)

9.4. A d.c. link comprises a line of loop resistance 5 Ω and is connected to transformers giving secondary voltage of 120 kV at each end. The bridge connected converters operate as follows:

$$\text{Rectifier:} \quad \alpha = 10° \qquad \text{Inverter:} \quad \delta_0 = 10°$$
$$X = 15 \, \Omega \qquad\qquad\qquad \text{Allow 5° margin}$$
$$\text{on } \delta_0 \text{ for } \delta$$
$$X = 15 \, \Omega$$

Calculate the direct current delivered if the inverter operates on constant δ control. If all parameters remain constant except α, calculate the maximum direct current transmittable.

(Answer: 610 A; 1104 A.)

9.5. The system in problem 9.4 is operated with $\alpha = 15°$ and on constant β control. Calculate the direct current for $\gamma = 15°$.

(Answer: 550 A.)

9.6. For a bridge arrangement, sketch the current waveforms in the valves and in the transformer windings and relate them in time to the anode voltages. Neglect delay and commutation times. Comment on the waveforms from the viewpoint of harmonics.

9.7. Draw a schematic diagram of an existing or proposed direct-current transmission scheme. Give vital parameters, e.g. voltage, rating of valves, etc.

9.8. A direct current transmission link connects two a.c. systems via converters, the line voltages at the transformer–converter junctions being 100 kV and 90 kV. At the 100 kV end the converter operates with a delay angle of 10°, and at the 90 kV end the converter operates with a δ of 15°. The effective reactance per phase of each converter is 15 Ω and the loop resistance of the link is 10 Ω. Determine the magnitude and direction of the power delivered if the inverter operates on constant-δ control. Both converters consist of six valves in bridge connexion. Calculate the percentage change required in the voltage of the transformer which was originally at 90 kV to produce a transmitted current of 800 A, other controls being unchanged. Comment on the reactive power requirements of the converters.

(Answer: 1.55 kA, 207 MW, 6.5%.)

9.9. A thyristor rectifier supplies direct current to a load. On no load the voltage across the load is 1.5 kV and the firing or delay angle 35°. The delay is controlled by a feedback system which holds the load voltage constant with changes in current. With this control the transformer reactance causes the ratio (voltage drop/no-load voltage) to be 0.08 p.u.

By considering only the fundamental components of the three-phase supply currents, plot a curve of volt-amperes reactive as a function of load current. Neglect the commutation angle.

Overvoltages and Insulation Requirements

10.1 Introduction

An area of critical importance in the design of power systems is the consideration of the insulation requirements for lines, cables, and stations. At first glance this may appear to be a sinple matter once the operating voltage of the system is decided, but unfortunately this if far from so. As well as the normal operating voltages, transients causing overvoltages occur in the system due to switching, lightning strokes, and other causes; the peak values of these can be much in excess of the operating voltage. Because of this devices must be provided to protect items of plant. The term extra high voltage (e.h.v.) has generally been accepted as describing systems of 230 kV up to 765 kV and for voltages above 765 kV the term ultra high voltage (u.h.v.) is applied; below 230kV, high voltage (h.v.) is in use.

Until recent years lightning has largely determined the insulation requirements, i.e. size of bushings, number of insulators per string, and tower clearances of the system, and the insulation of equipment tested with voltages of a waveform approximately that of a lightning surge. With the much higher operating voltages now in use and projected, the voltage transients or surges due to switching, i.e. the opening and closing of circuit breakers, have become the major consideration. In this context it is of interest to note that voltages of 500 kV and 765 kV are now in use and active discussion and research are taking place to decide the next voltage level which will be in the range 1000–1500 kV.

A factor of major importance is the *contamination* of insulator surfaces caused by atmospheric pollution. This considerably modifies the performance of insulation which becomes difficult to assess precisely. The presence of dirt, salt, etc., on the insulator discs or bushing surfaces results in these surfaces becoming slightly conducting and hence flashover occurs.

A few terms frequently used in high-voltage technology need definition. They are as follows.

1. *Basic impulse insulation level or basic insulation level (BIL)*—Reference levels expressed in impulse crest (peak) voltage with a standard wave not longer than a $1.2 \times 50 \ \mu s$ wave. Apparatus insulation as demonstrated by suitable tests

362

shall be equal or greater than the BIL. The two standard tests are the power frequency and 1.2/50 impulse wave withstand tests. The *withstand voltage* is the level the equipment will withstand for a given length of time or number of applications without disruptive discharge occurring, i.e. a failure of insulation resulting in a collapse of voltage and passage of current (sometimes termed 'sparkover' or 'flashover' when the discharge is on the external surface.) Normally several tests are performed and the number of flashovers noted. The BIL is usually expressed as a per unit of the peak (crest) value of the normal operating voltage to earth; e.g. for a maximum operating voltage of 362 kV,

$$1 \text{ p.u.} = \sqrt{2} \times \frac{362}{\sqrt{3}} = 300 \text{ kV}$$

so that a BIL of 2.7 p.u. = 810 kV.

2. *Critical Flashover Voltage (CFO)*—The peak voltage for a 50 per cent probability of flashover or disruptive discharge (sometimes denoted by V_{50}).

3. *Impuse ratio* (for flashover or puncture of insulation)—Impulse peak voltage divided by the crest value of power-frequency voltage to cause flashover or puncture.

Impulse tests are normally performed with a voltage wave which rises in 1.2 μs and falls to half the peak value in 50 μs (see Figure 10.1), this is known as a 1.2/50 μs wave and typifies the lightning surge. In most impulse generators the shape and duration of the wave may be modified. Basic impulse waves are shown in Figure 10.1. Switching surges consist of damped oscillatory waves, the frequency of which is determined by the system configuration and parameters,

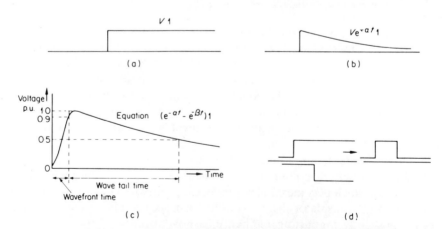

Figure 10.1   Basic impulse waveforms. (a) and (b) Frequently used in calculations. (c) Shape of lightning and switching surges; the former have a rise time of say 1.2 μs and a fall time of half maximum value of 50 μs (hence 1.2/50 wave). Switching surges are much longer, the duration times varying with situation, a typical wave is 175/3000 μs. Equation for 1/50 wave is $v = 1.036 \, (e^{-0.0146t} - e^{-2.56t})$. (d) Use of two unit functions to form a wave of finite duration.

they are normally of amplitude 2–2.8 p.u. although they can exceed 4 p.u. (per-unit values based on peak line to earth operating voltage as before).

10.2 Generation of Overvoltages

Lightning surges

A thundercloud is bipolar with positive charges at the top and negative at the bottom, usually separated by several kilometres. When the electric field strength exceeds the breakdown value a lightning discharge is initiated. The first discharge proceeds to the earth in steps (stepped leader stroke). When close to the earth a faster and luminous return stroke travels along the initial channel and several such leader and return strokes constitute a flash. The ratio of negative to positive strokes is about 5 to 1 in temperate regions. The magnitude of the return stroke can be as high as 200 kA, although an average value is of the order of 20 kA.

Following the initial stroke, after a very short interval, a second stroke to earth occurs, usually in the ionized path formed by the original. Again a return stroke follows. Usually several such subsequent strokes (known as dart leaders) occur, the average being between three and four. The complete sequence is known as a multiple-stroke lightning flash and a representation of the strokes at different time intervals is shown in Figure 10.2. Normally only the heavy current flowing over the first 50 μs is of importance and the current–time relationship has been shown to be of the form $i = i_{\text{peak}}(e^{-\alpha t} - e^{-\beta t})$.

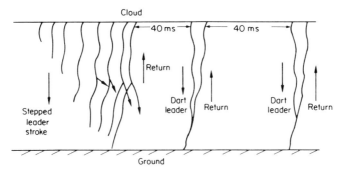

Figure 10.2   Sequence of strokes in a multiple lightning stroke.

When a stroke arrives on an overhead conductor equal current surges of the above waveform are propagated in both directions away from the point of impact. The magnitude of each voltage surge set up is therefore $\frac{1}{2} \cdot Z_0 \cdot i_{\text{peak}}(e^{-\alpha t} - e^{-\beta t})$, where Z_0 is the conductor surge impedance. For a current peak of 20 kA and a Z_0 of 350 Ω the voltage surges will have a peak value of $(350/2) \times 20 \times 10^3$, i.e. 3500 kV.

When a ground or earth wire exists over the overhead line a stroke arriving on a tower or on the wire itself sets up surges flowing in both directions along the wire. On reaching neighbouring towers they are partially reflected and transmitted further. This process continues over the length of the line as towers are encountered. If the towers are 300 metres apart the travel time between towers and back to the original tower is $(2 \times 300)/(3 \times 10^8)$, i.e. 2 μs, where the speed of propagation is 3×10^8 m/s. The voltage distribution may be obtained by means of the Bewley lattice diagram to be described in a later section.

An indirect stroke strikes the earth near a line and the induced current which is normally of positive polarity creates a voltage surge of the same waveshape which has an amplitude dependent on the distance from the ground. With a direct stroke the full lightning current flows into the line producing a surge travelling away from the point of impact in all directions. A direct stroke to a tower causes a *back flashover* due to the voltage set up across the tower inductance and footing resistance by the rapidly changing lightning current (typically 10 kA/μs); this appears as an overvoltage between the top of the tower and the conductors (which are at a lower voltage).

Switching surges—Interruption of short circuits and switching operations

When the arc between the circuit-breaker contacts breaks, the full system voltage (recovery voltage) suddenly appears across the open gap and hence across the R–L–C circuit comprising the system. The simplest form of single-phase equivalent circuit is shown in Figure 10.3(a) and (b). The resultant voltage appearing across the circuit is shown in Figure 10.3(c). It consists of a

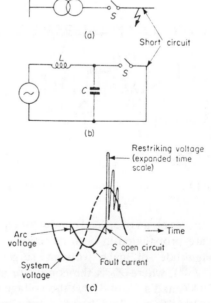

Figure 10.3 Restriking voltage set up on fault interruption. (a) System diagram. (b) Equivalent circuit. (c) Current and voltage waveforms.

high-frequency component superimposed on the normal system voltage, the total being known as the restriking voltage and constituting a switching surge. The equivalent circuit of Figure 10.3 may be analysed by means of the Laplace transform. When the circuit breaker opens the cessation of current may be simulated by the injection of an equal and opposite current into the system at time zero. Although this current is sinusoidal in form it may be approximated by a ramp function as shown in Figure 10.4. Let the fault current, $i = \sqrt{2}I \sin \omega t$.

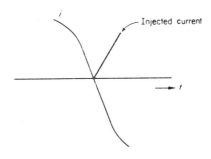

Figure 10.4   Application of Laplace transform to circuit of Figure 10.3. Ramp function of current.

The arc is extinguished at a current pause or zero (see Chapter 12) at which

$$\frac{di}{dt} = (\sqrt{2}I\omega \cos \omega t)_{t=0}$$

$$= \sqrt{2}I\omega$$

The short-circuit current $= I = V/\omega L$ (r.m.s. value). Hence the equation for the ramp function of injected current is

$$\left(\frac{di}{dt}\right)_{t=0} \cdot t = \sqrt{2}\left(\frac{V}{L}\right)t \tag{10.1}$$

The transform of the circuit after breaker opening is shown in Figure 10.5.

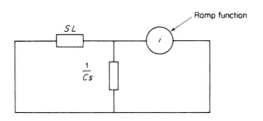

Figure 10.5   Equivalent circuit.

Transform of the ramp function

$$= i(s) = \frac{\sqrt{2}V}{Ls^2} \tag{10.2}$$

Voltage across the breaker contacts, i.e. across C,

$$v(s) = \frac{\sqrt{2}V}{Ls^2} \cdot \frac{L/C}{Ls + 1/Cs}$$

$$= \sqrt{2}V \frac{\omega_0^2}{s(s^2 + \omega_0^2)}$$

where $\omega_0^2 = 1/LC$.

$$\therefore \ v(s) = \sqrt{2}V \left(\frac{A}{s} + \frac{Bs + D}{s^2 + \omega_0^2} \right)$$

$$= \sqrt{2}V \left(\frac{1}{s} - \frac{s}{s^2 + \omega_0^2} \right)$$

From a table of transforms,

$$v(t) = \sqrt{2}V \left[1 - \cos\left(\sqrt{\frac{1}{LC}} t \right) \right] \tag{10.3}$$

Equation (10.3) assumes that the full system voltage V exists across the open switch, strictly the voltage across C is

$$V \left(\frac{X_c}{X_c - X_L} \right)$$

where X_c and X_L are the capacitive and inductive reactances so that (10.3) becomes

$$\frac{\sqrt{2}V}{[1 - (X_L/X_c)]} (1 - \cos \omega_0 t) \tag{10.4}$$

The difference in practice between these two expressions is small.

When series resistance (R ohms) is significant, equation (10.1) becomes,

$$\sqrt{(2)} \left(\frac{V\omega}{R + j\omega L} \right) t$$

and the ramp expression

$$i(s) = \frac{\sqrt{(2)}\omega V}{(R + j\omega L)s^2} \tag{10.5}$$

and $v(s) = i_s Z(s)$

The analysis of this case yields the expression,

$$v(t) = \omega IL (1 - e^{-\alpha t} \cos \omega_0 t) \tag{10.6}$$

where ω is the power frequency angular frequency, ω_0 the natural angular frequency of the circuit, and I the short-circuit current prior to the breaker opening,

$$\alpha = \frac{R}{2L} \quad \text{and} \quad (\alpha^2 + \omega_0^2) = \frac{1}{LC}$$

Note: $V \doteq I\omega L$.

From (10.6)

$$(dv/dt)_{\text{maximum}} \doteq \omega I \sqrt{\left(\frac{L}{C}\right)} e^{-\alpha t} \tag{10.7}$$

The restriking voltage (v) can thus rise to a maximum value of $2V$, where V is the peak value of the system recovery voltage. A similar expression is obtained when a line is suddenly energized by the system voltage.

If a resistance R_s be connected across the contacts of the circuit breaker the surge will be critically damped when $R_s = \frac{1}{2}\sqrt{(L/C)}$, and this offers an important method of reducing the severity of the transient. The initial rate of rise of the surge is very important as this determines whether the contact gap, which is highly polluted with arc products, breaks down again after the initial open circuit occurs. In the system shown in Figure 10.6(a) a double-frequency transient (Figure 10.6(b)) is set up, the two frequencies being determined by the circuit on each side of the switch, i.e.

$$\omega_1 = \frac{1}{\sqrt{(L_1 C_1)}} \quad \text{and} \quad \omega_2 = \frac{1}{\sqrt{(L_2 C_2)}}$$

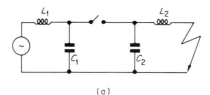

(a)

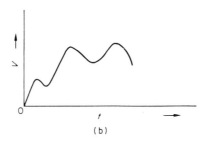

(b)

Figure 10.6(a)   System with double frequency restriking transient. (b) Typical waveform of double frequency, although initial amplitude is low, rate of rise is high.

Example 10.1   Determine the relative attenuation occurring in 5 cycles in the overvoltage surge set up on a 66 kV cable fed through an air-blast circuit breaker when the breaker opens on a system short circuit. The breaker incorporates resistance switching, i.e. an optimum resistance switched in across contact gap on opening. The network parameters are as follows:

$$R = 7.8 \, \Omega \qquad L = 6.4 \, \text{mH} \qquad C = 0.0495 \, \mu\text{F}$$

Solution   Switching resistance $= \frac{1}{2}\sqrt{(L/C)} = 180 \, \Omega$

$$\alpha = \frac{1}{2} \cdot \frac{R}{L} = 610$$

$$\omega = \frac{1}{\sqrt{LC}} = 5.61 \times 10^4 \, \text{rad/s}$$

i.e. transient frequency $= 8.93 \, \text{kHz}$

Hence,

$$5 \text{ cycles} = \frac{1}{8.93 \times 10^3} \times 5 \text{ s}$$

and

$$e^{-\alpha t} = \exp\left(-610 \times \frac{5}{8.93 \times 10^3}\right) = 0.712$$

The maximum theoretical voltage set up with the circuit breaker opening on a short-circuit fault would be

$$2 \times \frac{66000}{\sqrt{3}} \times \sqrt{2}$$

i.e. 107.5 kV, and in 560 μs this becomes $107.5 \times (1 - 0.712)$, i.e. 31 kV.

Note that resistance switching lowers the current to be broken and raises the power factor so that the voltage is not at peak value at opening.

Switching surges—interruption of capacitive circuits

The interruption of capacitive circuits is shown in Figure 10.7. The capacitance is left charged at the instant of arc interruption to value v_m but half a cycle later the system voltage is $-v_m$ giving a gap voltage of $2v_m$. If the gap breaks down an oscillatory transient is set up (as previously discussed) which can increase the gap voltage still further, as shown in the figure.

Current chopping   Current chopping arises with air-blast circuit breakers which operate on the same air pressure and velocity for all values of interrupted current. Hence on low-current interruption the breaker tends to open the

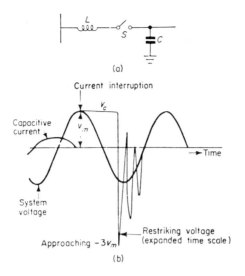

(a)

(b)

Figure 10.7   Voltage waveform when opening a capacitative circuit.

circuit before the current natural-zero and the electromagnetic energy present is rapidly converted to electrostatic energy, i.e.

$$\tfrac{1}{2}Li_0^2 = \tfrac{1}{2}Cv^2 \quad \text{and} \quad v = i_0\sqrt{\frac{L}{C}} \tag{10.8}$$

The voltage waveform is shown in Figure 10.8

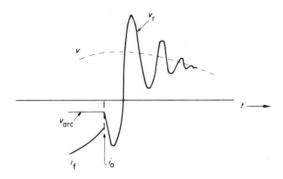

Figure 10.8   Voltage transient due to current chopping: i_f = fault current; v = system voltage, i_0 = current magnitude at chop.

An extension of (10.8) to include resistance and time yields

$$v = i_0\sqrt{\frac{L}{C}}\,e^{-\alpha t}\sin \omega_0 t \tag{10.9}$$

where, $\omega_0 = 1\sqrt{LC}$ and i_0 is the value of the current at the instant of chopping.

High transient voltages may be set up on opening a highly inductive circuit such as a transformer on no load.

Faults   Overvoltages may be produced by certain types of asymmetrical fault, mainly on systems with ungrounded neutrals. The voltages set up are of normal operating frequency. Consider the circuit with a three-phase earth fault as shown in Figure 10.9. If the circuit is not grounded the voltage across the first gap to open is 1.5 V_{phase}. With the system grounded the gap voltage is limited to the phase voltage. This has been discussed more fully in Chapter 7.

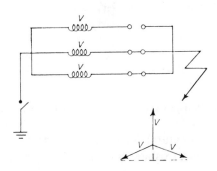

Figure 10.9   Three-phase system with neutral earthing.

Resonance   It is well known that in resonant circuits severe overvoltages occur dependent on the resistance present; the voltage at resonance across the capacitance can be high. Although it is unlikely that resonance in a supply network be obtained at normal supply frequencies it is possible to have this condition at harmonic frequencies. Resonance is normally associated with the capacitance to earth of items of plant and often brought about by an opened phase due to a broken conductor or a fuse operating.

In circuits containing windings with iron cores, e.g. transformers, a condition due to the shape of the magnetization curve known as *ferroresonance* is possible. This can produce resonance with overvoltages and also sudden changes from one condition to another.

A summary of important switching operations is given in Table 10.1 and means of reducing switching overvoltages summarized in Table 10.2.

10.3   Protection Against Overvoltages

Modification of transients

When considering the protection of a power system against overvoltages the transients may either be modified or even eliminated before reaching the substations, or if this is not possible to protect by various means the lines and substation equipment from flashover or insulation damage. By the use of

Table 10.1 Summary of the more important switching operations

Switching operation	System	Voltage across contacts
1. Terminal short circuit		
2. Short line fault		
3. Two out-of-phase systems—voltage depends on grounding conditions in systems		
4. Small inductive currents, current chopped (unloaded transformer)	Transformer	
5. Interrupting capacitive currents—capacitor banks, lines and cables on no load		
6. Evolving fault—e.g. flashover across transformer plus arc across contacts when interrupting transformer on no load	Transformer	
7. Switching—in unloaded e.h.v./u.h.v. line (trapped charge)		

Permission of *Brown Boveri Review*, December 1970.

overhead earth (ground) wires, phase conductors may be shielded from direct lightning strokes and the effects of induced surges from indirect strokes lessened. The shielding is not complete except perhaps for a phase conductor immediately below the earth wire. The effective amount of shielding is often described by an angle α as shown in Figure 10.10, a value of 35° appears to agree with practical experience. Obviously two earth wires horizontally separated provide much better shielding. Often for reasons of economy earth wires are installed over the last kilometre or so of line immediately before it enters a substation. It has already been seen that the switching-in of resistance across circuit-breaker contacts reduces the high overvoltages produced on opening, especially on capacitive or low-current inductive circuits.

An aspect of vital importance, quite apart from the prevention of damage, is the maintenance of supply, especially as most flashovers cause no permanent damage and therefore a complete and lasting removal of the circuit from

Table 10.2   Overvoltages in long e.h.v. and u.h.v. lines

Means of reducing switching overvoltages	Basic diagram
1. High-voltage shunt reactors connected to the line to reduce power-frequency overvoltage	
2. Eliminating or reducing trapped charge by:	
2.1 Line shunting after interruption	
2.2 Line discharge by magnetic potential transformers	
2.3 Low-voltage side disconnection of the line	
2.4 Opening resistors	
2.5 Single-phase reclosing	
2.6 Damping of line voltage oscillation after disconnecting a line equipped with h.v. reactors	
3. Damping the transient oscillation of the switching overvoltages	
3.1 Single-stage closing resistor insertion	
3.2 Multi-stage closing resistor insertion	
3.3 Closing resistor in-line between circuit breaker and shunt reactor	
3.4 Closing resistor in-line on the line side of the shunt reactor.	
3.5 Resonance circuit (surge absorber) connected to the line	
4. Switching at favourable switching moments:	
4.1 Synchronized closing	
4.2 Reclosing at voltage minimum of a beat across the breaker	
5. Simultaneous closing at both ends of the line	
6. Limitation by surge arresters when: energizing line at no-load (a) disconnecting reactor loaded transformers (b) disconnecting high-voltage reactors (c)	

Permission Brown Book Review, December 1970

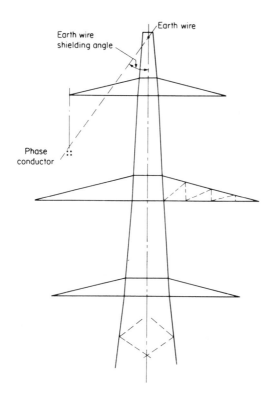

Figure 10.10 Single earth wire protection; shielding angle α normally 35°.

operation is not required. This may be achieved by the use of *auto-reclosing* circuit breakers.

A *surge modifier* may be produced by the connexion of a shunt capacitor between the line and earth or an inductor in series with the line; oscillatory effects may be reduced by the inclusion of damping resistors. It is uneconomical to attempt to modify or eliminate most overvoltages and means are required to protect the various items of power systems. Surge diverters are connected across the equipment and divert to earth the transient; surge modifiers are connected in series with the line entering the substation and attempt to reduce the steepness of the wave front—hence its severity.

Surge diverters

The basic requirements for diverters are that they should pass no current at normal voltage, interrupt the power frequency follow-on current after a flashover, and break down as quickly as possible after the abnormal voltage arrives. The simplest form is the *rod gap* as shown in Figure 12.2 for a circuit-breaker bushing. This may also take the form of rings (arcing rings) around the top and bottom of an insulator string. The breakdown voltage for a given gap is polarity dependent to some extent. It does not interrupt the post

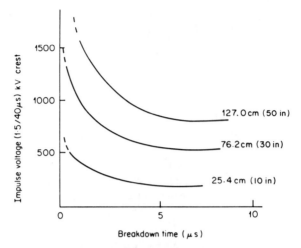

Figure 10.11   Breakdown characteristics of rod gaps.

flashover follow-on current (i.e. the power-frequency current which flows in the path created by the flashover) and hence the circuit protection operates, but it is by far the cheapest device for plant protection against surges. It is usually recommended that a rod gap be set to a breakdown voltage not less than 30 per cent below the voltage withstand level of the protected equipment. For a given gap the time for breakdown varies roughly inversely with the applied voltage; there exists, however, some dispersion of values. The times for positive voltages are lower than those for negative.

Typical curves relating the critical flashover voltage and time to breakdown for rod gaps of different spacings are shown in Figure 10.11. In Figure 10.12 the

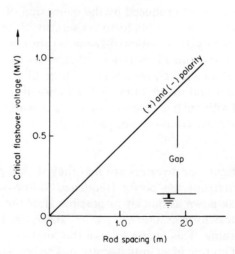

Figure 10.12   Effect of rod spacing on flashover of rod gaps—lower rod grounded.

flashover voltage-rod spacing relationship is shown for a rod-to-rod gap. The flashover voltage is to some extent dependent on the length of the lower (grounded) rod. For low values of this length there is a reasonable difference between positive (lower values) and negative flashover voltages. Usually a length of 1.5–2.0 times the gap spacing is adequate to diminish this effect to a reasonable amount.

Expulsion gaps or tubes (*protector tubes*) A disadvantage of the plain rod gap is the power-frequency current (follow-on current) that flows after breakdown which can only be extinguished by circuit-breaker operation. An improvement is the expulsion tube which consists of a spark gap in a fibre tube as shown in Figure 10.13. When a sparkover occurs between the electrodes the follow-on

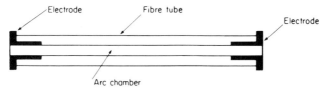

Figure 10.13   Expulsion tube.

current arc is contained in the relatively small fibre tube. The high temperature of the arc vaporizes some of the organic material of the tube wall, causing a high gas pressure to build up in the tube. This gas possesses considerable turbulence and it extinguishes the arc. The hot gas rapidly leaves the tube which is open at the ends. Very high currents have been interrupted in such tubes. The breakdown voltage is slightly lower than for plain rod gaps for the same spacing.

An improved but more expensive surge diverter is the *lightning arrester*. A porcelain bushing contains a number of spark gaps in series with silicon carbide discs, the latter possessing low resistance to high currents and high resistance to low currents, i.e. it obeys a law of the form $V = aI^{(0.2)}$, where a depends on the material and its size. The overvoltage breaks down the gaps and then the power-frequency current is determined by the discs and limited to such a value that the gaps can quickly interrupt it at the first current zero. The voltage–time characteristic of a lightning arrester is shown in Figure 10.14, the gaps break

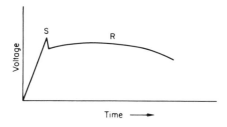

Figure 10.14   Characteristic of lightning arrester.

376

down at S and the characteristic after this is determined by the current and the discs; the maximum voltage at R should be in the same order as at S. For high voltages a stack of several such units are used. The basic arrangement of discs and gaps which are housed inside a porcelain bushing is shown in Figure 10.15.

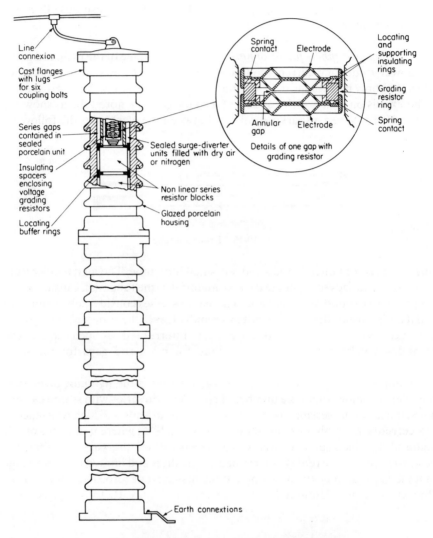

Figure 10.15   Diagram of lightning arrester—four-unit stack showing part section of one unit. (*Permission of the Electricity Council.*)

Although with multiple spark gaps diverters can withstand high rates of rise of recovery voltage (RRRV), the non-uniform voltage distribution between the gaps presents a problem. To overcome this, capacitors and non-linear resistors are connected in parallel across each gap. With the high-speed surge the voltage is mainly controlled by the gap capacitance and hence capacitive grading is used. A power-frequencies non-linear resistor provides effective voltage grading. The equivalent circuit is shown in Figure 10.16.

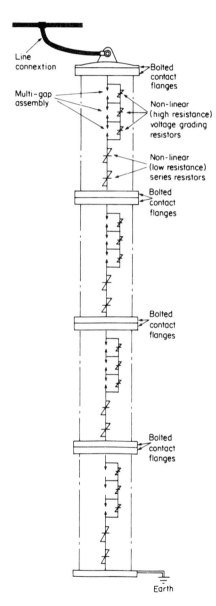

Figure 10.16   Equivalent circuit of components comprising a single-phase four-unit surge diverter stack. (*Permission of the Electricity Council.*)

10.4 Insulation Coordination

The equipment used in a power system comprises items having different breakdown or withstand voltages and different voltage–time characteristics. In order that all items of the system be adequately protected there is a need to consider the situation as a whole and not items of plant in isolation, i.e. the insulation protection must be coordinated. To assist this process standard insulation levels are recommended and are summarized in Tables 10.3 and 10.4. Reduced basic insulation impulse levels are used when considering switching surges and these are summarized in Table 10.4.

Table 10.3 British substation practice

Nominal voltage (kV)	Impulse (peak kV)	Power frequency withstand (peak kV)	Minimum clearance to ground (m)	Minimum clearance between phases (m)	Coord. gap setting (m)
11	100	29	0.2	0.25	2 × 0.03
132	550	300	1.1	1.25	0.66
400	1425	675	3.05	3.55	1.5–1.8

Table 10.4 Recommended BIL's at various operating voltages (United States practice)

Voltage class (kV)	15	23	34.5	46	69	92	115
BIL (kV)	110	150	200	250	350	450	550
Reduced BIL (kV)			125				450⎱ 350⎰

Voltage class (kV)	138	161	196	230	287	345	500
BIL (kV)	650	750	900	1050	1300	1550	1800
Reduced BIL (kV)	550⎱ 450⎰	650⎱ 550⎰		900⎱ 825⎬ 750⎰	1175⎱ 1050⎬ 900⎰	1425⎱ 1300⎬ 1050⎰	1675⎱ 1550⎬ 1300⎰

Coordination is rendered difficult by the different voltage–time characteristics of plant and protective devices, for example a gap may have an impulse ratio of 2 for a 20 μs front wave and 3 for a 5 μs wave. At the higher frequencies (shorter wavefronts) corona cannot form in time to relieve the stress concentration on the gap electrodes. With a lightning surge a higher voltage can be withstood because a discharge requires a certain discreet amount of energy as well as a minimum voltage and the applied voltage increases until the energy reaches this value. In Figure 10.17 the voltage–time characteristics for the system elements comprising a substation are shown and the protection of the weakest items by the arrester is illustrated.

Up to an operating voltage of 345 kV the insulation level is determined by lightning and the standard impulse tests suffice along with normal frequency tests. Above this value, however, the overvoltages resulting from switching are higher in magnitude and therefore decide the insulation. The characteristics of

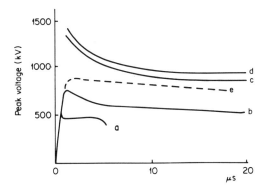

Figure 10.17 Insulation coordination in a HV substation. Voltage–
time characteristics of plant for a 1.5/40 μs wave. (a) Characteristic of
lightning arrester. (b) Transformer. (c) Line insulator string. (d) Busbar
insulation. (e) Maximum surge applied waveform.

air gaps and some solid insulations are different for switching surges than for
the standard impulse waves, and closer coordination of insulation is required
because of the lower attenuation of switching surges, although their
amplification by reflexion is less than with lightning. Recent work indicates that
for transformers the switching impulse strength is of the order of 0.95 of the
standard value while for oil-filled cables it is 0.7 to 0.8.

The design withstand level is selected by specifying the risk of flashover, e.g.
for 550 kV towers a 0.13 per cent probability has been used. At 345 kV, design
is carried out by accepting a switching impulse level of 2.7 p.u. which cor-

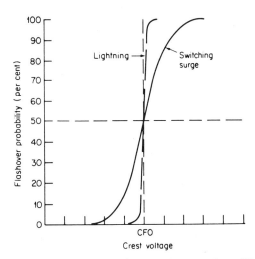

Figure 10.18 Flashover probability—peak surge voltage. Lightning
flashover is close-grouped near the critical flashover voltage (CFO),
whereas switching surge probability is more widely dispersed and
follows normal Gaussian distribution. (*Permission of the Westinghouse
Electrical Corporation, East Pittsburgh, Pennsylvania, U.S.A.*)

380

responds to the lightning level. At 500 kV, however, a 2.7 p.u. switching impulse would require 40 per cent more tower insulation than that governed by lightning. The tendency is therefore for the design switching impulse level to be forced lower with increasing system operating voltage and controlling the surges by the more widespread use of resistance switching in the circuit breakers. For example, for the 500 kV network the level is 2 p.u. and at 765 kV it is reduced to 1.7 p.u.; with further increases in system voltage it is hoped to decrease the level to 1.5 p.u.

The problem of switching surges is illustrated in Figure 10.18 in which flashover probability is plotted against peak (crest) voltage; *critical flashover voltage* (CFO) is the peak voltage for a particular tower design for which there is a 50 per cent probability of flashover. For lightning the probability of flashover below the CFO is slight, but with switching surges having much longer fronts the probability is higher, the curve following the normal Gaussian

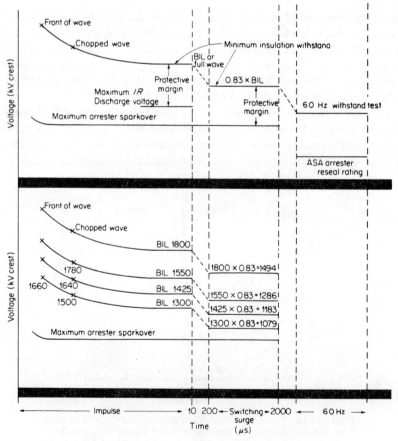

Figure 10.19 Coordination diagrams relating lightning arrester characteristics to system requirements. (*Permission of the Ohio Brass Co.*)

distribution, and the tower must be designed for a CFO much higher than the maximum transient expected.

An example of the application of lightning arrester characteristics to system requirements is illustrated in Figure 10.19. In this variation of crest voltage for lightning and switching surges with time is shown for an assumed BIL of the equipment to be protected. Over the impulse range (up to 10 μs) the front of wave and chopped wave peaks are indicated. Over the switching surge region the insulation withstand strength is assumed to be 0.83 of the BIL (based on transformer requirements). The arrester maximum sparkover voltage is shown along with the maximum voltage setup in the arrester by the follow-on current (10 or 20 per cent). A protective margin between the equipment withstand strength and the maximum arrester sparkover of 15 per cent is assumed.

10.5 Propagation of Surges

The basic differential equations for voltage and current in a distributed-constant line are as follows:

$$\frac{\partial^2 v}{\partial x^2} = LC \frac{\partial^2 v}{\partial t^2} \quad \text{and} \quad \frac{\partial^2 i}{\partial x^2} = LC \frac{\partial^2 i}{dt^2}$$

and these equations represent travelling waves. The solution for the voltage may be expressed in the form,

$$v = F_1(t - x\sqrt{LC}) + F_2(t + x\sqrt{LC})$$

i.e. one wave travels in the positive direction of x and the other in the negative. Also it may be shown that because $\partial v/\partial x = -L(\partial i/\partial t)$, the solution for current,

$$i = \sqrt{\frac{C}{L}}[F_1(t - x\sqrt{LC}) - F_2(t + x\sqrt{LC})] \quad \text{noting that} \sqrt{\frac{C}{L}} = \frac{1}{Z_0}$$

In more physical terms, if a voltage is injected into a line (Figure 10.20) a corresponding current i will flow and if conditions over a length dx are considered, the flux set up between the go and return wires is equal to iL dx, where L is the inductance per unit length. The induced back e.m.f. is $-d\Phi/dt$, i.e. $-Li(dx/dt)$ or $-iLU$, where U is the wave velocity. The applied voltage v must equal iLU. Also charge is stored in the capacitance over dx, i.e. $Q = i$ d$t = vC$ dx and $i = vCU$. Hence, $vi = viLCU^2$ and $U = 1/\sqrt{LC}$. Also, $i = v\sqrt{C/L} = v/Z_0$, where Z_0 is the characteristic or surge impedance. For single-circuit three-phase overhead lines (conductors not bundled) Z_0 lies in the range 400–600 Ω. U for overhead lines is 3×10^8 m/s, i.e. the speed of light and for cables

$$U = \frac{3 \times 10^8}{\sqrt{\varepsilon_r \mu_r}} \text{ m/s}$$

where ε_r is usually from 3 to 3.5, and $\mu_r = 1$.

382

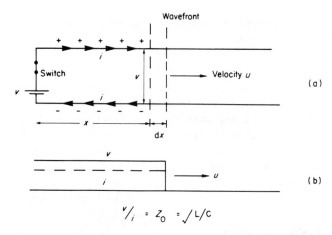

Figure 10.20 Distribution of charge and current as wave progresses along previously unenergized line. (a) Physical arrangement. (b) Symbolic representation.

From the above relations,

$$\tfrac{1}{2}Li^2 = \tfrac{1}{2}(iLU)\left(\frac{i}{U}\right)$$

$$= \tfrac{1}{2}\left(\frac{vi}{U}\right) = \tfrac{1}{2}Cv^2$$

The incident travelling waves of v_i and i_i, when they arrive at a junction or discontinuity, produce a reflected current i_r and a reflected voltage v_r which travel back along the line. The incident and reflected components of voltage and current are governed by the surge impedance Z_0, so that

$$v_i = Z_0 i_i \quad \text{and} \quad v_r = -Z_0 i_r$$

In the general case of a line of surge impedance Z_0 terminated in Z (Figure 10.21) the total voltage at Z is $v = v_i + v_r$ and the total current is $i = i_i + i_r$.
Also,

$$(v_i + v_r) = Z(i_r + i_i),$$

$$Z_0(i_i - i_r) = Z(i_r + i_i)$$

and

$$i_r = \left(\frac{Z_0 - Z}{Z_0 + Z}\right)i_i \qquad (10.10)$$

Again,

$$v_i + v_r = Z(i_i + i_r)$$

$$= Z\left(\frac{v_i - v_r}{Z_0}\right)$$

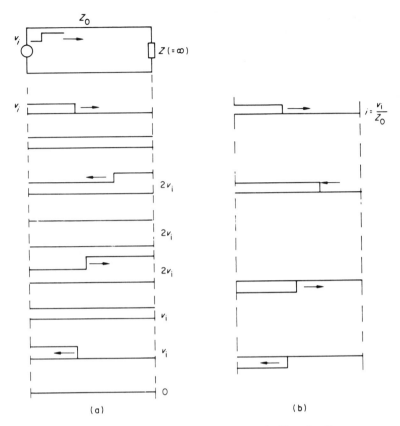

Figure 10.21 Application of voltage to unenergized loss-free line on open circuit. (a) Distribution of voltage. (b) Distribution of current. Voltage source is effective short circuit.

or,

$$v_r = \left(\frac{Z - Z_0}{Z + Z_0}\right)v_i = \alpha v_i \tag{10.11}$$

where α is the *coefficient of reflexion*
Hence,

$$v = \left(\frac{2Z}{Z + Z_0}\right)v_i \tag{10.12}$$

and

$$i = \left(\frac{2Z_0}{Z + Z_0}\right)i_i \tag{10.13}$$

From the above, if $Z \to \infty$, $v = 2v_i$ and $i = 0$. Also if $Z = Z_0$ (matched line) $\alpha = 0$, i.e. no reflexion. If $Z > Z_0$, then v_r is positive and i_r is negative, but if

$Z < Z_0$, v_r is negative and i_r is positive. The reflected waves will travel back and forth the line setting up in turn further reflected waves at the ends, and this process will continue indefinitely unless the waves are attenuated due to resistance and corona.

Summarizing, at an open circuit the reflected voltage is equal to the incident voltage and this wave along with a wave $(-i_i)$ travels back along the line; note that at the open circuit the total current is zero. Conversely, at a short circuit the reflected voltage wave is $(-v_i)$ in magnitude and the current reflected is (i_i), giving a total voltage at the short circuit of zero and a total current of $2i_i$. For other termination arrangements Thevenin's theorem may be applied to analyse the circuit. The voltage across the termination when it is open-circuited is seen to be $2v_i$ and the equivalent impedance looking in from the open-circuited termination is Z_0; the termination is then connected across the terminals of the Thevenin equivalent circuit (Figure 10.22).

(a) (b)

Figure 10.22 Analysis of travelling waves—use of Thevenin equivalent circuit. (a) System. (b) Equivalent circuit.

Consider two lines of different surge impedance in series. It is required to determine the voltage across the junction between them (Figure 10.23),

$$v_{AB} = \left(\frac{2v_i}{Z + Z_1}\right) Z_1 = \beta v_i \qquad (10.14)$$

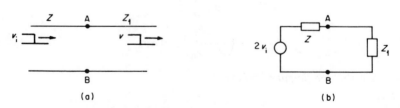

(a) (b)

Figure 10.23 Analysis of conditions at junction of two lines or cables of different surge impedance.

The wave entering the line Z_1 is the refracted wave and β is the *refraction coefficient*, i.e. the proportion of the incident voltage proceeding along the second line (Z_1).

$$v_r = v_i \alpha = v_i \left(\frac{Z_1 - Z}{Z_1 + Z}\right) \quad \text{and} \quad i = \frac{v_{AB}}{Z_1} = \frac{2v_1}{Z_1 + Z} = \text{refracted current}$$

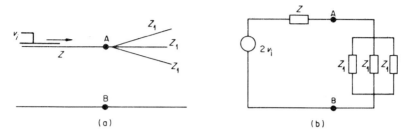

Figure 10.24 Junction of several lines. (a) System. (b) Equivalent circuit.

When several lines are joined to the line on which the surge originates (Figure 10.24) the treatment is similar, e.g. if the lines have equal surge impedances (Z_1) then,

$$i_A = \frac{2v_i}{(Z + Z_1/3)} \quad \text{and} \quad v_{AB} = \left(\frac{2v_i}{Z + Z_1/3}\right)\left(\frac{Z_1}{3}\right) \tag{10.15}$$

An important practical case is that of the clearance of a fault at the junction of two lines and the surges produced. The equivalent circuits are shown in Figure 10.25; the fault clearance is simulated by the insertion of an equal and

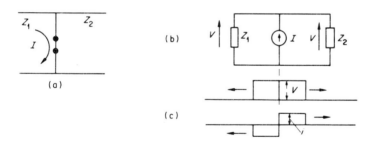

Figure 10.25 Surge set up by fault clearance. (a) Equal and opposite current (I) injected in fault path. (b) Equivalent circuit. (c) Voltage and current waves set up at a point of fault with direction of travel.

opposite current (I) at the point of the fault. From the equivalent circuit, the magnitude of the resulting voltage surges (v)

$$= I\left(\frac{Z_1 Z_2}{Z_1 + Z_2}\right)$$

and the currents entering the lines are

$$I\left(\frac{Z_1}{Z_1 + Z_2}\right) \quad \text{and} \quad I\left(\frac{Z_2}{Z_1 + Z_2}\right)$$

The directions are as shown in Figure 10.25(c).

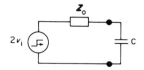

Figure 10.26   Termination of line (surge impedance $\mathbf{Z}_0$) in a capacitor
(C).

Termination in inductance and capacitance

(a) *Shunt capacitance*   Using the Thevenin equivalent circuit as shown in Figure 10.26 the voltage rise across the capacitor C, $v_c = 2v_i(1 - e^{-t/Z_0 C})$, where t is the time commencing with the arrival of the wave at C. The current through C is given by,

$$\left(\frac{2v_i}{Z_0}\right) e^{-t/Z_0 C}$$

The reflected wave,

$$v_r = v_c - v_i = 2v_i(1 - e^{-t/Z_0 C}) - v_i = v_i(1 - 2\,e^{-t/Z_0 C}) \qquad (10.16)$$

As to be expected, the capacitor acts initially as a short circuit and finally as an open circuit.

(b) *Shunt inductance*   Again from the equivalent circuit the voltage across the inductance,

$$v_L = 2v_i\,e^{-(Z_0/L)t}$$

and

$$v_r = v_L - v_i = v_i[2\,e^{-(Z_0/L)t} - 1] \qquad (10.17)$$

Here the inductance acts initially as an open circuit and finally as a short circuit.

(c) *Capacitance and resistance in parallel* (Figure 10.27(a))   Open circuit voltage across AB (Figure 10.27(b)),

$$= \left(\frac{2v_i}{Z_0 + R}\right) R$$

Figure 10.27   Two lines surge impedances $\mathbf{Z}_1$ and $\mathbf{Z}_2$ grounded at their junction through a capacitor C. (a) and (b) Equivalent circuits. (c) System diagram.

Equivalent Thevenin resistance

$$= \frac{RZ_0}{R + Z_0}$$

Voltage across R and C,

$$v = \frac{2v_i}{Z_0 + R}(1 - e^{-[t(R+Z_0)]/RZ_0C} \qquad (10.18)$$

This is the solution to the practical system shown in Figure 10.27(c), where C is used to modify the surge. The reflected wave is given by $(v - v_i)$.

Example 10.2 An overhead line of surge impedance 500 Ω is connected to a cable of surge impedance 50 Ω through a series resistor (Figure 10.28(a)).

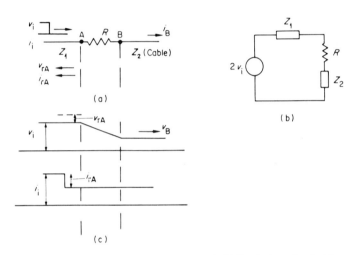

Figure 10.28   (a) System for Example 10.2. (b) Equivalent circuit. (c) Voltage and current surges.

Determine the magnitude of the resistor such that it absorbs maximum energy from a surge originating on the overhead line and travelling into the cable. Calculate:

(a) the voltage and current transients reflected back into the line, and
(b) those transmitted into the cable, in terms of the incident surge voltage; and
(c) the energies reflected back into the line and absorbed by the resistor.

Let the incident voltage and current be v_i and i_i. From the equivalent circuit (Figure 10.28(b)),

$$v_B = \frac{2v_i \cdot Z_2}{Z_1 + Z_2 + R}$$

and the reflected voltage at A

$$v_{rA} = \frac{2v_i(Z_2+R)}{Z_1+Z_2+R} - v_i = \frac{Z_2+R-Z_1}{Z_1+Z_2+R}$$

As

$$v_i = Z_1 i_i$$

$$i_B = \frac{2v_i}{Z_1+Z_2+R} = \frac{2Z_1 i_i}{Z_1+Z_2+R}$$

and

$$i_{rA} = i_i \frac{(Z_1-Z_2-R)}{Z_1+Z_2+R}$$

Power absorbed by the resistance

$$= R\left(i_i \frac{2Z_1}{Z_1+Z_2+R}\right)^2 \text{ watt}$$

This power is a maximum when

$$\frac{\mathrm{d}}{\mathrm{d}R}\left[\frac{R}{(Z_1+Z_2+R)^2}\right] = 0$$

if i_i is given and constant; from which $R = (Z_1+Z_2)$. With this resistance the maximum energy is absorbed from the surge.

Hence R should be $500+50 = 550 \, \Omega$. With this value of R,

$$i_B = \left(\frac{2\times500}{500+50+550}\right) \cdot i_i = 0.91 i_i$$

$$i_{rA} = -0.091 i_i$$

$$v_B = 0.091 v_i \quad \text{and} \quad v_{RA} = 0.091 v_i$$

Also the surge energy entering Z_2

$$= v_B i_B = 0.082 v_i i_i$$

The energy absorbed by $R = (0.91 i_i)^2 \cdot 550$

$$= 455\left(\frac{v_i}{Z_1}\right) i_i = 0.91 v_i i_i$$

and the energy reflected $= v_i i_i(1-0.082-0.91)$

$$= 0.008 v_i i_i$$

The waveforms are shown in Figure 10.28(c).

10.6 Determination of System Voltages Produced by Travelling Surges

In the previous section the basic laws of surge behaviour were discussed. The calculation of the voltages set up at any node or busbar in a system at a given instant in time is, however, much more complex than the previous section would suggest. When any surge reaches a discontinuity its reflected waves travel back and are in turn reflected so that each generation of waves sets up further waves which coexist with them in the system.

To describe completely the events at any node involves, therefore, an involved book-keeping exercise. Although many mathematical techniques are available and in fact used, the graphical method due to Bewley[6] indicates clearly the physical changes occurring in time, and this method will be explained in some detail.

Bewley lattice diagram

This is a graphical method of determining the voltages at any point in a transmission system and is an effective way of illustrating the multiple reflexions which take place. Two axes are established, a horizontal one scaled in distance along the system, and a vertical one scaled in time. Lines indicating the passage of surges are drawn such that their slopes give the times corresponding to distances travelled. At each point of change in impedance the reflected and transmitted waves are obtained by multiplying the incidence wave magnitude by the appropriate reflexion and refraction coefficients α and β.

The method is best illustrated by an example, and a loss-free system comprising a long overhead line (Z_1) in series with a cable (Z_2) will be considered. Typically, Z_1 is 500 Ω and Z_2 is 50 Ω and referring to Figure 10.29 the following coefficients apply:

Line to cable reflexion coefficient,

$$\alpha_1 = \frac{50-500}{50+500} = -0.818$$

Line to cable refraction coefficient,

$$\beta_1 = \frac{2 \times 50}{50+500} = 0.182$$

Cable to line,

$$\alpha_2 = \frac{500-50}{500+50} = 0.818$$

Cable to line,

$$\beta_2 = \frac{2 \times 500}{500+50} = 1.818$$

As the line is long, reflexions at its remote end will be neglected. The remote end of the cable is considered to be open-circuited, giving an α of 1 and a β of zero.

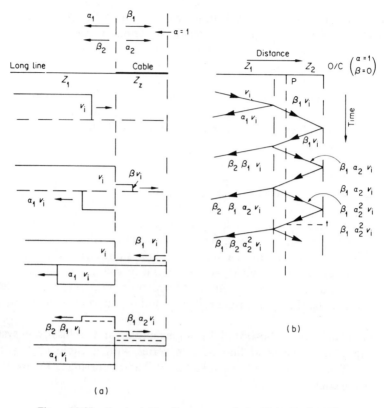

Figure 10.29 Bewley lattice diagram—analysis of long overhead line and cable in series. (a) Position of voltage surges at various instants over first complete cycle of events, i.e. up to second reflected wave travelling back along line. (b) Lattice diagram.

When the incident wave v_i (see Figure 10.29) originating in the line reaches the junction a reflected component travels back along the line $(\alpha_1 v_1)$, the refracted or transmitted wave $(\beta_1 v_1)$ traverses the cable and is reflected from the open-circuited end back to the junction $(1 \times \beta_1 v_1)$. This wave then produces a reflected wave back through the cable $(1 \times \beta_1 \alpha_2 v_1)$ and a transmitted wave $(1 \times \beta_2 \beta_1 v_1)$ through the line. The process continues and the waves multiply as indicated in Figure 10.29(b). The total voltage at a point P in the cable at a given time (t) will be the sum of the voltages at P up to time t; i.e. $v_i \beta_1 (2 + 2\alpha_2 + 2\alpha_2^2)$ and the voltage at infinite time will be, $2v_i \beta_1 (1 + \alpha_2 + \alpha_2^2 + \alpha_2^3 + \alpha_2^4 + \dots)$.

The voltages at other points are similarly obtained. The time scale may be determined from a knowledge of length and surge velocity, for the line the latter is of the order of 300 m per μs and for the cable 150 m/μs. For a surge 50 μs in duration and a cable 300 m in length there will be 25 cable lengths traversed and the terminal voltage will approach $2v_i$. If the graph of voltage at the cable open-circuited end is plotted against time an exponential rise curve will be obtained similar to that obtained for a capacitor.

The above treatment applies to a rectangular surge waveform, but may be modified readily to account for a waveform of the type illustrated in Figure 10.1(b) or (c). In this case the voltage change with time must also be allowed for and the process is more complicated.

Effects of line loss

Attenuation of travelling waves is caused mainly by corona which reduces the steepness of the wavefronts considerably as the waves travel along the line. Attenuation is also caused by series resistance and leakage resistance and these quantities are considerably larger than the power-frequency values. The determination of attenuation is usually empirical and use is made of the expression, $v_x = v_i e^{-\gamma z}$, where v_x is the magnitude of the surge at a distance x from the point of origination. If a value for γ is assumed then the wave magnitude of the voltage may be modified to include attenuation for various positions in the lattice diagram. For example, in Figure 10.29(b), if $e^{-\gamma x}$ is equal to a_L for the length of line traversed and a_c for the cable, then the magnitude of the first reflexion from the open circuit is $a_L v_i a_c \beta_1$ and the voltages at subsequent times will be similarly modified.

Considering the power and losses over a length dx of a line of resistance and shunt conductance per unit length R ohms and G ohms^{-1}, the power loss

$$dp = i^2 R\ dx + v^2 G\ dx \quad \text{(watts)}$$

also,

$$p = vi = i^2 Z_0 \quad \text{and} \quad dp = 2iZ_0\ di.$$

As dp is a loss it is considered negative and

$$-2iZ_0\ di = (i^2 R + v^2 G)\ dx$$

hence,

$$\frac{di}{i} = -\tfrac{1}{2}\left(\frac{R + Z_0^2 G}{Z_0}\right) dx$$

giving

$$i = i_i \exp\left[-\frac{1}{2}\left(\frac{R}{Z_0} + GZ_0\right)x\right] \tag{10.19}$$

where

$$i_i = \text{surge amplitude (amperes)}$$

Also it may be shown that

$$v = v_i \exp\left[-\frac{1}{2}\left(\frac{R}{Z_0} + GZ_0\right)x\right] \tag{10.20}$$

and the power at x

$$v_i i_i = vi = v_i i_i \, e^{-[(R/Z_0)+GZ_0]x} \qquad (10.21)$$

If R and G are realistically assessed (including corona effect) attenuation may be included in the travelling wave analysis.

Digital methods

The lattice diagram becomes very cumbersome for large systems and normally either analogue or digital methods are applied. Digital methods may use strictly mathematical methods, i.e. the solution of the differential equations or the use of Fourier or Laplace transforms. These methods are capable of high accuracy, but require large amounts of data and long computation times. The general principles of the graphical approach described above may also be used to develop a computer program which is very applicable to large systems. Such a method will now be discussed in more detail and an example considered.

A rectangular wave is used, of infinite duration. The theory developed in the previous sections is applicable. The major role of the program is to scan the nodes of the system at each time interval and compute the voltages. In Figure 10.30 a particular system (single-phase representation) is shown and will be used to illustrate the method.

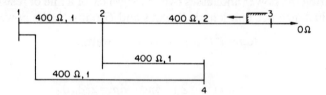

Figure 10.30 Application of digital method—network analysed. Each line labelled with surge impedance and surge travel time (multiples of basic unit), e.g. 400 Ω, 1.

The relevant data describing the system are given in tabular form (see Table 10.5). Branches are listed both ways in ascending order of the first nodal number and are referred to under the name BRANCH. The time taken for a wave to travel along a branch is recorded in terms of a positive integer (referred to as PERIOD) which converts the basic time unit into actual travel time. Reflexion coefficients are stored and referred to as REFLECT and the corresponding refraction coefficients obtained, i.e. $(1 + \alpha_{ij})$. Elements of time as multiples of the basic integer are also shown in Table 10.5.

The method is illustrated by examining the system after the arrival of the rectangular wave at node 3 at TIME (0). This voltage (magnitude 1 p.u.) is entered in the BRANCH (3, 2), TIME (0) element of the BRANCH–TIME matrix. On arrival at node 2 at time equal to zero plus PERIOD (3, 2) two waves are generated, on BRANCH (2, 1) and BRANCH (2, 4) both of magnitude, $1(1 + \alpha_{32})$, i.e. 2/3. A reflected wave is also generated on BRANCH (2, 3) of magnitude

Table 10.5

BRANCH i j	1 2	1 4	2 1	2 3	2 4	3 2	4 1	4 2
PERIOD	1	1	1	2	1	2	1	1
α_{ij}	$-\frac{1}{3}$	0	0	-1	0	$-\frac{1}{3}$	0	$-\frac{1}{3}$
$1+\alpha_{ij}$	$\frac{2}{3}$	1	1	0	1	$\frac{2}{3}$	1	$\frac{2}{3}$
TIME 0						1		
1								
2			$\frac{2}{3}$	$-\frac{1}{3}$	$\frac{2}{3}$			
3	0	$\frac{2}{3}$						
4								

$1 \times \alpha_{32}$, i.e. $-1/3$. These voltages are entered in the appropriate BRANCH in the TIME (2) row of Table 10.5. On reaching node 1, TIME (3), a refracted wave of magnitude $\frac{2}{3}(1+\alpha_{21})$, i.e. 2/3, is generated on BRANCH (1, 4) and a reflected wave $\frac{2}{3} \times \alpha_{21}$, i.e. 0 is generated on BRANCH (1, 2). This process is continued until a specified time is reached. All transmitted waves for a given node are placed in

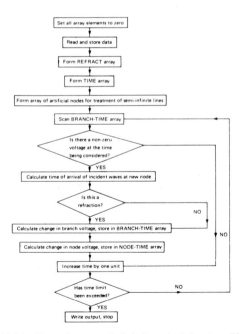

Figure 10.31 Flow diagram of digital method for travelling-wave analysis.

a separate node–time array, a transmitted wave is considered only once even though it could be entered into several BRANCH–TIME elements. Current waves are obtained by dividing the voltage by the surge impedance of the particular branch. The flow diagram for the digital solution is shown in Figure 10.31.

It is necessary for the programs to cater for semi-infinite lines (i.e. lines so long that waves reflected from the remote end may be neglected) and also inductive/capacitive terminations. *Semi-infinite* lines require the use of artificial nodes labelled (say) 0. For example, in Figure 10.32 node 2 is

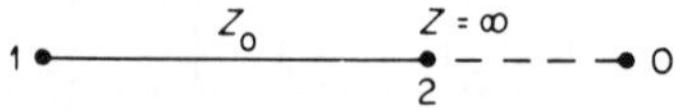

Figure 10.32   Treatment of terminations—use of artificial node 0 and line of infinite impedance to represent open circuit at node 2.

open-circuited and this is accounted for by introducing a line of infinite surge impedance between nodes 2 and 0, and hence, if

$$Z \to \infty \qquad \alpha_{12} = \frac{Z - Z_0}{Z + Z_0} \to 1$$

Although the scanning will not find a refraction through node 2 it is necessary for the computer to consider there being one to calculate the voltage at node 2. Lines with short-circuited nodes may be treated in a similar fashion with an artificial line of zero impedance.

Inductive/capacitive terminations may be simulated by *stub lines*. An inductance (L henrys) is represented by a stub transmission line short-circuited at the far end and of surge impedance, $Z_L = L/t$, where t is the travel time of the stub. Similarly, a capacitance C (farads) is represented by a line open-circuited and of surge impedance, $Z_c = t/C$. For the representation to be exact t must be small and it is found necessary for Z_L to be of the order of 10 times and Z_c to be one-tenth of the combined surge impedance of the other lines connected to the node.

For example, consider the termination shown in   Figure 10.33(a). The equivalent stub line circuit is shown in Figure 10.33(b). Stub travel times are chosen to be short compared with a quarter cycle of the natural frequency, i.e.

$$\frac{2\pi\sqrt{LC}}{4} = \frac{2\pi}{4}\sqrt{0.01 \times 4 \times 10^{-8}} = 0.315 \times 10^{-4}\,\text{s}$$

Let $t = 5 \times 10^{-6}$ s for both L and C stubs (corresponding to the total stub length of 1524 m). Hence,

$$Z_c = \frac{t}{C} = \frac{2.5 \times 10^{-6}}{4 \times 10^{-8}} = 67.5\,\Omega$$

and

$$Z_L = L/t = \frac{0.01}{2.5 \times 10^{-6}} = 4000\,\Omega$$

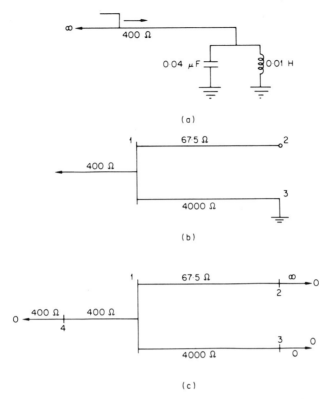

Figure 10.33 Representation of line terminated by L–C circuit by means of stub lines. (a) Original system. (b) Equivalent stub lines. (c) Use of artificial nodes to represent open and short-circuited ends of stub lines.

The configuration in a form acceptable for the computer program is shown in Figure 10.33(c). Refinements to the program to incorporate attenuation, waveshapes, and non-linear resistors may be made without changing its basic form.

A typical application is the analysis of the nodal voltages for the system shown in Figure 10.30. This system has been previously analysed by a similar method by Barthold and Carter[17] and good agreement found. The printout of nodal voltages for the first 20 μs is shown in Table 10.6, and in Figure 10.34 the voltage plot for nodes 1 and 4 is compared with a transient analyser solution obtained by Barthold and Carter.[17]

Program for determination of voltage magnitudes as busbars (in Fortran).

Input data: IPROB = number of time steps,

   I No = number of time units for each step or increment, in the example 1.

   NB = number of time intervals to be considered.

Table 10.6   Digital computer printout. Node voltages (system of Figure 10.30)

Time (μs)	Node 1	Node 2	Node 3	Node 4
0	0.0000E–01	0.0000E–01	1.0000E 00	0.0000E–01
1	0.0000E–01	0.0000E–01	1.0000E 00	0.0000E–01
2	0.0000E–01	6.6670E–01	1.0000E 00	0.0000E–01
3	6.6670E–01	6.6670E–01	1.0000E 00	6.6670E–01
4	1.3334E 00	6.6670E–01	1.0000E 00	1.3334E 00
5	1.3334E 00	1.5557E 00	1.0000E 00	1.3334E 00
6	1.5557E 00	1.7779E 00	1.0000E 00	1.5557E 00
7	2.0002E 00	1.7779E 00	1.0000E 00	2.0002E 00
8	2.2224E 00	2.0743E 00	1.0000E 00	2.2224E 00
9	2.2965E 00	1.7779E 00	1.0000E 00	2.2965E 00
10	1.8520E 00	1.8520E 00	1.0000E 00	1.8520E 00
11	1.4075E 00	1.9508E 00	1.0000E 00	1.4075E 00
12	1.5062E 00	1.0617E 00	1.0000E 00	1.5062E 00
13	1.1604E 00	7.6534E–01	1.0000E 00	1.1604E 00
14	4.1955E–01	8.2297E–01	1.0000E 00	4.1955E–01
15	8.2086E–02	2.6307E–01	1.0000E 00	8.2086E–02
16	–7.4396E–02	1.6435E–01	1.0000E 00	–7.4396E–02
17	7.8720E–03	1.0732E–02	1.0000E 00	7.8720E–03
18	9.3000E–02	–2.5556E–01	1.0000E 00	9.3000E–02
19	–1.7043E–01	4.1410E–01	1.0000E 00	–1.7043E–01
20	1.5067E–01	6.2634E–01	1.0000E 00	1.5067E–01

NN = number of nodes.

N REFR = artificial nodes used – when none equal to zero.

Read (1,101) BRANCH (1,K), BRANCH (2,K), PERIOD (K), REFLECT (K).

BRANCH (1,K) refers to the starting-end node (i.e. first row of Table 10.5), BRANCH (2,K) refers to the other-end node of the corresponding branches (i.e. row two of Table 10.5).

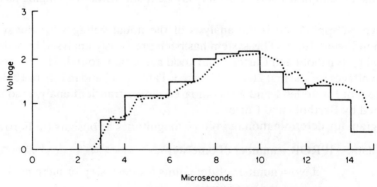

Figure 10.34   Voltage–time relationship at nodes 1 and 4 of system in Figure 10.30. ———— = computer solution: ..... = transient analyser.
(*Permission of the Institute of Electrical and Electronic Engineers.*)

Each card will contain the start and end nodes of a particular branch, time period for the wave to transverse that line, and the reflection coefficient scan by the wave travelling from the start node to the branch end.

For example, from Table 10.5, the K = 1 card will read,

$$1 \quad 2 \quad 1 \quad -0.33$$

and the K = 8 card, 4   2   1   0.0

IN NODE = node on which surge originates (in this case 3).

VOLT IN = magnitude of surge (1 p.u.).

TIME IN = time of arrival of surge (in example at time zero).

The last three quantities are applied singly, each node requiring a separate card. In the application discussed the basic time unit is 1 (i.e. 1 μs).

Other terms used are as follows:

REFRACT (J)—contains refraction coefficients.

TIME (Q)—contains consecutive positive integers representing multiples of basic time units.

VOLT A (Q, J)—contains incident voltage wave on the respective branch at time $(Q-1)$ units.

VOLT A (Q1, J)—contains the reflected and refracted voltages on a branch at time $(Q1-1)$ time units, refers to BRANCH–TIME array.

VOLT N (Q1, J)—contains the node voltages for time $(Q1-1)$, refers to NODE–TIME array.

NELSE (I)— contains the nodes with a semi-infinite line connected to them.

```
 1              PROGRAM VERYGEN
 2              INTEGERQ,P,X,Z,BRANCH,PERIOD,TIME,NB,NT,NV,T,NODE,Q1,C,TIMEIN
 3              DIMENSION BRANCH(2,50),PERIOD(50),REFLECT(50),REFRACT(50), VOLTA(1
 4              100,21),TIME(100),JTIME(100),VOLTN(100,21),NELSE(10)
 5      C              READ IN DATA
 6              READ(1,100) IPROB
 7      100     FORMAT(I5)
 8      2       READ(1,1) INO
 9      1       FORMAT(I5)
10              IF(INO.GT.IPROB) GO TO 8
11              READ(1,4)NB,NT,NN,NREFR
12      4       FORMAT(4I5)
13              DO 10 K=1,NB
14              READ(1,101)BRANCH(1,K),BRANCH(2,K),PERIOD(K),REFLECT(K)
15      101     FORMAT(3I5,F7.3)
16      C              FORM REFRACTION COEFFICIENT ARRAY
17              REFRACT(K)=REFLECT(K)+1.0
18              DO 60 L=1,NT
19      C              CLEAR VOLTAGES ARRAYS
20              VOLTA(L,K)=0.0
21              VOLTN(L,K)=0.0
22      C              FORM TIME ARRAY
23              TIME(L)=L-1
24      60      CONTINUE
25      10      CONTINUE
26              READ(1,5) INNODE,VOLTIN,TIMEIN
27      5       FORMAT(I5,F7.3,I5)
28              MT=TIMEIN+1
29              DO 6 L=MT,NT
30      C              SET INPUT VOLTAGES IN ARRAYS
```

398

```
31              VOLTN(L,INNODE)=VOLTIN
32              DO 7 Z=1,NB
33              IF(BRANCH(1,Z).NE.INNODE) GO TO 7
34              VOLTA(L,Z)=VOLTIN
35      7       CONTINUE
36      6       CONTINUE
37      C               FORM TERMINATION ARRAY
38              IF(NREFR.EQ.0)GO TO 79
39              DO 31 L=1,NB
40              IF(BRANCH(2,L).NE.0)GO TO 31
41              DO 32 I=1,NREFR
42              NELSE(I)=BRANCH(1,L)
43      32      CONTINUE
44      31      CONTINUE
45      79      DO 50 Q=1,NT
46              DO 40 L=1,NB
47      C               FIND NON-ZERO VOLTAGE
48              IF(VOLTA(Q,L).EQ.0)GO TO40
49              IC=1
50              DO 20 J=1,NB
51              IF(BRANCH(1,J).NE.BRANCH(2,L))GO TO 20
52              Q1=Q+PERIOD(L)
53      C               REFLECTION OR REFRACTION?
54              IF(BRANCH(2,J).EQ.BRANCH(1,L)) GO TO 30
55      C               CALCULATE REFRACTED & NODE VOLTAGE
56              VOLTA(Q1,J)=VOLTA(Q,L)*REFRACT(L)+VOLTA(Q1,J)
57              DO 810 I=1,NN
58              IF(IC.NE.1) GO TO 810
59              IF(I.NE.BRANCH(2,L)) GO TO 810
60              VOLTN(Q1,I)=VOLTA(Q,L)*REFRACT(L)+VOLTN(Q1,I)
61              IC=IC+1
62      810     CONTINUE
63              GO TO 20
64      C               CALCULATE REFLECTED & NODE VOLTAGES
65      30      VOLTA(Q1,J)=VOLTA(Q,L)*REFLECT(L)+VOLTA(Q1,J)
66              IF(NREFR.EQ.0) GO TO 20
67              IF(IC.NE.1)GO TO 20
68              DO 33 IJ=1,NREFR
69              I=NELSE(IJ)
70              IF(BRANCH(1,J).NE.NELSE(IJ)) GO TO 33
71              VOLTN(Q1,I)=VOLTA(Q,L)*REFRACT(L)+VOLTN(Q1,I)
72      33      CONTINUE
73      20      CONTINUE
74      40      CONTINUE
75      50      CONTINUE
76              WRITE(2,3)INO
77      3       FORMAT(1H1,12H PROBLEM NO.,I2)
78              WRITE(2,209)
79      209     FORMAT(15X,13HNODE VOLTAGES//)
80              WRITE(2,210)
81      210     FORMAT(5H TIME,10X,1H1,14X,1H2,14X,1H3,14X,1H4//)
82              DO 800 Q=1,NT
83              IF(INO.EQ.2) NN=4
84              WRITE(2,211)TIME(Q),(VOLTN(Q,I),I=1,NN)
85      211     FORMAT(1H0,1X,I2,6X,1P5E14.4)
86      800     CONTINUE
87              IF(INO.LE.IPROB) GO TO 2
88      8       STOP
89              END
```

Transient analysers

The main alternative to digital solutions is the equivalent-circuit modelling of systems. These circuits are physically realized and pulses applied by means of pulse generators and voltages measured by cathode ray oscilloscopes. The modelling of large systems is expensive and time-consuming. The exact representation of plant such as generators and transformers is complex because of the distributed capacitance presented by the windings. A typical equivalent circuit representation of a rotating machine is shown in Figure 10.35. The

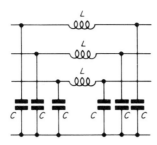

Figure 10.35 Equivalent circuit representation of rotating machines. $C = 1/3$ of total capacitance to ground. $L = 0.65$ of subtransient inductance.

difficulty in representing items of plant other than overhead lines holds equally for any digital method.

Three-phase analysis

The single-phase analysis of a system as presented in this chapter neglects the mutual effects which exist between the three phases of a line, transformer, etc. The transient voltages due to energization may be further increased by this mutual coupling and also by the three contacts of a circuit breaker not closing at the same instant. The difference resulting between the use of three-phase and single-phase representation has been discussed by Bickford and Doepel.[20]

10.7 Electromagnetic Transient Program (EMTP)

The digital method previously described is very limited in scope. A much more powerful method has been developed by the Bonneville Power Administration known as EMTP. This is widely used especially in the USA.

It is assumed that the variables of interest are known at the previous time step $t - \Delta t$ where Δt is the time step. Δt must be small enough to give reasonable accuracy with a finite difference method.

Lumped element modelling

Consider an inductance L, Figure 10.36(a) and the voltage–time curve in 10.36(b).

$$v = L \, di/dt$$

and

$$i = \int \frac{v}{L} \, dt$$

$$\tfrac{1}{2} \frac{\Delta t}{L} \left[v(t) - v(t - \Delta t) \right]$$

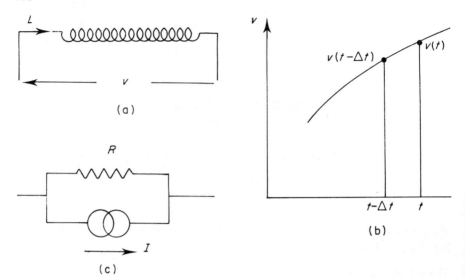

Figure 10.36

Hence the current at time interval (t)

$$i(t) = i(t - \Delta t) + \frac{\Delta t[v(t) - v(t - \Delta t)]}{2L}$$

or

$$i(t) = v(t)\frac{\Delta t}{2L} + i(t - \Delta t) + v(t - \Delta t)\frac{\Delta t}{2L} = \frac{v(t)}{R} + I \qquad (10.22)$$

where

$$R = \frac{2L}{\Delta t} \quad \text{and} \quad I = i(t - \Delta t) + v(t - \Delta t)/R.$$

Here R is constant and I varies with time. The equivalent circuit is shown in Figure 10.36(c).

A similar treatment applies to capacitance (C) see Figure 10.37(a). Here

$$v = \int \frac{i}{C} dt$$

and over the interval Δt,

$$i(t) = i(t - \Delta t) + \frac{2C}{\Delta t}[v(t) - v(t - \Delta t)]$$

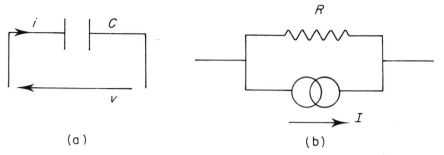

(a)   (b)

Figure 10.37 Equivalent circuits for transient analysis.

Again, if

$$R = \Delta t/2C$$

$$I = -i(t - \Delta t) - \frac{v(t - \Delta t)}{R} \qquad (10.23)$$

giving the equivalent circuit of Figure 10.37(b).

Resistance is represented directly (Figure 10.38(a)).

An application is shown in Figure 10.39.

The procedure is as follows:

(a) From initial conditions determine $i(t - \Delta t) = i(0)$
and $v(t - \Delta t) = v(0)$

(b) Solve for $i(t)$ and $v(t)$. Increase time step by Δt and calculate new values of i and v and so on. The analysis is done by use of the nodal admittance matrix and Gaussian elimination. If mutual coupling between elements exists then the representation becomes very complex.

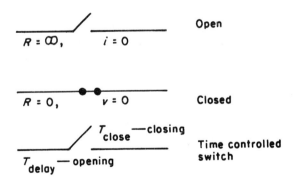

Figure 10.38 Representation of switch.

Switching

The various representations are shown in Figure 10.38. A switching operation changes the topology of the network and hence the $[Y]$ matrix. If $[Y]$ is formed

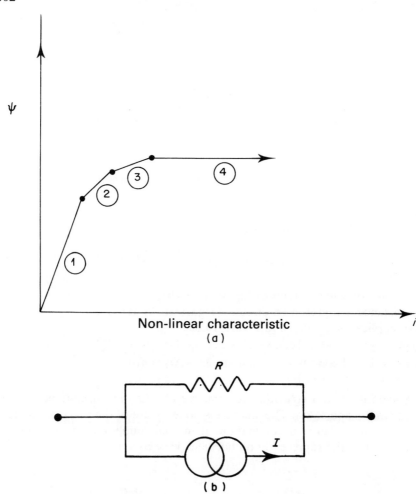

Figure 10.39 Representing non-linear characteristics.

with all the switches open then the closure of a switch is obtained by the adding of the two rows and columns of $[Y]$ together and the associated rows of $[i]$.

Another area where switching is used is to account for non-linear $\psi - i$ characteristics of transformers, reactors, etc. The representation is shown in Figure 10.39(a) and (b) in which

$$R = 2b_k/\Delta t$$

and (10.24)

$$I = v(t - \Delta t)/R + i(t - \Delta t)$$

where b_k = incremental inductance.

If ψ is outside of the limits of sequent K the operation is switched to either $k-1$ or $k+1$. This changes the $[Y]$ matrix.

Because of the random nature of certain events, e.g. switching time or lightning incidence, Monte Carlo (Statistical) methods are sometimes used. A useful account of this approach is given in Reference 7.

Travelling wave approach

Lines and cables would require a large number of Pi circuits for accurate representation. An alternative would be the use of travelling wave theory.

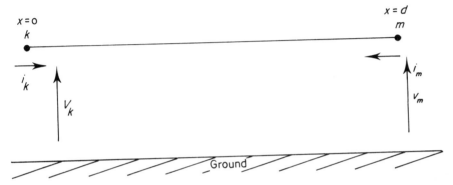

Figure 10.40   Transmission line with earth return.

Consider Figure 10.40

$$i(x, t) = f_1(x - st) + f_2(x + st)$$
$$v(x, t) = z_0 f_1(x - st) + z_0 f_2(x + st)$$

where s = speed of propagation and z_0 = characteristic impedance. At node k,

$$i_k(t) = \frac{v_k(t)}{z_0} + I_k \qquad (10.25)$$

where

$$I_k = -i_m(t - \tau) - v_m(t - \tau)/z_0$$

and

$$\tau = \frac{d}{s} = \text{travel time}$$

The equivalent circuit is shown in Figure 10.41. I_k depends on the current and voltage at the other end of the line τ seconds previously. An advantage of this method is that the two ends of the line are decoupled. I_k depends on

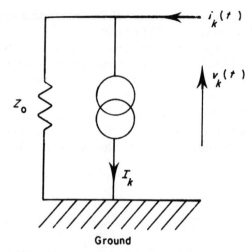

Figure 10.41   Equivalent circuit for single-phase line.

the situation some time previously, e.g. if $\tau = 0.36$ ms and $\Delta t = 100$ μs storage of four previous times is required.

Detailed models for synchronous machines and hvdc converter systems are given in Reference **6**.

Example   The equivalent circuit of a network is shown in Figure 10.42. Determine the network which simulates this network for transients using the EMTP method after the first time step of the transient of 5 μs.

Solution

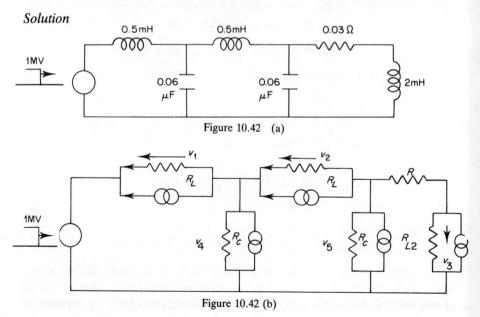

Figure 10.42   (a)

Figure 10.42 (b)

$$R_L = \frac{2 \times 0.0005}{5 \times 10^{-6}} = 200\,\Omega$$

$$R_c = \frac{5 \times 10^{-6}}{2 \times 0.06 \times 10^{-6}} = 41.7\,\Omega$$

$$R_{L2} = \frac{2 \times 0.002}{5 \times 10^{-6}} = 800\,\Omega$$

At $t = 0$,
note current
sources are zero

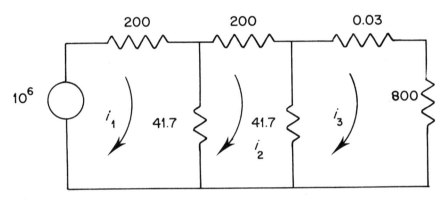

Figure 10.42 (c)

$$\begin{bmatrix} 10^6 \\ 0 \\ 0 \end{bmatrix} = \begin{bmatrix} 241.7 & -41.7 & 0 \\ -41.7 & 241.7 & -41.7 \\ 0 & -41.7 & 800 \end{bmatrix} \begin{bmatrix} i_1 \\ i_2 \\ i_3 \end{bmatrix}$$

Invert

$$\begin{bmatrix} i_1 \\ i_2 \\ i_3 \end{bmatrix} = \begin{bmatrix} 0.0043 & 0.0043 & 0.0007 \\ 3.8 \times 10^{-5} & 0.0007 & 0.0043 \\ 0.0002 & 3.8 \times 10^{-5} & 0.0002 \end{bmatrix} \begin{bmatrix} 10^6 \\ 0 \\ 0 \end{bmatrix}$$

$$i_1 = 4300\ \text{A}$$

$$i_2 = -38,\ i_3 = -200$$

Equivalent circuit after 5 μs is shown in Figure 10.42(d)

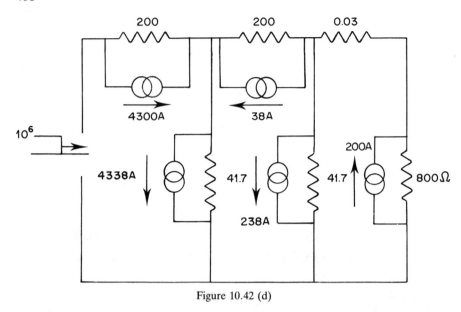

Figure 10.42 (d)

The process is then repeated for the next 5 μs step using Figure 10.42(d) as the starting condition.

10.8 Ultra-high-voltage Transmission

In this section arrangements for extra high voltage (e.h.v.) lines will be reviewed and factors influencing designs for u.h.v. discussed. At the present time for e.h.v. lines 'V' type insulator strings are used to support bundle conductors in horizontal formation on steel lattice towers with two earth (ground) wires; a typical structure is shown in Figure 10.43. In this figure dimensions of interest are labelled and some critical distances are shown. Span lengths (i.e. distances between towers) are of the order of 400–500 m at present and are expected to continue in this range. The number of subconductors in each bundle is determined by radio-interference and audible noise levels stipulated along the route.

Tower dimensions are determined by insulation requirements and the importance of the switching surge in this connexion has already been discussed. The tower surge–withstand voltage is set below the critical flashover (i.e. 50 per cent probability) voltage to give a reasonable value of flashover probability. When designing a tower the insulation string length, insulator surface creep distance, tower grounding, and strike distances must be decided upon. Tower strike distance is the distance from the conductor to the tower structure. These dimensions are chosen to make the insulation strength such that the applied surge results in an acceptable surge flashover rate.

An example of laboratory assessment of the strength of tower insulation (V-string insulators) is shown in Figure 10.44. It has been shown (Figure 10.18)

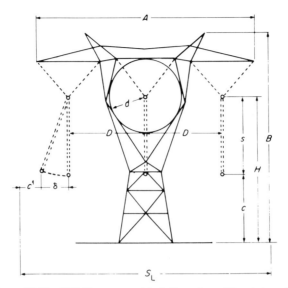

Figure 10.43   U.H.V. tower—critical dimensions. (*Permission of the Institute of Electrical and Electronics Engineers.*)

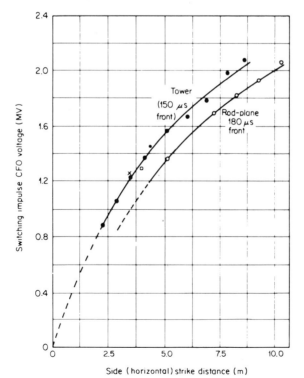

Figure 10.44   Maximum obtainable switching impulse CFO voltage for a specific side strike distance. (*Permission of the Westinghouse Electric Corporation.*)

that for a given insulation the flashover voltage follows a Gaussian cumulative distribution curve to at least four standard deviations (σ_f) below the CFO; σ_f is about 4.6 per cent to 5 per cent of the CFO. The procedure is to equate the switching surge voltage to the withstand strength of the insulation where withstand is defined as that voltage which results in a 0.13 per cent flashover probability, i.e. $3\sigma_f$ below the CFO. The CFO is determined from a knowledge of σ_f and the withstand voltage and from it the strike distances obtained. This process is illustrated in Figure 10.45 for CFOs under wet and dry conditions.

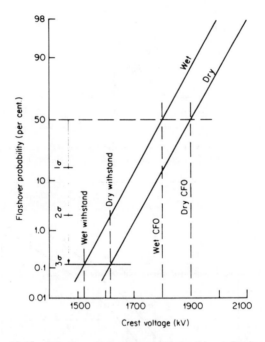

Figure 10.45   Selection of tower surge withstand level. Design withstand level for a tower is established by specifying some acceptable risk of flashover, in this case three standard deviations below the 50 per cent probability CFO voltage. (*Permission of the Westinghouse Electric Corporation.*)

Contamination requirements are expressed in terms of the shortest distance along the insulator disc from cap to pin (creep distance); the creep distance required per kilovolt for flashover-free operation with normal practice at present is between 0.83 in (2.1 cm) and 1 in (2.54 cm) per kilovolt. The various factors involving flashover and insulation requirements are summarized in Figure 10.46, in which system operating voltage is related to the minimum distance for flashover between the phase conductor and tower sides (strike distance) for a 'V' formation of string insulators. The switching–surge curves refer to various per-unit peak values of switching surge, control over which may be exercised by circuit-breaker resistance switching. It is seen that increases in

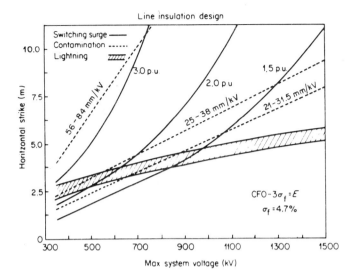

Figure 10.46   Estimates of o.h.v.–u.h.v. tower insulation requirements. (*Permission of the Westinghouse Electric Corporation.*)

system voltages require progressively larger increases in the strike distance
(conductor-to-tower minimum distance) and the tower dimensions would
become intolerable from both economic and appearance standpoints unless the
per-unit value of the surge is reduced.

The method illustrated in this section has been questioned for the u.h.v.
region. It does not produce an estimate of the switching surge flashover rate
and only matches the withstand voltage to the maximum switching surge. In
fact, two probability distributions are involved, one for surge magnitude, the
other for insulation strength. The exact form of the surge magnitude probability curves for systems is not as yet known to an acceptable degree of
accuracy for the calculation of flashover probability. This complete probability
approach is being studied and revised versions of the curves shown in Figure
10.46 obtained.

Paris[21] has suggested various criteria for the design of u.h.v. towers, e.g.
dimensions are chosen in such a way as to make the cost of the inactive
components a constant ratio (about 0.8) of the active components (i.e. the
conductors). The tower size is defined by two parameters: the 'line size' giving
the width of the 'right of way' (width of the strip of ground required underneath
the line) and the 'tower size' defined as the product of tower width, height, and
number of towers per kilometre. As the height of the conductors will not
increase proportionally with the voltage the magnitude of the electric field at
the ground surface becomes a critical factor. Paris's analysis suggests that
towers for voltages up to 1500 kV are feasible and some of his suggested towers
are shown in Figure 10.47 based on line characteristics given in Table 10.7 and
the criteria mentioned above. The guyed structures already in use lead to the

410

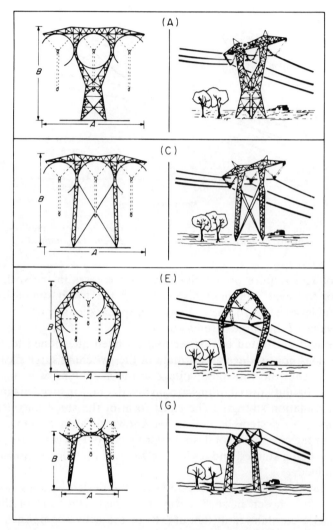

Figure 10.47   Proposed types of towers for 1500 kV lines.[21] Proposed new tower types, C to H. Traditional tower types, A and B. (*Permission of the Institute of Electrical and Electronics Engineers.*)

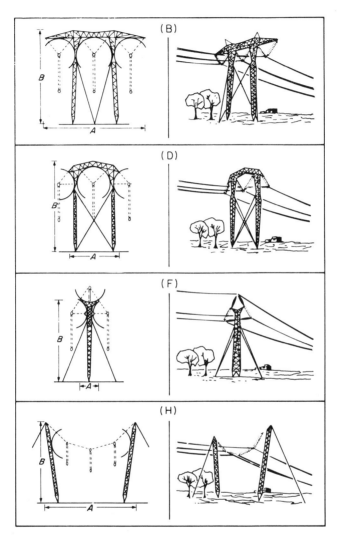

Figure 10.47 —*continued*

Table 10.7   Characteristics of lines at various system voltages

Highest system voltage (V_m)(kV):	420	525	765	1000	1300	1500
Overall aluminium section per phase, S (mm^2)	1240	1660	2680	3780	5250	6300
Number of subconductors per phase, n	2	3	4	6	8	8
Subconductor diameter, ϕ (mm)	34.5	32.4	35.8	34.7	35.5	38.8
Conductor-tower clearance, d (m)	3.00	3.90	5.60	7.20	8.50	9.40
Switching impulse 50% discharge voltage of tower insulation, $V_{50\%}$ (per unit)	3.2	2.95	2.60	2.25	1.95	1.80
Conductor-ground clearance at midspan, C (m)	7.2	8.45	10.8	13.1	15.0	16.2
Span length, L (m)	400	420	445	475	500	515
Midspan sag, s (m)	12	13.5	15	17	19	20
Conductor height at the tower, H (m)	19.2	21.7	25.8	30.1	34.0	36.2
Interphase distance, D (m)	7.30	9.20	12.8	16.1	19.0	20.8
Tower width, A (m)	20.0	25.4	35.6	45.2	53.3	58.4
Tower height, B (m)	24.6	28.2	35.5	42.25	47.9	51.5
Line-size parameter (right of way), S_L (m)	35.5	42.3	52.0	62.5	72.0	76.5
Tower-size parameter, $S_T = 1000AB/L$ (m^2/km)	1230	1700	2840	4020	5110	5840
RI limit gradient of lateral phase conductor, (kV/cm)	15.8	15.7	15.35	15.5	15.25	14.85
Voltage gradient at ground, G (kV/m)	7.35	9.50	11.4	13.1	16.55	17.55
Surge impedance, Z_s (Ω)	284	268	264	249	240	245
Surge impedance loading, P_s (MW)	560	925	1970	3615	6335	8265

Permission I.E.E.E.—reference 21.

possible use of towers to withstand vertical and transverse forces only, with just a small number of special towers to withstand tthe longitudinal forces as well, say one special tower to five light two-dimensional towers. Other possibilities for reducing size at high voltages include the use of insulated cross-arms of the form shown in Figure 10.48.

10.9   Design of Insulation by Digital Computer

The design of the complete system involves many parameters and variables and the design-analysis procedure is complex. Some factors, e.g. loss, RI, and the swinging of insulator strings and conductors, depend on the weather conditions. Hence, some form of statistical analysis dependent on the geographical position of the line is important to obtain an economical design. Although the presentation here is of an introductory nature, flow diagrams of typical programs will be presented mainly to indicate the general manner in which design studies are developing at the present time. In particular such programs endeavour to give a realistic account of weather conditions and to perform computations to cover a long time-period, i.e. 10–20 years, to obtain reasonable statistical distributions. As well as weather conditions it is also necessary statistically to account for switching-surge magnitudes and flashover occurrences. A general critical-path diagram covering the steps in the design of an e.h.v. line is shown in Figure 10.49.

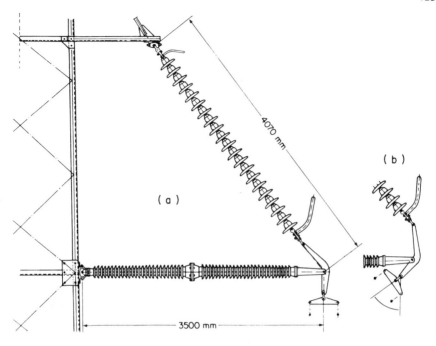

Figure 10.48   Insulating cross-arm for a double-circuit 420 kV line
(Italian). (a) Normal conditions. (b) Windy conditions.

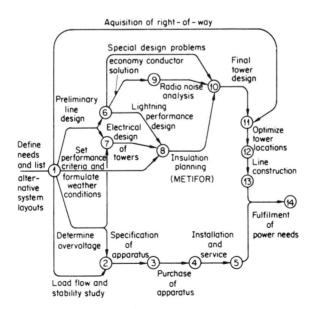

Figure 10.49   Critical path diagram for steps in the design of an e.h.v.
line. (*Permission of the Edison Electric Institute.*)

414

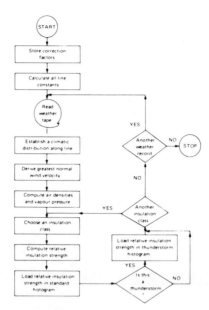

Figure 10.49 Flow chart for digital computation of relative insulation strength—METIFOR 1. (Permission of the Institute of Electrical and Electronics Engineers.)

Programs have been developed[3,9] called METIFOR (METeorologically Integrated FORecasting) and will now be briefly described. METIFOR 1 determines insulation strength and a flow diagram is shown in Figure 10.50. Having previously obtained the weather variables the following calculations are made.

(a) Using a set of velocity–force tables previously stored, the wind swings of the conductors are computed and from these all reduced air-gap dimensions obtained.

(b) Flashover strengths of the above gaps (from practical tests) are used and modified to account for precipitation, air density, humidity, and fog.

(c) The weakest of the three insulation forms, i.e. of vertical and horizontal air-gaps and insulator strings, is compared with the strength of the configuration under standard dry conditions with zero wind. The ratio of the two is called the relative insulation strength and is stored; it represents the per-unit strength of the tower configurations.

The output of METIFOR 1 is fed into another program called METIFOR 2 which evaluates the performances of alternative designs based on the following:

(a) the probability of experiencing a surge of given magnitude;
(b) the probability that an insulator string will be swung to a certain position;
(c) the probability that weather factors (precipitation, etc.) will produce a certain flashover strength of insulators and gaps;

(d) the probability that a gap or string will flashover above or below its critical strength.

The flow chart of METIFOR 2 is shown in Figure 10.51. The output is a tabulation of the flashover paths and the frequency of flashover for each path at specified insulator swing angles. A large number of surges are conceptually applied and all gaps and insulator strings examined for flashovers. The final result is the probability of flashover occurring following a circuit-breaker operation for the whole line.

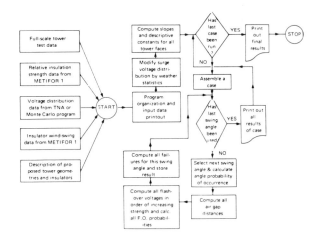

Figure 10.51   Flow chart for insulation performance program—
METIFOR 2. (*Permission of the Institute of Electrical and Electronic Engineers.*)

References

BOOKS

1. *Electrical Transmission and Distribution Reference Book*, Westinghouse Electric Corp., East Pittsburgh, Pennsylvania.
2. Rüdenburg, R., *Electric Shock Waves in Power Systems*, Harvard University Press, Cambridge, Mass., 1968.
3. *E.H.V. Transmission Line Reference Book*, Edison Electric Institute, New York, 1968.
4. *Transmission Line Reference Book, 345 kV and Above*, Electric Power Research Institute, Palo Alto, U.S.A., 1975.
5. Greenwood, A., *Electrical Transients in Power Systems*, Wiley, New York, 1971.
6. Bewley, L. V., *Travelling Waves on Transmission Systems*, Dover Books, New York, 1961.
7. *Digital Simulation of Electrical Transient Phenomena*, IEEE Course Text 81 EHO 173-5-PWR.

PAPERS

8. Paris, L., 'Influence of air-gap characteristics on line-to-ground switching surge strength', *Trans. I.E.E.E., P.A.S.*, **PAS-86** (1967), 936.

416

9. Anderson, J. G., and L. O. Barthold, 'METIFOR—a statistical method for insulation design of E.H.V. lines', *Trans. I.E.E.E., P.A.S.*, **83** (1964), 271.
10. Anderson, J. G., 'Monte Carlo calculation of transmission-line lightning performance', *A.I.E.E. Trans., P.A.S.*, **80** (1961), 414.
11. Young, F. S., J. M. Clayton, and A. R. Hilemann, 'Shielding of transmission lines', *Trans. I.E.E.E., P.A.S.*, **83** (1963), 132.
12. Armstrong, H. R., and E. R. Whitehead, 'Field and analytical studies of transmission line shielding', *Trans. I.E.E.E., P.A.S.*, **PAS-87** (1968), 270.
13. Billings, M. J., and J. T. Storey, 'Consideration of the effect of pollution on the potential distribution of insulator systems', *Proc. I.E.E.*, **115** (1968) 1661.
14. Ely, C. M. A., and W. J. Roberts, 'Switching-impulse flashover of air gaps and insulators in an artificially polluted atmosphere', *Proc. I.E.E.*, **115** (1968), 1667.
15. Garrard, C. J. O., 'High voltage switchgear— a review of progress', *Proc. I.E.E.*, **113** (1966), 1523.
16. I.E.E.E. Committee, 'Switching surges Part IV Control and reduction on a.c. transmission lines', *I.E.E.E. Trans.*, **PAS-101** (1982), 2694.
17. Barthold, L. O., and G. K. Carter, 'Digital travelling wave solutions, 1—single phase equivalents', *Trans. A.I.E.E.*, **80** (1961).
18. Uram, R., and R. W. Miller, 'Mathematical analysis of transmission line transients: Part 1. Theory', *Trans. I.E.E.E., P.A.S.*, **83** (1964).
19. McElroy, A. J., and R. M. Porter, 'Digital computer calculation of transients in electric networks', *Trans. I.E.E.E., P.A.S.*, **82** (1963).
20. Bickford, J. P., and P. S. Doepel, 'Calculation of switching transients with particular reference to line energisation', *Proc. I.E.E.*, **114** (1967).
21. Paris, L., 'The future of U.H.V. transmission lines', *I.E.E.E. Spectrum*, **44** (1969).
22. I.E.E.E. Committee Reports on *Switching Surges, Power Apparatus and Systems*, 1948, p. 912; 1961, p. 240; 1966, p. 1091; 1970, p. 173.
23. Swift, G., 'An analytical approach to ferroresonance', *Trans. I.E.E.E.*, **PAS-88** (1969), 42.
24. Kawai, M., 'Research at project UHV on the performance of contaminated insulators', *Trans. I.E.E.E.*, **PAS-92**, Pt. II (1973), 1111–1120.
25. Annestrand, S., 'Bonneville Power Administration—Prototype, 1100/1200 kV line', *Trans. I.E.E.E.*, **PAS-96** (1977), 357–366.

Problems

10.1. A 345 kV, 60 Hz system has a fault current of 40 kA. The capacitance of a busbar to which a circuit breaker is connected is 25,000 p.F. Calculate the surge impedance of the busbar and the frequency of the restriking (recovery) voltage on opening.
(Answer: 674 Ω, 875 Hz.)

10.2. A highly capacitive circuit of capacitance per phase 100 μF is disconnected by a circuit breaker, the source inductance being 1 mH. The breaker gap breaks down when the voltage across it reaches twice the system peak line-to-neutral voltage of 38 kV. Calculate the current flowing with the breakdown and its frequency and compare it with the normal charging current of the circuit.
(Answer: 34 kA, 503 Hz; note $\hat{I} = 2\ V_p/Z_0$.)

10.3. A 10 kV, 64.5 mm^2 cable has a fault 9.6 km from a circuit breaker on the supply side of it. Calculate the frequency of the restriking voltage and the maximum voltage of the surge after 2 cycles of the transient. The cable parameters are (per km), capacitance per phase = 1.14 μF, resistance = 5.37 Ω, inductance per phase = 1.72 mH. The fault resistance is 6 Ω.
(Answer: 374 Hz; 16 kV.)

10.4. The effective inductance and capacitance of a faulted system as viewed by the contacts of a circuit breaker are 2 mH and 500 $\mu\mu$F, respectively. The circuit breaker chops the fault current when it has an instantaneous value of 100 A. Calculate the restriking voltage set up across the circuit breaker. Neglect resistance.
(Answer: 200 kV.)

10.5. A 132 kV circuit breaker interrupts the fault current flowing into a symmetrical three-phase to earth fault at current zero. The fault infeed is 2500 MVA and the shunt capacitance, C, on the source side is 0.03 μF. The system frequency is 50 Hz. Calculate the maximum voltage across the circuit breaker and the restriking-voltage frequency.

If the fault current is prematurely chopped at 50 A, estimate the maximum voltage across the circuit breaker on the first current chop.
(Answer: 215.5 kV; 6.17 kHz; 45 kV.)

10.6. Repeat Example 10.2 but with the surge travelling from the cable into the overhead line.
(Answer: Current into line = 0.091 × incident surge current; current reflected back into cable = 0.91 × incident current, reflected energy = 0.83 × incident surge energy.)

10.7. Repeat Example 10.2 but with zero resistance between the line and cable.
(Answer: Energy reflected back to line = 0.67 × incident surge energy.)

10.8. A cable of inductance 0.188 mH per phase and capacitance per phase 0.4 μF is connected to a line of inductance 0.94 mH per phase and capacitance 0.0075 μF per phase. All quantities are per km. A surge of 1 p.u. magnitude travels along the cable towards the line. Determine the voltage set up at the junction of the line and cable.
(Answer: 1.85 p.u.)

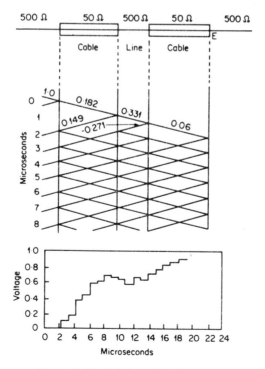

Figure 10.52   Solution of Problem 10.13.

10.9. A long overhead line has a surge impedance of 500 Ω and an effective resistance at the frequency of the surge of 7 Ω/km. If a surge of magnitude 500 kV enters the line at a certain point, calculate the magnitude of this surge after it has traversed 100 km and calculate the resistive power loss of the wave over this distance. The wave velocity is 3×10^5 km/s.

(Answer: 250 kV; 375 MW.)

10.10. A rectangular surge of 2 μs duration and magnitude 2 p.u. travels along a line of surge impedance 350 Ω The latter is connected to another line of equal impedance through an inductor of 800 μH. Calculate the value of the surge transmitted to the second line.

(Answer: $v = v_i[1 - e^{-(2Z_0/L)t}]$ (i.e. 1.67 p.u.))

10.11. A lightning arrester employs a thyrite material possessing a resistance characteristic described by $R = (72 \times 10^3)/(I^{0.75})$. An overhead line of surge impedance 500 Ω is terminated by the arrester. Determine the voltage across the end of the line when a rectangular travelling surge of magnitude 500 kV travels along the line and arrives at the termination. (A graphical method using the voltage–current characteristics is useful.)

(Answer: 375 kV.)

10.12. A rectangular surge of 1 p.u. magnitude strikes an earth (ground) wire at the centre of its span between two towers of effective resistance to ground of 200 Ω and 50 Ω. The ground wire has a surge impedance of 500 Ω. Determine the voltages transmitted beyond the towers to the earth wires outside the span.

(Answer: $0.44v_i$ from 200 Ω tower and $0.18v_i$ from 50 Ω tower)

10.13. A system consists of the following elements in series; a long line of surge impedance 500 Ω, a cable (Z_0 of 50 Ω), a short line (Z_0 of 500 Ω), a cable (Z_0 of 50 Ω), a long line (Z_0 of 500 Ω). A surge takes 1 μs to traverse each cable (they are of equal length) and 0.5 μs to traverse the short line connecting the cables. The short line is half the length of each cable. Determine by means of a lattice diagram the p.u. voltage of the junction of the cable and long line if the surge originates in the remote long line.

(Answer: see Figure 10.52)

II

Overhead Lines and Underground Cables

Brief outlines of the nature of lines and cables are given in Chapter 3 where the emphasis is on parameters for use in analysis. Rotating machines and transformers are normally treated in prerequisite machine courses, but as lines and cables are not included in such courses some aspects of design will be discussed here. The design of insulation will be considered and also thermal design for current ratings. The latter topic will be less familiar to electrical specialists but is of great importance. Normally the insulation provided must be adequate to withstand overvoltages in the system in the form of lightning strikes (see Figure 10.1) and switching surges which have a slower (e.g. 200 μs) rise time.

OVERHEAD LINES

11.1 Introduction

Overhead lines are used whenever possible on grounds of economy. The major part of the insulation is provided by air which at the time of writing is still relatively inexpensive. The form of the insulators used is given in Chapter 3. At subtransmission voltages and above, suspension-type insulators are used made of toughened glass in Europe and porcelain in the U.S.A. The use of plastic materials for insulators is under active development.

Regulations exist governing the minimum heights of line conductors above the ground. Hence it is necessary to calculate the sag in the line and its variation with temperature. Although the power transmitted by a line may be limited by electrical considerations, e.g. stability, often and especially for shorter lines around cities—it is the current which is limited, this being decided by allowable temperature. Hence the normal current rating will depend on ambient temperatures and summer, normal, and winter ratings are used. In addition the thermal time-constants of the conductors are utilized to give short-term and long-term emergency ratings. Daily currents of a cyclic nature attain the rated value for only a few hours per day. However, because of the relatively short temperature time-constant (up to 20 min) the line must be rated for the full-load current.

Line route selection

Because of the visual impact of towers and lines the selection of routes is of great importance. Attempts are made to modify the designs of towers to give reduced weights and dimensions and by the use of guyed structures. Route design must consider natural landscapes and the geology, flora, and fauna of the land. Also industrial and residential constraints must be met. Computer programs now exist which develop 'suitability' maps with suggested right-of-way (wayleave) corridors based on minimal environmental impact.

The effective use of a right of way by a given line or lines may be expressed by an index defined as follows:

(line loading)/(width of right of way)(tower height)

Values of this index for lines in the U.S.A. are given in Table 11.1

Table 11.1   Transmission route parameters

Voltage (kV)	Loading (MVA)	Right of way (m)	Tower Height (m)	Index value
345	500	45.8	27.4	0.037
500	1200	61.0	36.6	0.050
765	2500	76.2	41.2	0.074
1200*	7500	91.5	50.4	0.152

* Proposed design.

On one single 1200 kV right of way the following number of lines (circuits) are required to transmit 7500 MW. 345 kV, 15 (index 0.058); 765, 3 (index 0.085). The improvement in corridor use by increase in voltage is very evident.

D.C. lines   The reduced insulation requirements for direct voltages and the reduced number of conductors result in a smaller, compact, and more economic line for a given power transfer. Other factors are different from a.c., e.g. corona appears to be reduced in wet weather, audible noise levels are relatively low, and radio interference levels are less (but RI emanating from converter stations must be considered). A comparison of a.c. and d.c. line capabilities based on economic grounds is given in Table 11.2.

Table 11.2   Economic comparison—HVAC and HVDC

(Approximate economic Loading)		Equivalent d.c. based on equal insulation to ground		Equivalent d.c. based on equal right of way	
kV	MW	kV	MW	kV	MW
230	240	±200	400	±300	900
345	580	±300	900	±500	2500
500	1280	±400	1600	±700	4500
765	2700	±600	3600	±1000	8000

11.2 Mechanical Design

As stated in Chapter 3 for higher voltage lines, i.e. subtransmission and above the conductor (ACSR) is formed of strands of aluminium wire wound around a core of steel strands, the latter providing the mechanical strength. At lower voltages copper, cadmium copper, aluminiun, and aluminium alloy, all hard drawn, are used. A typical cross-section of a ACSR conductor is shown in Figure 11.1. An alternative to the ACSR (aluminium conductor steel rein-forced) conductor is one using aluminium alloy (instead of steel) as the core

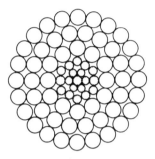

Figure 11.1 Cross-section of ACSR overhead line conductor—19 strands steel, 59 strands aluminium.

with an aluminium conductor ACAR). Advantages claimed are better cor-rosion resistance, lighter weight, and simpler splices. To minimize corona in lines using one conductor per phase an expanded (larger diameter) conductor has been developed. In one method non-metallic strands of fibrous material are inserted between the steel core and aluminium conductor wires. Another incorporates air pockets formed by wrapping aluminium strands helically around the steel core.

The usual way to reduce line corona with its attendant problems of audible and radio noise is to use more than one conductor per phase forming a *bundle*. The effect of having two or more subconductors per phase is to reduce the maximum surface electric field and hence the magnitude of corona discharge. A further effect of bundled conductors is reduced inductive reactance. The subconductors are spaced apart by steel spacers at frequent intervals along the line. Three or four subconductor bundles are frequently used in the 400–765 kV range.

The calculation of tension and sag in a span of conductor between towers must account for regulations often stipulating ice and wind loading. For example British regulations call for ice of radial thickness 3.75 mm($\frac{3}{8}$ in) (temperature $-5°C$) in conjunction with a wind of 80 kmh^{-1} (50 mph) at right angles to the line such that the pressure is 37.5 N/m^2 (8 lbf/ft^2) over the projected area. The resultant force or loading is used in calculations.

Consider the line shown in Figure 11.2. Here, l is the length of the complete span, s the length along the line conductor, w the effective weight per unit

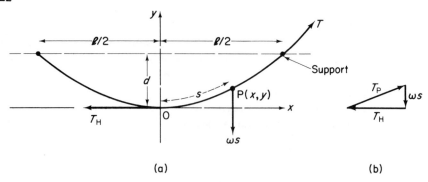

Figure 11.2 (a) Overhead line—sag. (b) Forces on an overhead line.

length and d the maximum sag. The horizontal component of tension T_H is constant along the line. The relation between T_H, the actual tension T, and the weight of line (ws) relative to point P is shown in Figure 11.2(b). At point P (x, y),

$$\frac{dy}{dx} = \frac{ws}{T_H} \qquad (11.1)$$

Also

$$ds = (dx^2 + dy^2)^{\frac{1}{2}}$$

Hence

$$\frac{ds}{dx} = \left[1 + \left(\frac{dy}{dx}\right)^2\right]^{\frac{1}{2}} = \left[1 + \left(\frac{w^2 s^2}{T_H^2}\right)\right]^{\frac{1}{2}} \qquad (11.2)$$

Integrating and substituting $s = 0$ when $x = 0$,

$$s = \frac{T_H}{w} \sinh\left(\frac{wx}{T_H}\right) \qquad (11.3)$$

Substituting for s in 11.1 and integrating for y,

$$y = \frac{T_H}{w}\left[\cosh\left(\frac{wx}{T_H}\right) - 1\right] \qquad (11.4)$$

which is a catenary. The maximum sag d at $x = l/2$

$$= \frac{T_H}{w}\left[\cosh\left(\frac{wl}{2T_H}\right) - 1\right]$$

$$= l\left[\frac{1}{8}\left(\frac{wl}{T_H}\right) + \frac{1}{384}\left(\frac{wl}{T_H}\right)^3 + \dots\right] \qquad (11.5)$$

At the supports the tension

$$T = T_H\left[\cosh\left(\frac{wl}{2T_H}\right)\right]$$

The length of the line

$$= \frac{2T_H}{w} \sinh\left(\frac{wl}{2T_H}\right)$$

$$= \frac{2T_H}{w}\left(\frac{wl}{2T_H} + \frac{w^3l^3}{T_H^3 48} + \cdots\right) \doteq l + \frac{w^2l^3}{24T_H^2} \tag{11.6}$$

Normally, given l and w, it is required to calculate the sag for a given maximum tension. Note that,

$$T/T_H = \cosh\frac{wl}{2T_H} \doteq 1 + \frac{w^2l^2}{8T_H^2} \quad \text{and if} \quad \frac{w^2l^2}{8T_H^2} \ll 1 \quad T \doteq T_H$$

Typical tensile strengths are:

for aluminium $1.6{-}1.9 \times 10^8$ N/m^2 ($23{-}28 \times 10^3$ lb/in^2)
for steel 13.8×10^8 N/m^2 (200×10^3 lb/in^2)

The line will be strung for weather conditions different to those in service and the relationships between parameters relevant to the two situations which will be suffixed 1 and 2 are required. Let the line effective cross-section (steel) be 'a' square metres. If the mechanical stress originally f_1 increases to f_2 then the increase in length of the line over the original length $2s$,

$$= 2s_1(f_2 - f_1)/E$$

where E is the modulus of elasticity. Similarly, for a temperature increase from θ_1 to θ_2, the increase in length

$$= \alpha 2s_1(\theta_2 - \theta_1)$$

where α is the temperature coefficient of expansion. Hence the new half-length of conductor,

$$s_2 = s_1 + s_1(f_2 - f_1)E + \alpha s_1(\theta_2 - \theta_1)$$

The expression for the complete length $2s_2$ may be expanded and the approximate value, $l + w^2l^3/24T_H^2$, i.e. $l + w^2l^3/24a^2f^2$ substituted to give

$$f_2^2(f_2 - M) = w_2^2l^2E/24a^2 \tag{11.7}$$

where,

$$M = f_1 - \frac{w_1^2l^2E}{24T_1^2} - \alpha E(\theta_2 - \theta_1)$$

The new sag d_2 is then found by substituting f_1 (i.e. $T_1/a = w_1l^2/8ad_1$) and f_2 (i.e. $w_2l^2/8ad_2$) into equation (11.7). The equation for both f_2 and d_2 are cubic and not readily solved. Iterative procedures involving an initial informed guess are best employed.

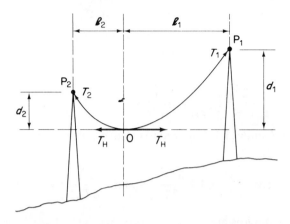

Figure 11.3   Line with supports at different heights.

Occasionally the support points are at different levels as shown in Figure 11.3. OP_1 and OP_2 are both catenaries. and using equation (11.4),

$$d_1 = \frac{T_H}{w}\left[\cosh\left(\frac{wl_1}{T_H}\right) - 1\right]$$

and

$$d_2 = \frac{T_H}{w}\left[\cosh\left(\frac{wl_2}{T_H}\right) - 1\right]$$

The difference in height of the support points is

$$d_1 - d_2 = \left[\cosh\frac{wl_1}{T_H} - \cosh\frac{wl_2}{T_H}\right]$$

Also, $l_1 + l_2 = $ total span l. The maximum stress $= T_1 = T_H \cosh(wl_1/T_H)$. As $(d_1 - d_2)$, l, and T_1 are specified the unknowns T_H, l_1, l_2, and hence the sag and tension can be found. Again iterative methods are advisable.

With steel-cored conductors spans of about 300–400 m are normally used, with corresponding values of sag between 8 and 13 m. When towers are in a straight line the conductor tensions on each side balance and there is no overturning force. When the route direction changes by an angle α then the tower at the change experiences a side for of $2T\sin(\alpha/2)$, where T is the total tension of all the conductors.

11.3   Electrical Design

The electric breakdown of air-gaps was discussed in Chapter 10. Arrangements of insulators are shown in Figure 11.4; the less widely used insulated cross-arm was shown in Figure 10.40. The length of insulator string is roughly the same as the air clearance for Figure 11.4(a), but the conductor is free to swing and the clearance must be increased accordingly. With the V-string of

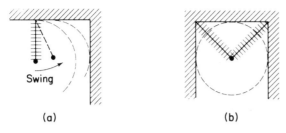

Figure 11.4 Suspension-type insulators. (a) Single string. (b) V-string.

Figure 11.4(b) there is no swinging and the tower width is reduced. At the higher voltages, e.g. above 500 kV, the cost saving on the tower with V-strings appears to outweigh the extra cost of insulation. At angle towers the insulators must withstand the full conductor tension.

The strength of rod-plane gaps is higher for negative impulse voltages (e.g. mean stress of 0.75 MV/m at 4 m gap length) than for positive impulse (0.3 MV/m for 4 m gap). The configuration of the air insulation is very influential; the various geometries encountered are shown in Figure 11.5.

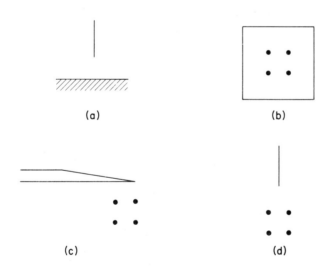

Figure 11.5 Tower–electrode gap configurations.

Obviously the performance with positive surges is of crucial importance. An approach to the problem for positive polarity switching surges, particularly for large gaps, is as follows. An empirical factor 'k', is used to define the particular tower–conductor configuration as follows.

The breakdown voltage, $(V_{50}) = 500kd^{1.6}$, where V is in kilovolts and d is the clearance (m). In Figure 11.5 for the gap shown in (a) $k = 1$, i.e. this is the reference situation, for (b) $k = 1.2$, for (c) $k = 1.55$, and for (d) $k = 1.9$. The inclusion of insulators between gaps under dry conditions with positive surges creates little change in V_{50}.

The minimum distances between the conductors and the tower structure are determined experimentally by high-voltage breakdown tests for the various geometries involved. On the other hand, the electric field on the surface of line conductors may be calculated and this is necessary to determine corona onset. Various methods have been used for this purpose, e.g. Maxwell's potential coefficients, conformal transforms, filament conductors, images, charge simulation technique, and the use of Laplace's equation. In these methods the following assumptions are made:— infinitely long, parallel conductors; horizontal ground plane modelled by image conductors; smooth conductor surfaces.

Maxwell's potential coefficients   For a multiconductor system with m conductors;

$$[V] = [P][Q]$$

where

$[V]$ = column matrix of line to ground voltages (r.m.s.);

$[Q]$ = conductor charges, complex (a.c.) or real (d.c.);

$[P]$ = Maxwell's potential coefficients;

$$P_{ii} = \frac{1}{2\pi\varepsilon_0} \ln\left(\frac{2h_i}{r_i}\right) \text{ m/F}$$

$$P_{ij} = \frac{1}{2\pi\varepsilon_0} \ln\frac{d'_{ij}}{d_{ij}} \text{ m/F}$$

h_i = height of conductor i(m);

r_i = radius (equivalent radius of bundle conductors) of conductor i(m);

d'_{ij} = distance between centre conductor i and image of conductor j;

d_{ij} = distance between conductors i and j.

Note. $P_{ii}Q_i$ is the change of potential of i due to the acquisition of charge Q_i per metre, P_{ji} is the potential on j due to Q_i on i, and so on.

For a single conductor (of voltage V and radius r) at height h above the ground,

$$V = \frac{Q}{2\pi\varepsilon_0} \ln\left(\frac{2h}{r}\right)$$

The stress at x above ground vertically below the conductor,

$$E_x = \frac{Q}{2\pi\varepsilon_0(h-x)} - \frac{-Q}{2\pi\varepsilon_0(h+x)}$$

$$= \frac{2Q}{2\pi\varepsilon_0}\left(\frac{1}{h-x} + \frac{1}{h+x}\right)$$

Substituting for Q,

$$E_x = \frac{V 2\pi\varepsilon_0}{\ln(2h/r)}\left(\frac{1}{h-x}+\frac{1}{h+x}\right) \tag{11.8}$$

For a three-phase system as shown in Figure 11.6:

$$V_1 = \frac{Q_1}{2\varepsilon_0}\left[\ln\left(\frac{2h_1}{r}\right)+\ln\left(\frac{d_{12}^1}{d_{12}}\right)+\ln\left(\frac{d_{13}^1}{d_{13}}\right)\right]$$

$$V_2 = \frac{Q_2}{2\pi\varepsilon_0}\left[\ln\left(\frac{d_{21}^1}{d_{21}}\right)+\ln\left(\frac{2h_2}{r}\right)+\ln\left(\frac{d_{23}^1}{d_{23}}\right)\right] \tag{11.9}$$

$$V_3 = \frac{Q_3}{2\pi\varepsilon_0}\left[\ln\left(\frac{d_{31}^1}{d_{31}}\right)+\ln\left(\frac{d_{32}^1}{d_{32}}\right)+\ln\left(\frac{2h_3}{r}\right)\right]$$

Hence the capacitance of each conductor can be determined $(C = Q/V)$. The electric field at a point below three-phase lines may be determined as for the single-conductor case, but accounting for the components of stress at the point due to all three conductors and their images.

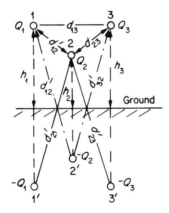

Figure 11.6 Three-phase line—calculation of stress by means of images—charges on conductors Q_1, Q_2, Q_3; conductor radius r.

Maximum stress on conductor surface of three-phase systems

Consider three conductors in horizontal formation with conductor to conductor spacing of s as shown in Figure 11.7. When the line to neutral voltage of (a),

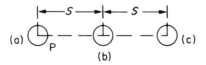

Figure 11.7 Three-phase line—horizontal formation—conductor radius r.

v_{an}, is at a maximum the *instantaneous* line-to-line voltage (a)–(b) is $(\sqrt{3}/2)\hat{V}_{ab}$. If the charge on (a) is $\hat{Q}$ then the charges on (b) and (c) are each $Q/2$. When considering the conductor surface stress the effects of the ground plane will be ignored. $\hat{V}_{ab}$ is the peak voltage between a and b.

The stress (E_p) on the conductor surface at point P on (a) is

$$E_p = \frac{\hat{Q}}{2\pi\varepsilon_0}\left[\frac{1}{r}+\frac{1}{2(s-r)}+\frac{1}{2(2s-r)}\right]$$

Also

$$\frac{\sqrt{3}}{2}\hat{V}_{ab} = \int_{r}^{s-r} E(x)\,dx$$

$$= \frac{Q}{4\pi\varepsilon_0}\left[3\ln\left(\frac{s-2}{r}\right)+\ln\left(\frac{2s-r}{s+r}\right)\right]$$

Hence E_p is obtained in terms of $\hat{V}_{ab}$.

Bundle Conductors

The advantages compared with a single conductor are as follows:

(a) the maximum surface electric field is reduced and in the limit this may be seen by assuming the number of subconductors to be large enough to form a continuous cylinder of large diameter;

(b) the surface field required for corona to form is greater for the bundle conductor.

The calculations for the surface electric fields for bundle conductors are complex and only general findings will be given here. The effects of the number and diameter of the subconductors are shown in Figure 11.8. The optimum subconductor spacing (with respect to surface stresses) is about 8–14 times the subconductor diameter for bundles of 2, 3, and 4. Surface electric stresses are not constant around the periphery of subconductors due to the presence of the radial (self) field and that due to other subconductors. This variation can be approximated by the law,

$$E_\theta = E_{average}\left[1+\frac{r}{R}(n-1)\cos\theta\right]$$

see Figures 11.9 and 11.10

Corona

Air at normal atmospheric pressure and temperature breaks down at 30 kV/cm (peak or crest value). For smooth cylinders this stress may be determined from the expression,

$$\frac{V}{r\ln(d/r)}\,\text{V/cm}$$

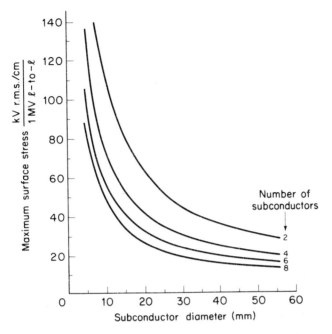

Figure 11.8 Variation of maximum surface stress with subconductor diameter and number of subconductors. Phase spacing 7.6 m; subconductor spacing 0.305 m. (*Permission of the Institution of Electrical Engineers.*)

where

V = the voltage to neutral;
d = spacing between lines (cm);
r = radius of conductor (cm).

The discharge which occurs in the air surrounding a conductor subject to a stress above this value is known as corona; it is easily detected by a hissing

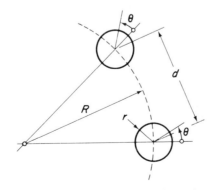

Figure 11.9 Configuration of bundle conductor.

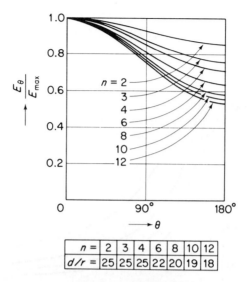

$n =$	2	3	4	6	8	10	12
$d/r =$	25	25	25	22	20	19	18

Figure 11.10   Variation of stress round conductor. (*Permission of the Institute of Electrical and Electronics Engineers.*)

sound and at night by a blue glow around the conductors. Corona is established at a stress $\mathscr{E}_v$ called the visual critical stress (corresponding to a voltage V_v) which is greater than the basic breakdown value $\mathscr{E}_0$. For smooth cylinders,

$$\mathscr{E}_v = \mathscr{E}_0 \left(1 + \frac{0.3}{\sqrt{r}}\right) \quad \text{and as} \quad \mathscr{E}_v = \frac{V_v}{r \ln (d/r)}$$

and air breaks down at 21.1 kV (r.m.s.) per cm, it follows that for fair weather conditions,

$$V_v = 21.1 r \left(1 + \frac{0.3}{\sqrt{r}}\right) \ln \left(\frac{d}{r}\right) \text{ kV (r.m.s. to earth)}$$

This voltage is dependent on air density and the nature of the conductor surface. The following formula due to Peek[1] incorporates these features:

$$V_v = 21.1 \, m\delta r \left(1 + \frac{0.3}{\sqrt{\delta r}}\right) \ln \left(\frac{d}{r}\right) \text{ kV (r.m.s.)} \qquad (11.10)$$

where

δ = air density factor = $3.92b/(273 + T)$;
b = barometric pressure in centimetres of mercury;
T = temperature in degC;
m = is typically 0.9 for weathered stranded conductors.

The power loss (kW per km of conductor) due to corona under fair-weather conditions (Peterson—Discussion *A.I.E.E. Transactions*, **52**, 1933) is given by)

$$p = \frac{0.000021}{[\log_{10}(2s/d)]^2} fV^2 F \qquad (11.11)$$

where

f = frequency (Hz);
V = line-to-ground voltage;
s = spacing between conductors;
d = conductor diameter;
F = corona factor determined by test.

With fair weather conditions the line loss can approach the same order of magnitude as the insulator leakage losses. These losses have little economic or technical significance for e.h.v. lines. Typical measured values in kW for all three conductors per km (reference 2) are 1–5 for a 500 kV line of conductor diameter 3.7 cm (1.465 in) and 3–19 for the same conductor at 700 kV (average peak gradient 23.13 kV/cm). The loss for an insulator string (24 discs) at 500 kV is 95 W and at 735 kV (32 discs) 150 W.

Foul-weather corona losses are much more significant. Tests summarized in reference 2 indicate that the conductor loss is proportional to the following: logarithm of the rate of rain fall (rain rate), number of conductors, and approximately to the product of the operating voltage and the fifth power of the voltage gradient on the underside of the conductor surface where the loss is greatest due to the accumulation of water. The following formula applies:

total loss per three-phase km in kW
= total fair-weather loss per three-phase km (kW)

$$+ \left[\frac{V}{\sqrt{3}} Jr^2 \ln(1 + KR) \sum_1^n (E^5) \right]$$

where

J = loss-current constant (approx. 3.35×10^{-10} at 700 kV and 500 kV and 4.4×10^{-10} at 400 kV);
r = conductor radius (cm);
n = number of conductors per bundle $\times 3$;
E = voltage gradient on underside of conductor kV (peak)/cm;
K = a wetting coefficient = 10 for R in mm/h;
R = rain rate (mm/h).

Typical measured values in kW per three-phase mile (reference 2) are as follows: 700 kV, 14 m (45.5 ft) spacing, 190 at 2.5 mm/h, 260 at 12.5 mm/h; 500 kV, 6.1 m (20 ft) spacing, 150 at 2.5 mm/h and 215 at 12.5 mm.

Radio interference (RI) or radio noise

Although the presence of corona results in a power loss, a more important effect is that the discharge causes radiations to be propagated in the frequency bands used by radio and television. The most effective way to reduce or avoid corona and radio interference is the use of bundle conductors, i.e. several conductors per phase suspended from common insulators and separated mechanically by spacers of various designs. On some systems four-conductor bundles are in use. The configuration of conductors forming a bundle modifies the surface electric field such that the maximum stress is lower than with a single conductor, also the current rating of the circuit is increased. Troubles with bundle conductor lines have been experienced owing to aerodynamic instability resulting in mechanical oscillations along the line.

The corona discharges occur at discontinuities on the conductor surface and a random generation of pulses occurs. In wet weather water droplets form on the underside of the conductor which deform under the stress. Each droplet elongates and a sharp point is formed providing a strong source of RI and loss. Humidity has a marked effect on the corona and RI phenomena, increase in humidity causing a marked increase in these quantities. Radio interference is also caused by discharges from insulators and bad contacts.

Contamination of insulator surfaces

The nature of the contamination varies according to location and often consists of soot, fly-ash cements, etc. In coastal districts surface films of salt are formed. The surface conductivity (σ_s) of a conducting film of thickness t centimetres is defined by $\sigma_s = \sigma t$, where σ is the specific (volume) conductivity of the film. For an axisymmetric insulator surface with a uniform film of contaminating material of surface resistivity ρ_s (i.e. $1/\sigma_s$), the total electrode to electrode resistance,

$$R = \int_0^L \frac{\rho_s dl}{\pi D_1}$$

where

$$L = \text{length of leakage path};$$
$$dl = \text{element of leakage path};$$
$$D_1 = \text{diameter of surface at } dl.$$

The small leakage current of a dry insulator is determined by the electrostatic field alone. When the pollution layer present becomes moist due to rain, fog, or dew it becomes conductive and a surface leakage current of a much higher magnitude than the dry value flows; the electric field becomes very distorted. Where the current density is greatest the heat generated increases the film temperature to boiling point and evaporation of the moisture occurs and the formation of *dry bands*. These bands around the insulation have a high resistance and support almost all the voltage. This results in the breakdown of the air and an arc is formed across the dry band. The arc roots burn upon the moisture film at the dry band boundary and dry out more film under the roots. Hence the arc extends further into the moist area. If the moisture precipitation is low the dry band widens until the arc, when it reaches a certain length, extinguishes. The stress across the dry band is then just below the air break-down value, and should a local increase in precipitation occur breakdown results and the arc reappears. If the rate of precipitation balances the loss of moisture due to the arc then a stable condition is reached and the discharge may continue for a considerable time, see Figure 11.11.

Figure 11.11   Formation of dry bands around polluted insulators.

Various methods are used to prevent flashover due to pollution. The leakage length of the insulator is increased although difficulty is experienced in produc-ing insulators with leakage paths much greater than 2.54 cm (1 in) per kV. Greasing of the entire insulator surface prevents the formation of a continuous film and is very effective. However, a fresh application of grease is required periodically, e.g. 1–2 years. The washing of surfaces with water from hoses when the equipment is live is also effective.

11.4   Current Rating

The voltage rating of overhead lines is a function of insulation design and hence physical size. The current capability is decided as in all electric plant by thermal, i.e. temperature rise, considerations. Because of electrical limitations long lines are not usually operated close to their thermal rating especially with bundled conductors. The conductor operating temperature is limited (70°C is often used) by mechanical aspects such as allowable mid-span sag, creep in the conductor, and long-term mechanical effects. Perhaps the most limiting aspect on lines is the temperature of joints and clamps.

Heat is dissipated from overhead lines by radiation and convection. Heat loss by radiation is given by

$$q = 5.7 \times 10^{-8} \varepsilon (\pi d)(T_c^4 - T_a^4) \text{ watts/metre length}$$

where

ε (emissivity) = 0.2 for newly installed conductors rising to
0.8 when oxidized in service;

T_c and T_a = absolute temperatures (K) of the conductor
and ambient;

d = conductor diameter (m).

For example, for a line of $d = 20$ mm at 65°C with an ambient of 27°C:

$$T_c = (273 + 65)\text{K} \qquad T_a = 300 \text{ K} \qquad \varepsilon = 0.8 \quad \text{and} \quad q = 14.18 \text{ W/m}$$

As the temperature difference is 38°C, the effective thermal resistance per
metre = 38/14.18 = 2.7 degC/W. Often a wind velocity of 0.5 m/s is assumed
and the following empirical formula used (air flow at right angles to the line):
heat transfer coefficient (W/m^2K)

$$= \frac{1}{d \cdot g} (0.35 + 0.47Re^{0.52})0.7^{0.3} \tag{11.12}$$

where

g = thermal resistivity of air = 36K m/W (normal temperature and pressure)

$$Re = \text{Reynolds number} = \frac{d \times \text{velocity}}{\text{kinematic viscosity}} = \frac{d \times \text{velocity}}{1.568 \times 10^{-5}}$$

This gives a value of thermal resistance per metre length in the order of
0.9 degC m/W for wind convective cooling for average size lines. These two
thermal resistances are effectively in parallel (heat is removed by each mode)
and the total effective value = 0.68 degC m/W. If the area of conductor
(aluminium) in an ACSR line is say 20 mm^2 and of resistance 1.8×10^{-4} Ω/m,
for a conductor temperature of 65°C and an ambient of 27°C the heat evolved
per metre

$$= \frac{38}{0.68} = 55.9 \text{ W}$$

$$I^2R = I^2 \times 1.8 \times 10^{-4} = 55.9$$

$$\therefore \ I = 560 \text{ A}$$

This ignores the heat input from solar radiation which is given by

$$\alpha sd \text{ (W/m)}$$

where

α = absorbtivity (1 for black surface is maximum);
s = radiation (say 800 W/m^2);
d = line diameter (m).

The short-circuit temperature rise ($\Delta\theta$) is calculated by assuming that all the
heat is stored in the conductor. Hence,

$$I_F^2 R \Delta t = \text{mass} \times c_p \times \Delta\theta$$

where

R = mean conductor resistance over temperature range Ω/m;
Δt = fault duration (s);
c_p = specific heat = 0.9 kJ/kg degC (for aluminium).

The current rating of bundle conductors is increased because of the cooling of the individual subconductors. The ratings for a given number of subconductors for a British 400 kV line of *total* conductor cross-section of 800 mm^2 are as follows: one subconductor (1.3 kA), three sub conductors (1.7 kA), 4 sub conductors (2.3 kA).

11.5 Environmental Criteria

Although the visual impact of lines comes immediately to mind, other aspects have become of importance for e.h.v. lines and will become more so for u.h.v. These are:

(a) audible noise emission;
(b) emission of ozone and oxides of nitrogen;
(c) radio and television interference;
(d) safety and comfort problems caused by electrostatic fields.

Criteria used in the design of lines must meet various central and local government agencies with jurisdiction over the land over which the line is constructed.

Table 11.3   Recommended electrical environmental criteria for e.h.v. a.c. lines

Criteria	Magnitude	Justification
Ground gradient	Gradients not exceeding 13 kV/m	5 mA let-go-current criteria in forest areas and areas non-accessible by large vehicles
	Maximum 7.5 kV/m	Farmlands and areas accessible to heavy machinery. 5 mA let-go-current criteria
	Around 2 kV/m at edge of right-of-way	Most lines meet this criterion
Radio interference Fair weather	37 dB above 1 μV/m at the edge of right-of-way in residential areas or as legally required	I.E.E.E. recommendations of 27 db S/N ratio, for Grade B quality reception of Class 1 stations
Audible noise	52 dB(A) at edge of right-of-way or as legally required for residential areas	Bonneville Power Administration data based on public response to a survey
Ozone	0.080 ppm maximum 1 h concentration	U.S. Environmental Protection Agency recommendation

Permission *Energy International (Miller Freeman Publications)*, July 1975.

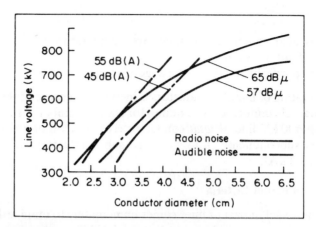

Figure 11.12 Conductor size required at a given voltage to meet foul-weather levels for radio interference and audible noise at 15 m from outermost conductor.

Audible, radio noise, and ozone emission are caused by corona and are reduced by any measures which will reduce the field strength at conductor surface. Electric fields from e.h.v. lines induce charge on well-insulated objects, e.g. cars, trucks, etc. close to the line. The national safety code (U.S.) recommends a limit of 5 mA for the maximum short-circuit current from the

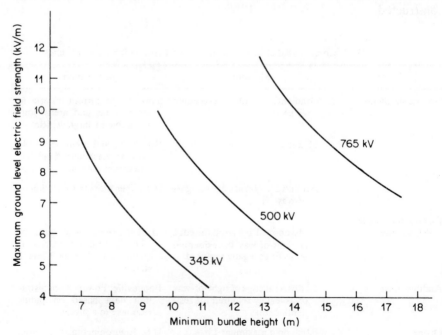

Figure 11.13 Variation of ground-level electric field with minimum height of conductor bundle. (*Permission Energy International (U.S.A.) July 1975.*)

largest anticipated vehicle. Recommended criteria for e.h.v., a.c. lines are given in Table 11.3

The line design requires correct selection of subconductor size, phase-to-ground and phase-to-phase clearances and mechanical loading. The size of bundle and subconductor are most crucial for the effects indicated above. The effects of conductor diameter on audible and radio noise are shown in Figure 11.12 and the variation of ground level electric field with minimum bundle conductor height in Figure 11.13.

UNDERGROUND CABLES

11.6 Introduction

A condiderable amount of transmission and distribution, especially in urban areas is carried out by means of underground cables. At the present time, growing public pressure to preserve the amenities of both town and countryside is forcing electricity supply authorities to consider the undergrounding of many circuits which they would prefer on economic grounds to place overhead. Even in sparsely populated regions, high-voltage bulk-transmission circuits have been placed underground where areas of outstanding natural beauty exist.

Underground transmission is more expensive than the overhead alternative; at 345 kV a figure of 10 times the cost of an equivalent overhead line is quoted for average suburban areas in the U.S.A., this ratio decreasing with lower voltages. A breakdown of costs for a 345 kV pipe-type cable is shown in Figure 11.14. A major problem in present-day technology is the development of

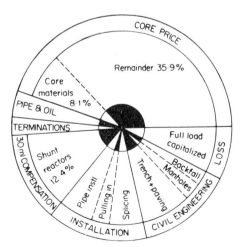

345 kV pipe-type cable

Figure 11.14 Cost components of conventional 345 kV pipe-type cable—U.S. practice. (*Permission of the Institute of Electrical and Electronics Engineers.*)

caples which are not only economically more attractive but physically able to carry the very large powers in use and envisaged. The major cause of the lack of current-carrying capacity in cables is the restriction on the temperature rise of the insulating material used; with overhead lines this is by no means so severe a problem. If the cable is to remain flexible there exists an upper limit to its overall diameter and hence to conductor size.

In spite of earlier hopes for plastic insulations, the combination of paper and oil still forms the most effective dielectric at the higher voltages. The difficulty of manufacturing extruded polyethylene cables without the creation of voids in the insulation has limited their use to relatively low voltages although such cables are becoming available for voltages above 100 kV. The temperature limitation of these cables is 70°C, above which the conductor tends to sink in the hot plastic, becoming offset and creating higher voltage stresses. This limit has been raised to 90°C by the use of cross-linked polythene.

In low-voltage distribution the use of plastic cables is more widespread.

A considerable development in recent years has been the widespread use of aluminium as both a conductor and sheath material. It has largely superseded lead as a sheathing material and is quickly gaining ground as a conductor. A major problem with aluminium conductors has been the making of effective joints which are in practice subject to thermal cycling and creep. Many techniques have been used and today much more confidence is placed in the available jointing techniques. The main reason for the use of aluminium is its price stability compared with copper which has severe price fluctuations. A comparison of possible conductor materials is given in Table 11.4.

Table 11.4   Conductor materials for power transmission

Material	Resistivity σ ($\mu\Omega$ cm)	Density ρ (g/cm^3)	$\rho\sigma$	Cost per pound weight $C(\$)$	$C\sigma\rho$
Copper	1.72	8.9	15.3	0.43	658
Aluminium	2.82	2.7	7.62	0.31	236
Silver	1.62	10.5	17.0	19.00	32300
Sodium	4.3	0.97	4.17	0.17	71.0

If l = length, A = area and R = permitted resistance of conductor, $A = \sigma l/R$ and total cost = $C\rho Al = C\rho l^2\sigma/R$ = const. $\times$ $C\rho\sigma$.

At lower voltages oil-impregnated paper-insulated cables (solid type) are used often with the three conductors contained in a single sheath. The three conductors are stranded and insulated separately and then laid up spirally together. The space between and around the conductors is packed with paper or jute to form a circular surface which is then further wrapped with insulation. This is called the *belted* type of cable (see Figure 11.15) and it may have steel-wire armouring over the sheath because no induced eddy currents are

439

produced in the armouring wires, whereas with single conductor cables severe loss and increased impedance would result. With the three-core cable electric stresses are set up tangentially to the paper insulation surfaces in which direction the insulation strength is weakest. To overcome this each core is wrapped with a conducting layer of metallized paper which converts the cable electrically into three single-core cables with the electric stress completely in the radial direction. This form of construction was originated by Hochstadter and is known as the 'H' type. As system voltages increased above 33 kV the solid-type cable became prone to breakdown because of the voids formed (small pockets of air or gas) in the insulation when constituent parts of the cable expanded and contracted to different extents with the heat evolved on load cycles. The stress across these voids is high and local discharges occur, which impair the paper. Eventually complete breakdown results.

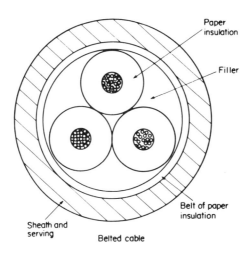

Figure 11.15   Belted-type construction, three-core cable.

In many countries, for voltages above 33 kV, the type of cable system in most common use is the oil-filled cable with paper/oil insulation, due to Emanueli.[7] In the oil-filled cable the hollow centre of the conductor is filled with insulating oil maintained under pressure by reservoirs feeding the cable along the route. As the cable heats on load, oil is driven from the cable into the reservoirs and vice versa, hence the creation of voids is avoided. In gas-pressure installations nitrogen at several atmospheres maintains a constant pressure on the inner sheath, compressing the dielectric, and so preventing the formation of voids. In the U.S.A. the paper/oil-insulated conductors are installed in a rigid pipe containing the insulating oil and a conductor oil-duct is not required. At the highest voltages, at present of the order of 500 kV, the oil-filled cable has a disign working stress of 15 kV/mm. A 275 kV oil-filled cable and joint are shown in Figure 11.16 (a) and (b) and a pipe-type cable in Figure 3.35.

440

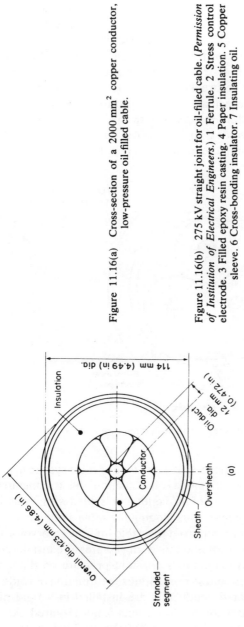

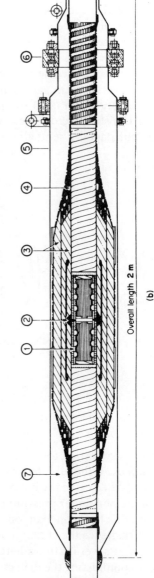

Figure 11.16(a) Cross-section of a 2000 mm² copper conductor, low-pressure oil-filled cable.

Figure 11.16(b) 275 kV straight joint for oil-filled cable. (*Permission of Institution of Electrical Engineers.*) 1 Ferrule. 2 Stress control electrode. 3 Filled epoxy resin casting. 4 Paper insulation. 5 Copper sleeve. 6 Cross-bonding insulator. 7 Insulating oil.

Insulation

114 mm (4.49 in) dia.

12 mm dia. (0.472 in)

Oil duct

Conductor

Overall dia. 123 mm (4.86 in)

Sheath

Overshealth

Stranded segment

(a)

Overall length 2 m

(b)

Methods of laying

The four main methods of installing cables are as follows:

(a) Direct in the soil—the cable is laid in a trench which is refilled with a backfill consisting of either the original soil or imported material of lower or more stable thermal resistivity.
(b) In ducts or troughs usually of earthenware or concrete.
(c) In circular ducts or pipes through which cables are drawn—this has the advantage of further cables being installed without excavation.
(d) In air, e.g. installed in tunnels built for other purposes.

Current developments include a gas-filled cable consisting of a conductor supported in a rigid external pipe which is filled with a gas under pressure, usually sulphur hexafluoride (SF_6) at 3 atm pressure. One advantage is much better heat dissipation by natural convection in the gas. Perhaps more revolutionary are cables incorporating cryogenic coolants which will be described in more detail in a further section.

The possibility of transmitting large amounts of power through *wave guides* has been considered. These would use oversized circular pipes and work in the TE_{01} mode at frequencies from 3 to 10 GHz. Attenuation of the TE_{01} mode in a circular wave guide decreases with increasing radii of pipe and increasing frequency. The variation of the specific attenuation (α) per kilometre and the maximum power transmittable with radius are shown in Figure 11.17. For 4 GW (4000 MW) of power to be transmitted over 1000 km with the same loss as with a 50 Hz e.h.v., a.c., line (about 1 dB) the wave guide will be at least 1.15 m at 3 GHz and 0.65 m at 10 GHz, giving power transfers of 10.6 and 3.4 GW, respectively. The relative merits of such a system have been discussed

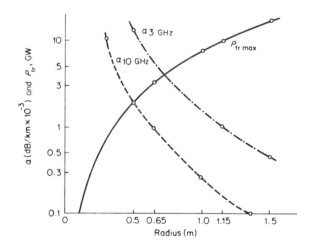

Figure 11.17   Attenuation (α) and power handling capacity (P_{tr}) of a circular waveguide. (*Permission of Institution of Electrical Engineers*; *Reference 39.*)

442

by Paul.[39] At the present time this application is remote and receiving small attention.

Conductor cross-sectional areas are described in different units in various countries. In the U.S.A. the circular mil is used (area enclosed by a circle 0.001 in. diameter) and in Europe the square millimetre. Relevant conversion factors are:

To convert mm^2 to circular mils multiply by 1970
To convert in^2 to circular mils multiply by 1.273239 × 10^6

11.7 Electrical Characteristics of Cable Systems

Joints

The maximum length between joints is determined by the reel size and for most cable circuits several lengths must be joined. At high voltages the design of the joint is complex and critical, especially as the paper tapes are applied by hand *in situ* and hence do not possess the electric strength of the cable tapes because of the presence of moisture. A low thermal resistance straight-through joint (connecting two lengths) is shown in Figure 11.16(b). Oil-filled systems need special joints to terminate the circuit ends and joints to allow oil from the reservoirs to enter the cable (feed joints). Both from an electrical and thermal standpoint, the joint represents the limiting part of a cable system.

Stresses in coaxial conductor systems.

In a.c. cables three stresses are considered, i.e. impulse, a.c., and inception stress for partial discharges. Normally cable systems operate discharge free and the design is based on the impulse stress with the a.c. stress perhaps becoming critical at very high voltages. The following impulse (1/50 μs waveform) and switching (200/300 μs-waveform) withstand levels are relevant to cables.

System voltage (kV)	Impulse level (MV)	Switching level (MV)
138	0.65	—
345	1.175	0.95
500	1.425/1.550	1.175
765	2.1	1.675
1200	2.4	1.95

To date the highest working a.c. stress for oil-filled cable insulation is 15 kV/mm with the prospect of further increase with increase in oil pressure as indicated in Figure 11.18. The following treatment applies to 'capacitance-determined' stress distributions, i.e. a.c. and surge. For a concentric (conductor-sheath) configuration let the charge per metre length on the conductor (radius r) be Q coulombs. The stress at radius x,

$$E_x = \frac{Q}{\varepsilon 2\pi x}$$

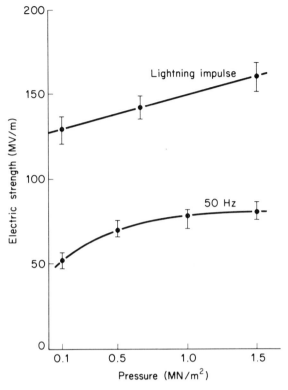

Figure 11.18 Impulse and a.c. breakdown strength of oil-impregnated paper tape as a function of oil pressure. (*Permission of Institution
of Electrical Engineers.*)

where $\varepsilon = \varepsilon_0 \cdot \varepsilon_r$. The voltage between the conductor and outer dielectric
screen or sheath (radius R),

$$V = \int_r^R E_x \, \mathrm{d}x = \frac{Q}{2\pi\varepsilon} \ln\left(\frac{R}{r}\right)$$

and as

$$Q = CV$$

$$C = \frac{2\pi\varepsilon}{\ln(R/r)} \text{ farads/metre}$$

Hence,

$$E_x = \frac{V}{x \ln (R/r)} \text{ volts/metre} \qquad (11.13)$$

and

$$E_r = \frac{V}{r \ln (R/r)} \text{ volts/metre} \tag{11.14}$$

The stress which is a maximum at the surface of the conductor is further increased by the effects of conductor stranding by an amount dependent on the strand diameter, but in the range 15–25 per cent. This increase is overcome by the use of taped screens of metallized paper or carbon-black paper, or in the case of extruded cables an extruded screen of semiconducting plastic. The minimum value of E_r is found from, $dE_r/dr = 0$, and occurs when, $\ln (R/r) = 1$, i.e. $R = 2.718r$, when $E_r = V/r$. This optimum relationship is often overridden by other considerations for conductor radius.

Because of the variation of stress through the dielectric economy may be obtained in taped dielectrics by the use of thinner tapes near the conductor and thicker tapes as the outer shield is approached. Because of limitations in manufacture the outer diameter of the dielectric is limited to 150 mm in oil-filled, self-contained systems. The determination of stress in three-core cables, i.e. three conductors in one sheath, is more complex and two-dimensional field plots are required.

Great care must be taken when reviewing working and breakdown stresses in the literature. It is American practice to quote the *average value* of stress in the insulation, whereas most other countries quote the *maximum* stress. In taped dielectrics it is the maximum stress that determines the dielectric performance and design and this would seem to be the preferable quantity to use.

A cross-section of impregnated taped insulation is given in Figure 11.19 for the 65/35 registration, i.e. effective angle of lay. To allow the cable to bend

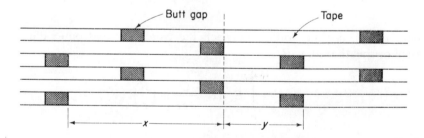

Figure 11.19   Cross-section of impregnated taped insulation. Registration $[x/(x+y)]$.

without adjacent tapes riding over each other (causing creasing and tearing) gaps are provided between tapes known as butt gaps. The butt gaps are filled with the impregnant and largely determine the electrical performance of the insulation. Because of the lower permittivity in the gaps the stress is higher than in the tape material. In Figure 11.19 it is evident that the 65/35 registration gives two tape thicknesses between gaps and is hence often used. The electric strength of taped insulation increases with decrease in tape thickness, and often a thinner tape is used close to the conductor and sheath with a thicker tape for the bulk of the insulation.

Internal partial discharges   In taped insulation gas is evolved in the butt gaps due to chemical processes, and in extruded (solid wall) insulation small cavities or voids exist along with impurities as a result of the manufacturing process. Breakdown of the gas occurs regularly in the a.c. cycle and randomly on d.c., giving discharges known as partial discharges. These erode the surrounding solid insulation and can cause breakdown. In extruded insulation the erosion is often in the form of a propagation channel and discharge paths take the general shape of trees. In paper–oil cables the working a.c. stress is made considerably below the inception stress for discharges. With extruded polyethylene cables in attempts to avoid discharges and 'treeing' the working stress is made low, typically an average value of about 3 kV/mm.

Life.   Experience over many years on samples and real cables has indicated that the life of a cable at constant temperature is governed by the empirical equation, $t \cdot E^n = $ constant. This law is tested by maintaining constant stress on the dielectric and measuring time to failure. Life under service conditions is obtained by extrapolating the straight line resulting from the plot of log E against log t. This assumes that the law which has shown to be valid over relatively short test times holds for a value of t of say 30 years or more. There is also the problem, especially for extruded insulation, of whether to use the maximum or average stress in the cable dielectric.

Cable losses

The $I^2 R_{a.c.}$ loss of the conductor represents the largest heat source. The alternating-current resistance of the conductor $R_{a.c.}$ is the direct-current resistance $R_{d.c.}$, modified to account for the skin and proximity effects. Skin effect, even at power frequency, is significant and increases with conductor cross-section. With large conductors e.g. 2000 mm^2 the increase in $R_{a.c.}$ due to this cause is of the order of 20 per cent.[26] It is minimized by the use of stranded conductors which in large conductors are formed into segments which are transposed.

Proximity effects include the eddy currents induced in the conductor and sheath of a cable in a circuit comprising three separate single-conductor cables by the conductor fluxes of the neighbouring cables. This loss decreases with increased separation between cables. Further losses are induced into steel-wire armouring wound around some cables (especially submarine) for mechanical protection.

The dielectric loss, due to leakage and hysteresis effects in the dielectric, is usually expressed in terms of the loss angle δ; $\delta = 90 - \phi_d$, where ϕ_d is the dielectric power-factor angle. The dielectric loss $= \omega C V^2 \tan \delta$ where $C = $ capacitance to neutral per metre and $V = $ phase voltage. For paper–oil $\tan \delta$

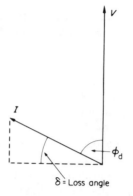

Figure 11.20   Loss angle in dielectric. Loss $= VI \cos \phi_d = VI \sin \delta = \omega CV^2 \sin \delta$.

lies between 0.002 and 0.003, but this value increases rapidly with temperature above 80°C in oil-filled cables. In low-voltage cables this loss is negligible, but is very appreciable in cables at 275 kV and above. A phasor diagram illustrating the nature of the loss angle is shown in Figure 11.20.

Currents and voltages in sheaths—self-contained cables

Two cables will be considered as shown in Figure 11.21a. If the sheaths are isolated,.i.e. not grounded at the ends, the fluxes set up by one cable, cut the

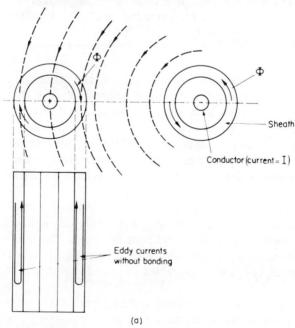

(a)

Figure 11.21(a)   Currents and fluxes in sheaths without bonding of ends of sheaths.

sheath of the other and induce eddy currents as shown in Figure 11.21(a)
Although the loss in this condition is very small except when the cables are
close or touching, voltages are set up along the sheath which may become
excessive. To avoid this the sheaths are bonded at the ends forming the
end-connexions indicated in Figure 11.21(b).

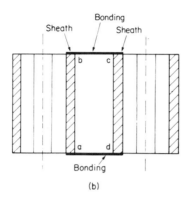

Figure 11.21(b) Two cables with sheaths bonded. Spacing between
centres of conductors = s metres. r and r_s = radius of conductor (solid)
and mean radius of sheath (metres).

The flux through the loop abcd forms a mutual inductance M between sheath
and conductor equal to

$$4 \times 10^{-7} \ln \left(\frac{s}{r_s} \right) H/m$$

where

$$s = \text{spacing between sheath centres};$$
$$r_s = \text{mean radius of sheath};$$

and for a three-phase system with equilateral spacing the effective value of M is

$$\frac{2}{10^7} \ln \left(\frac{s}{r_s} \right) H/m$$

The self-inductance of the sheath L_s is approximately equal to the mutual
inductance M, and letting the resistance of the current path be R_s, the sheath
current,

$$I_s = \frac{I\omega M}{\sqrt{R_s^2 + (\omega M)^2}}$$

and the sheath loss per phase

$$= I_s^2 R_s = R_s \left(\frac{I^2 \omega^2 M^2}{R_s^2 + \omega^2 M^2} \right) \qquad (11.15)$$

Hence total effective resistance, $R_{\text{a.c.}}^* = R_{\text{d.c.}}\,(1+\text{skin effect factor}+\text{proximity}$ effect factor)

$$+\left(\frac{\omega^2 M^2}{R_s^2+\omega^2 M^2}\right) R_s \quad \text{or} \quad R_{\text{a.c.}}^* = R_{\text{a.c.}}+\left(\frac{\omega^2 M^2}{R_s^2+\omega^2 M^2}\right) R_s$$

The equivalent circuit is shown in Figure 11.22, the conductor and sheath form a coupled circuit.

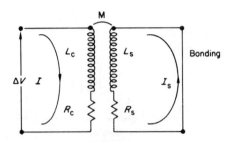

Figure 11.22   Equivalent circuit formed by conductor and bonded sheaths.

Consider a metre length of the circuit, the voltage drop per metre

$$= \Delta V = I(R_c+j\omega L_c)+I_s j\omega M$$

and in the sheath

$$0 = I_s(R_s+j\omega L_s)+I j\omega M$$

From these two equations

$$Z = \frac{\Delta V}{I} = \left[R_c+R_s\left(\frac{\omega^2 M^2}{R_s^2+\omega^2 M^2}\right)\right]+j\left(\omega L_c-\frac{\omega^3 M^3}{R_s^2+\omega^2 M^2}\right) \quad (11.16)$$

$$= R_{\text{effective}}+jX_{\text{effective}}$$

where $L_s = M$. Although the equivalent circuit shown is for a single-phase circuit equation (11.16) applies also for a three-phase, three-cable system, and

$$\omega L_c = X_c = 2\omega\left(2\ln\frac{s}{r}+\frac{1}{2}\right)\times 10^{-7}\ \Omega/\text{m}$$

and

$$X_s = \omega M = 2\omega\left(2\ln\frac{s}{r_s}\right)\times 10^{-7}\ \Omega/\text{m}$$

where $s = $ spacing between conductor and sheath centres, and r and $r_s = $ radius of conductor (solid) and mean radius of sheath.

The losses occurring with sheaths short-circuited have resulted in the widespread use of a system of sheath transposition known as *crossbonding*

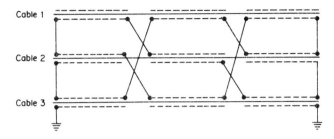

Figure 11.23   Cross-bonding of single-conductor cables.

which is illustrated in Figure 11.23. At each joint along the route the sheath is interrupted and transposed, and after three lengths have been traversed the sheaths are bonded to ground as shown. The effect is to add the sheath voltages in the adjacent sections and for balanced loads the sum will be zero. High sheath currents are avoided and induced voltages kept in reasonable proportions.

Example 11.1   Determine the effective electrical parameters of a three-phase, 66 kV, 50 Hz underground circuit, 65 km in length and comprising three single-core 200 mm^2 cables each of conductor radius 0.915 cm. The internal and external radii of the lead sheath are 2.5 and 2.8 cm respectively. The cables are in touching equilateral formation and the sheaths are bonded to ground at several points. The conductor a.c. resistance per kilometre at 15°C is 0.0875 Ω/km and the resistivity of lead at the operating temperature may be assumed to be 23.2×10^{-6} Ω-cm.

Solution   Conductor resistance $= 0.0875(1 + 0.004 \times 50) = 0.105$ Ω/km (conductor-operating temperature assumed to be 65°C and temperature coefficient 0.004). For the whole length,

$$R_{\text{d.c.}} = 0.105 \times 65 = 6.8 \ \Omega$$

Resistance of sheath

$$= \frac{23.2 \times 10^{-6} \times 65 \times 10^5}{\pi(2.8^2 - 2.5^2)} = 30 \ \Omega$$

In a three-phase system the equivalent phase to neutral reactance is considered, and in this case the conductor to sheath mutual inductive reactance,

$$X_m = \omega M = 2\pi \times 50 \times 2 \ln\left(\frac{d}{r_s}\right) \times 10^{-7} \ \Omega/\text{m}$$

where r_s = mean radius of sheath and d = the axis to axis spacing (6.1 cm).

$$\therefore \ X_m = 2\pi \times 50 \times 2 \times \ln\left(\frac{6.1}{\frac{1}{2}(2.5 + 2.8)}\right) \times 10^{-7} \times 65,000$$

$$= 3.4 \ \Omega \text{ for 65 km length}$$

Effective a.c. resistance of conductor $R^*_{\text{a.c.}}$.

$$= R_{\text{a.c.}} + \left(\frac{X^2_m R_s}{R^2_s + X^2_m}\right) = 6.8 + \frac{3.8^2 \times 30}{30^2 + 3.8^2} = 7.2 \ \Omega$$

Effective reactance per cable

$$= X - \left(\frac{X^3_m}{R^2_s + X^2_m}\right)$$

(where X = reactance with sheaths open circuit)

$$= \omega \times 2 \ln \left(\frac{d}{r_c}\right) \times 10^{-7} \times 65,000$$

$$= 314 \times 2 \ln \left(\frac{6.1}{0.915}\right) \times 10^{-7} \times 65,000$$

$$= 7.707 \ \Omega$$

Total effective reactance

$$= 7.75 - \frac{3.4^3}{30^2 + 3.4^2}$$

$$= 7.737 \ \Omega$$

$$\frac{\text{sheath loss}}{\text{conductor loss}} = \left(\frac{I^2 R_s X^2_m}{R^2_s + X^2_m}\right) \frac{1}{I^2 R}$$

$$= \left(\frac{30 \times 3.4^2}{30^2 + 3.4^2}\right) \frac{1}{6.8}$$

$$= 0.0568$$

For a current of 400 A the e.m.f. induced without bonding (per sheath) = $X_m I = 1.36$ kV.

11.8 System Operating Problems with Underground Cables

The determination of cable electrical parameters has been discussed in Chapter 3. It has been seen that the capacitance of cable systems is high because of their physical form. Table 11.5 shows a comparison of the charging reactive power requirements for typical overhead lines and underground cables in the United States. As the charging current flows in the cable conductor, a severe decrease in the value of load current transmittable (de-rating) occurs if the thermal rating is not to be exceeded; in the higher voltage range lengths of the order of 32 km (20 miles) create a need for drastic derating. A further current reduction is caused by the appreciable magnitude of dielectric losses at higher voltages.

Table 11.5 Comparison of charging MVA requirements for typical
overhead and underground a.c. transmission lines (60 Hz)
(pipe-type cables)

Voltage	Overhead three-phase charging MVA/km	Underground three-phase charging MVA/km
69	0.0155	1.18
138	0.066	3.04
230	0.188	5.465
345	0.53	10.56
500	1.00	18.82

The charging current (amperes per metre)

$$I_c = 2\pi f C V / \sqrt{3}$$

where C is the capacitance (farads) and V the line voltage (volts). When $I_c = I_{rated}$, the cable length is termed critical. Also, $C = 2\pi\varepsilon_0\varepsilon_r/\ln(R/r)$. The total three-phase reactive power

$$= \sqrt{3}\,VI_c = \sqrt{3}\,V2\pi f\left(\frac{V}{\sqrt{3}}\right) \cdot \left(\frac{2\pi\varepsilon_0\varepsilon_r}{\ln\,(R/r)}\right)$$

$$= 4\pi^2 f V^2 \varepsilon_0\varepsilon_r 10^{-6}/\ln\,(R/r)\text{MVAr/metre}$$

A 345–kV cable with approximately 25.3 mm (1 in) thickness of insulation and an ε_r of 3.5 has a critical length of 26.5 miles (42 km), the corresponding MVAr requirement is about 10.6 MVAr per km.

Apart from the reduction in transmission capability the presence of underground cables in a system presents further problems. Troubles arise when switching cables into circuit owing to their high capacitance (see Chapter 10) and also control of system voltage is difficult. Examination of the simple equivalent circuit of a cable indicates that on light or no load an appreciable voltage rise occurs at the load end (originally called the Ferranti effect). In a system containing a reasonable proportion of h.v. cables such as in large cities, significant voltage changes occur as the load varies between its maximum and minimum value. It is difficult to control these voltage excursions by normal system equipment and it is necessary to install shunt inductive reactors in the cable circuit. These may be connected at the ends of the circuit, and although technical advantage would result in connecting, say, one-third of the required amount of vars at each end and one-third at the middle, the cost of connexions at places other than the ends is relatively very high.

A further aspect of voltage control which is becoming more serious as the capacity of generators increases is the limited ability of large generators to operate at leading power factors. This requirement occurs especially at night when the load is low. It has been shown in section 3.5 of Chapter 3 by means of a performance chart that the ability to generate leading MVArs is limited by stability considerations. In particular as the synchronous reactance becomes larger the leading MVAr capability decreases and in high-capacity modern generators the reactance is larger than in smaller machines. These problems are avoided when cables are used in direct-current transmission schemes. Against the saving in the cost of compensating reactors must be set the large cost of the d.c. conversion equipment. It will be appreciated that there is a considerable incentive to develop cable materials of low permittivity, i.e. use of plastics.

11.9   Steady-state Thermal Performance of Cable Systems

In this section it is proposed to consider the thermal characteristics of the components of a cable system from a steady state and transient viewpoint. The transient aspect can be due to short circuits or the consideration of temperature–time characteristics when a steady load is changed under control to another value; this is electrically a steady-state condition as opposed to conditions on sudden short circuits.

In a transmission link, the component with the lowest limit will decide the current-carrying capacity; differences in thermal time-constant may, however, have a marked influence on this. An aspect of great economical importance is the possibility of short-time overloading.

The ambient temperature, depending as it does on the seasons and weather conditions, influences thermal limits; this operates favourably as the low winter ambient temperatures coincide with high load demands. The thermal limitations to transmission capacity are therefore not rigidly fixed and must be determined scientifically to obtain the optimum utilization of plant.

The thermal limit on an item of plant is that value of current which produces a maximum or 'hot-spot' temperature which should not be exceeded. Care should be taken to ascertain the difference between the 'hot-spot' and mean temperatures. The critical temperature rise is usually that in the insulation adjacent to the conductor.

Depending upon the class of insulation used the maximum allowable temperature rise is determined by a mixture of scientific analysis and practical experience. Difficulty arises in the assessment of allowable temperatures as to what precisely the criterion of plant failure or length of life should be. A transformer, for example, will continue to function with insulation which from the point of view of many physical tests is unsound. Failure usually occurs where some undue stress is applied due to a fault condition, when high mechanical and thermal stresses are set up. Factors which influence the maximum permissible temperature rise in cables are as follows.

(a) The differential expansion between the insulation, the surrounding sheath, and the conductor.
(b) Changes in the electrical properties of the insulation, especially in the dielectric loss.
(c) Changes in mechanical and chemical properties which result in electrical changes.

The following law relates the probable *life of insulation* with temperature $L = A e^{-m\theta'}$, where L is the insulation life, θ' the temperature, and A and m constants depending on the type of insulation.

Recommended steady-state working temperatures of the conductors of paper/oil cables are as follows:

armoured cables buried direct in soil—65°C;
oil-filled or gas pressure cables direct in soil—85°C.

The above values assume an earth temperature of 15°C; the range of earth temperature to be expected depends upon the depth and geographical location.

Thermal resistance of a single-core cable

Let the thermal resistivity of the dielectric material be g degC m/W. Then considering an annulus of width dx for a 1 m length of cable (Figure 11.24) the thermal resistance

$$= g \cdot \frac{dx}{2\pi x \times 1} \text{ degC/W}$$

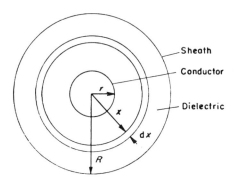

Figure 11.24   Thermal resistance of a single-core cable.

Total thermal resistance of the dielectric is

$$G_d = \int_r^R \frac{g \, dx}{2\pi x} = \frac{g \ln R/r}{2\pi} \text{ degC/W} \tag{11.17}$$

The thermal resistance between a core and the sheath in a three-core cable is difficult to calculate analytically. In practice a factor usually obtained by computer tests is used to modify equation (11.17).

Thermal resistance of the cable environment

The cables are assumed to be directly buried with the trench refilled with the original soil, i.e. they are situated in a semi-infinite homogeneous medium. The ground surface is assumed to be an isothermal plane. The thermal resistivity (g) of soil varies widely depending on the nature of the soil (clay, sand, etc.), and its moisture content. Values for g frequently taken are 1.2 degC m/W and 0.9 degC m/W. The value used in different countries varies somewhat, but is in the above region. By far the best policy when specifying a cable is to carry out a thermal resistivity survey of the proposed route. This will highlight the point of highest g, which value should be used in the calculation of the current rating. The mean value of g taken over the route is of little use as it is the hottest section of the cable that will fail; little heat is conducted longitudinally from the sections of highest temperature.

The external thermal resistance of a single cable may be found readily by considering the image of the line heat source in the thermal field as shown in Figure 11.25. If q watts is the total loss in 1 m of the cable, the temperature

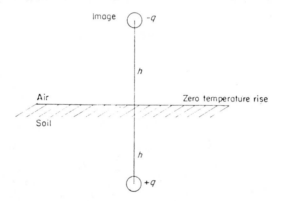

Figure 11.25   Image method for determining thermal resistance due to surrounding soil—one cable (radius r).

difference between the external surface of the cable and the ground surface is

$$\frac{q \cdot g}{2\pi} \ln \frac{2h}{r}.$$

h is normally in the order of 1 m. The external thermal resistance is therefore

$$\frac{\theta}{q} = \frac{g}{2\pi} \ln \frac{2h}{r} \text{ degC/W per metre of cable} \qquad (11.18)$$

Modifications to (11.8) have been made in the past to allow for the difference in the actual value of g and that of the soil sample tested in the laboratory. Nowadays the value of g is found directly *in situ* by probe methods and the unmodified expression (11.18) is used.

With several equally loaded, similar cables, in a group, which do not touch each other, the method of images is again used. Consider Figure 11.26. The temperature rise of a cable in the group is the sum of its own independent rise

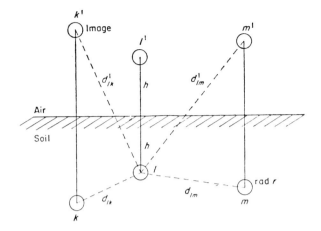

Figure 11.26   Image method for determining external thermal resistance for group of buried cables.

and that due to each of the other cables. The temperature rise at the surface of the l cable due to a loss q watts per centimetre in cable k is

$$\theta_{kl} = \frac{g}{2\pi} q \ln \frac{d'_{lk}}{d_{lk}}$$

Also, the temperature rise of l due to a loss q in m

$$= \frac{qg}{2\pi} \ln \frac{d'_{lm}}{d_{lm}}$$

Hence the temperature rise of l due to both k and m

$$= \frac{qg}{2\pi} \ln \frac{d'_{lm} d'_{lk}}{d_{lm} d_{lk}}$$

The total temperature rise of cable l therefore will be

$$\frac{qg}{2\pi} \ln \frac{2h}{r} + \frac{qg}{2\pi} \ln \left(\frac{d'_{lm} d^{d_l}_{lk}}{d_{lm} d_{lk}} \right)$$

For the popular configuration of three cables laid in horizontal formation with a spacing a between adjacent cable centres and a burial depth of h,

$$d_{kl} = a, \ d'_{kl} = \sqrt{(2h)^2 + a^2}$$

$$d_{lm} = a, \ d'_{lm} = \sqrt{(2h)^2 + a^2}$$

$$\therefore \ \theta_l = \frac{qg}{2\pi} \left[\ln \frac{2h}{r} + \ln \left(\frac{4h^2 + a^2}{a^2} \right) \right] \tag{11.19}$$

The effective thermal resistance of the centre cable of the group is $(\theta_l / q)\,°C/W$.

For cables drawn into ducts the external thermal resistance consists of the sum of the duct ground-surface resistance, the duct-wall resistance, and the cable surface to inner duct-wall resistance.

Moisture migration[28] An important aspect of cable environment is the increase in the resistivity of soil when it dries out. This has become important owing to the less cyclic nature of the load on some high-voltage cables now installed. When these cables form interconnexions to high-efficiency generating stations they carry full load continuously. The moisture in the surrounding soil migrates, leaving air in the interstices previously occupied by water and the thermal resistivity increases. This in turn increases the cable temperature which increases the dielectric loss and to a lesser extent the copper loss, thus causing increased moisture migration. The result is *thermal instability* and the electrical failure of the insulation.

Pipe-type cables (HPOF)

The effective thermal resistance from cable screens to pipe is complex, involving natural convection in the oil space. Similarly, the corresponding resistance for self-contained cables in air ducts must be evaluated. For values of the equivalent diameter (D_e) of the insulated cores between 75 and 125 mm, the effective thermal resistance, cable screen to pipe wall, is given by the empirical relation

$$\frac{100A}{1+(B+C\theta_m)D_e} \text{ degC/W per cm} \tag{11.20}$$

For three cores in a pipe-type cable, $D_e = 2.15 \times$ outside diameter of core (cm), for two cores $D_e = 1.65 \times$ core outside diameter (cm); θ_m is the mean temperature (°C) of the filling medium (e.g. oil), and values of A, B, and C are given in Table 11.6. An assumed value is initially used for θ_m in conjunction with an iteration process.

Table 11.6   Value of constants A, B, and C

Installation	A	B	C
In metallic conduit	5.2	1.4	0.011
In fibre duct in air	5.2	0.83	0.006
In fibre duct in concrete	5.2	0.91	0.010
In asbestos cement duct in air	5.2	1.2	0.006
In asbestos cement duct in concrete	5.2	1.1	0.011
Oil pressure pipe-type cable	0.26	0.0	0.0026

(Data from IEC Report 287, 1969)

Calculation of cable steady-state temperature rises

Methods for the calculation of the relevant thermal resistances and cable losses have been discussed and it now remains to outline the methods of calculation of the temperature rises. As with other items of electrical plant it is desirable to obtain an equivalent lumped-constant thermal circuit. This is shown for a cable of a directly buried, naturally cooled single-core system, in Figure 11.27. It is

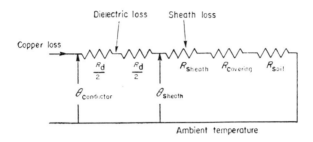

Figure 11.27 Equivalent thermal circuit for single-core, naturally cooled, buried cable. R_d = radial thermal resistance of dielectric.

seen that the dielectric loss is injected at the midpoint of the dielectric resistance; this is an approximation and the dielectric should be further subdivided for more accurate representation. Often the sheath thermal resistance is negligible. Although much attention has been devoted to cable heating it has now become apparent that with artificially cooled schemes the joint will rise to a higher temperature than the cable. Thermal analysis of the joint is difficult compared with cables and requires computer analyses.

Example 11.2 In this example the traditional empirical approach to the current rating of a pipe-type cable will be given.

Determine the rating of a 500 kV, 60 Hz, HPOF pipe-type cable. Data— diameter over conductor 41.4 mm; thickness of insulation 34 mm; diameter of pipe 300 mm; depth of pipe centre 1 m; soil g 0.9 degC m/W; ambient 25°C; conductor $R_{ac} = 0.044$ Ω/ph per km; dielectric g 5.0 degC m/W; tan δ 0.002, ε_r 3.5; pipe loss 0.28 × conductor loss; operating temperature 80°C.

Solution Thermal resistance cable surface to pipe (R_{c-p})

$$= \frac{100A}{1 + (B + C\theta_m)D_e}$$

For this system, $A = 0.26$;

$\qquad B = 0$;

$\qquad C = 0.0026$;

$\qquad D_e = 2.15(41.4 + 68) = 235.2 \text{ mm} = 23.52 \text{ cm}.$

Assume mean oil temperature to be 55°C,

$$R_{c-p} = \frac{100 \times 0.26}{1 + (0.0026 \times 55)(23.52)}$$

$$= 5.95 \text{ degC/W per cm length}$$

$$R_{\text{dielectric}} = \frac{500}{2\pi} \ln \left(\frac{235.2}{41.4}\right)$$

$$= 77.32 \text{ degC/W per cm length}$$

$$R_{\text{soil}} = \frac{90}{2\pi} \ln \frac{2 \times 100}{15.0}$$

$$= 36.9 \text{ degC/W per cm}$$

Conductor loss

$$= I^2 0.044 \times 10^{-5} \text{ W/cm}$$

Pipe loss

$$= 3 \times 0.28 I^2 \times 0.044 \times 10^{-5} \text{ W/cm}$$

Capacitance per core

$$= \frac{2\pi \times 8.85 \times 10^{-12} \times 3.5}{\ln (109.4/41.4)} \text{ F/m}$$

$$= 200 \ \mu\mu\text{F/m}$$

Dielectric loss = 15.21 W/m = 0.152 W/cm per core; temperature difference conductor to ambient = 55 degC

$$= 0.44 \times 10^{-6} I^2 (77.32) + 3 \times 0.44 \times 10^{-6} I^2 (5.96 + 36.9) + 0.152 \left(\frac{77.32}{2}\right)$$

$$+ 3 \times 0.152 (5.96 + 36.9) + 3 \times 0.28 \times 0.44 I^2 \times 10^{-6} \times 36.9$$

Hence $I = 530$ A. With this value of current the mean oil temperature is 64°C. As the initial guess for θ_m was 55°C it is worth while recalculating I with θ_m taken as, say, 62°C.

The new value of R_{c-d}

$$= \frac{100 \times 0.26}{1 + (0.0026 \times 62)(23.52)} \text{ degC/W per cm}$$

$$= 5.43 \text{ degC/W}$$

In the context of the whole chain of thermal resistances this is a small change and the current will not be charged to any significant extent. It is left to the reader to ascertain the resulting change in current.

11.10 Transient Thermal Performance

The following regimes of transient heating occur.

(a) *Short-circuit.* All the heat evolved is assumed to be stored in the cable. A temperature limit of 120°C is often assumed.

(b) *Cyclic loading.* Calculations are often based on an idealized daily load curve of full-load current for 8 h and no-load for 16 h. This results in a lower temperature rise than when permanently on full load which is allowed for in practice by the use of cyclic rating factors. For directly buried cables of conductor area less than 64.5 mm^2 (0.1 in^2) this factor is 1.09, and for conductor areas from 64.5–645 mm^2 (0.1–1.0 in^2) it is taken as 1.13, i.e. 1.13 times the rated full-load current can be passed on cyclic loading.

(c) *Short-time emergency loading.* An overload may be sustained for perhaps several hours without excessive temperatures due to the long thermal time-constant of the cable.

A lumped-constant thermal network may be used for transient calculations by the connexion of appropriate thermal capacitances between the nodes of the steady-state network and the ambient or reference line. For a cable if uniformly radial heat flow is assumed, i.e. the sheath is isothermal and the thermal circuit is an R–C ladder network. The number of nodes depends on the number of annular cylinders into which the dielectric is divided. The conductor and sheath temperatures may be obtained from a simple model such as suggested by Wormer[31] and shown in Figure 11.28. Reasonable results over a few hours of transient are obtained for pipe-type cables by making C_3 in Figure 11.28 the thermal capacity of the pipe oil.

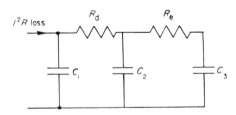

Figure 11.28 Simple model for transient thermal response of buried cable: C_1 = thermal capacitance of conductor + (p × capacitance of insulation); C_2 = thermal capacitance of sheath + $(1-p)$×(capacitance of insulation); C_3 = effective capacitance of soil; R_i = thermal resistance of insulation, R_e = effective thermal resistance of soil, $p = 1/[2 \ln (R/r)] - 1/(R^2/r^2 - 1)$, where R = outside radius of cable and r = conductor radius.

In transients of short-time interest the surrounding soil may be represented by a few rings with the assumption of radial flow. For an accurate simulation the temperature rise at the cable surface due to the soil alone is described by the equation

$$\theta(t) = \frac{q(t)g_{\text{soil}}}{4\pi}\left[\left\{-Ei\left(-\frac{d^2}{16\alpha t}\right) + Ei\left(-\frac{l^2}{\alpha t}\right)\right\}\right.$$

$$\left. + \sum_{i=1}^{i=N}\left\{-Ei\left(-\frac{r_{1,i}^2}{4\alpha t}\right) + Ei\left(-\frac{r_{2,t}^2}{4\alpha t}\right)\right\}\right] \tag{11.21}$$

where i = number of cable, $r_{1,i}$;
$\quad r_{2,i}$ = distances of cable i and its image from reference cable;
$\quad l$ = depth;
$\quad d$ = cable diameter;
$\quad \alpha$ = thermal diffusivity;
$\quad Ei$ = exponential integral

If $q(t)$, the value of the heat transmitted from the cable to the soil, is known at the instant t, then the temperature of the cable surface is known and the cable thermal network may be solved. However, $q(t)$ varies from zero at $t = 0$ to q (steady state) at $t = \infty$. A reasonably accurate formula for obtaining $q(t)$ is

$$\frac{q(t)}{q_{\text{steady state}}} = \left(\frac{\text{temperature across cable at } t}{\text{steady-state temperature rise across cable}}\right)$$

(cable surface node short-circuited)

The simplest analysis results from a single lumped constant representation, i.e. a single R–C network. The response of such a network to a time-varying load is illustrated in Figure 11.29 in which τ is the thermal time constant ($R \times C$) and θ_{m1}, θ_{m2}, and θ_{m3} are steady-state temperatures.

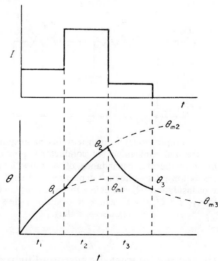

Figure 11.29 Temperature rise–time curves, varying load: $\theta_1 = \theta_m$
$(1 - e^{-t1/\tau})$;　　$(\theta_2 - \theta_1) = (\theta_{m2} - \theta_1)(1 - e^{-t2/\tau})$;　　$(\theta_3 - \theta_{m3}) =$
$(\theta_2 - \theta_{m3})\,e^{-t3/\tau}$.

Most items of power equipment possess temperature rise-time characteristics having more than one exponential term. This fact is important, as often when carrying out heat runs the temperature will appear to no longer rise with the time scale being used, usually a few hours. It is quite possible, however, for the temperature to continue rising appreciably over a considerable time. In buried cables, for example, many hundred hours elapse before steady conditions are reached.

In all the above equations the losses have been assumed constant. In practice the I^2R loss increases with increase in temperature due to the temperature coefficient of resistance of copper (0.4 per cent per °C). The losses will normally consist of a constant plus a variable portion, i.e. $q_0 + a\theta$.

11.11   Artificial Cooling of Underground Cables

In recent years progress has been made in schemes for reducing the external thermal resistance of cables by artificial cooling. These methods take two main forms.

(a) *Internal cooling.*   The pumping of water or insulating oil along a central duct formed in the cable conductor.

(b) *External cooling.*   Water is pumped through plastic pipes laid beside the cables as illustrated in Figure 11.30(a). The normal thermal resistance presented to the cable losses is shunted by the thermal resistance into the water. The water increases in temperature along the route so that there is a limit to the individual lengths which can be cooled in this manner. The pipe-to-soil resistance may be calculated by the method of images assuming an infinite homogeneous medium or by field plots using appropriate techniques. From an

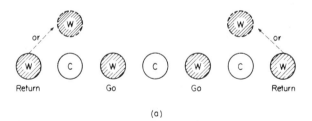

(a)

Figure 11.30(a)   Cooling pipes laid along buried cables—two pipes take outward flow and the other two the return flow.

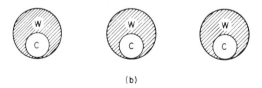

(b)

Figure 11.30(b)   Integral (surface) cooling.

462

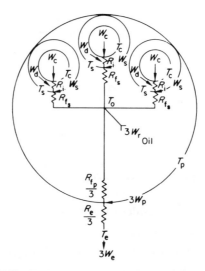

Figure 11.30(c) Equivalent thermal circuit of U.S. pipe-type cable system with forced cooling by passage of coil. Three cables are installed per pipe. (*Permission of the Institute of Electrical and Electronics Engineers.*) W_c = conductor loss, R_i = insulation thermal resistance, T_c = conductor temperature, T_s = sheath temperature, T_o = oil temperature, T_e = earth temperature, T_p = temperature of pipe, R_{fs} = thermal resistance sheath to oil, R_{fp} = heat-transfer resistance oil to pipe walls, R_f = thermal resistance of pipe, R_e = effective thermal resistance of soil.

engineering standpoint this scheme is much simpler than (a). Greatly increased currents are achievable by the use of large pipes of diameter, say, 150 mm.

(c) A more intensive water-cooled system than the use of external pipes involves the placing of each cable in a rigid plastic pipe of approximately 20 cm diameter and pumping water through the pipe. It is more expensive than method (b), but with a 1935 mm^2 conductor cable currents of 3200 A are achieved. This method is known as integral or surface cooling, see Figure 11.30. In American pipe-type cables the insulating oil may be pumped, thus providing enhanced heat transfer from the cable sheaths. The thermal network is shown in Figure 11.30(c). The necessity for artificial cooling if transmitted powers are to be increased at higher voltages is illustrated in Figure 11.31. It is seen that for each mode of cooling there exists a maximum power at a certain dielectric stress after which, owing to increased dielectric losses, the power falls. With normal burial in soil of thermal resistivity 1.2 degC m/W, the limit is reached at a stress of 15 MV/m. The rating of cables laid in tunnels or large ducts may be increased by the provision of air flow, either artificially induced or by natural ventilation.

Calculation of coolant temperature rise

In artificially cooled schemes the reference temperature is that of the coolant and the temperature rise of this is a limiting parameter; it may be calculated as

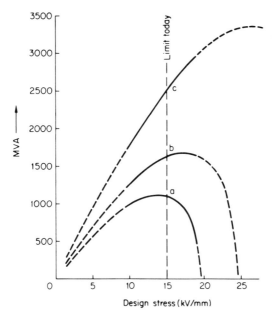

Figure 11.31   Relation between transmittable MVA and maximum working stress on cable dielectric (buried, self-contained, LPOF 2000 mm² conductor cable): (a) natural cooling (normally buried); (b) use of selective backfill (normally buried); (c) artificial cooling.

follows. Consider a pipe, tunnel or duct traversed by a coolant at W cubic metres/second entering at θ_0°C. The cables emit a loss in the steady state of q watts/metre and the thermal resistance through the containing walls and surrounding soil to ambient is R degC/watts per metre. At a distance x from the entrance the coolant temperature is θ (see Figure 11.32) and the following equation applies:

$$\frac{\theta}{R/\mathrm{d}x} + CW\,\mathrm{d}\theta = \mathrm{d}x\,q$$

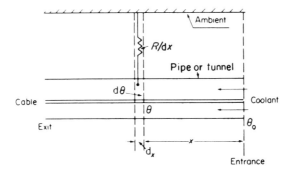

Figure 11.32   Temperature rise of cooling medium in a tunnel or duct.

where C is the volumetric specific heat of the coolant. Hence,

$$\frac{d\theta}{Rq - \theta} = \frac{dx}{CWR}$$

integrating and putting $CWR = \lambda$,

$$\left(\frac{Rq - \theta}{A}\right) = e^{-x/\lambda}$$

where A is the integration constant. When $x = 0$, $\theta = \theta_0$, and $Rq - \theta_0 = A$. Also with zero coolant velocity the temperature rise, $\theta_m = qR$ degC. So that

$$\theta = \theta_m - (\theta_m - \theta_0)\, e^{-x/\lambda} \qquad (11.22)$$

Hence the steady-state coolant temperature at any distance may be calculated. The actual temperature of the cable conductor is obtained by adding to the coolant rise the temperature drop across the cable dielectric and covering plus the drop across the cable-to-air heat-transfer resistance.

11.12 Cables for Direct Current Transmission

Owing to the absence of periodic charging current with direct voltage, high-voltage cables will play an increasingly important role in direct-current transmission links. For inland schemes direct current will be used because of the greater utility of underground cables. In an alternating current cable a power factor of 0.003 can be represented by a loss resistance of 3×10^{12} Ω-cm. The direct-current resistivity of the same dielectric would be greater than 10^{14} Ω-cm, hence the loss in the dielectric on direct current is only about 3 per cent of that on alternating current. Whereas the electric stress distribution in alternating-current cables is determined by the dielectric capacitance, in direct-current cables it is determined by the electric resistance of the dielectric. The electrical resistivity of the conventional dielectrics is very temperature dependent; for oil-impregnated paper for example the resistivity at 20°C is 100 times that at 60°C. In direct-current cables thermal considerations not only determine the rating but also influence the electric stress distribution in the dielectric. The electrical resistivity also varies with the electric stress. Instead of the electric stress decreasing through the dielectric from the conductor to the sheath, in direct-current cables the stress increases and can be larger at the sheath, as shown in Figure 11.33. This is known as stress inversion and can lead to troubles at terminations and joints where longitudinal stresses are created.

Example 11.3 An underground cable system is cooled by water pipes laid along the route. On normal operation with the water cooling in action the cable conductors attain a steady temperature rise of 70°C; this value increases to 120°C without water cooling. The temperature–time relation without water cooling is an exponential with a time constant of 7 h, while with water cooling the time constant is 0.65 h. If the water cooling is out of action for 7 h calculate

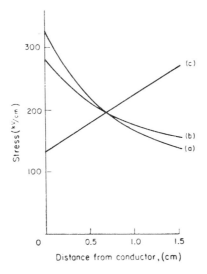

Figure 11.33   Distribution of electric stress in a direct-current cable.
Curve (a) on a.c., curve (b) on d.c. (no load), curve (c) on d.c. with load,
working voltage 300 kV. (*Permission of Electrical Journal.*)

the time that will elapse after the cooling is restored for the conductor temperature rise to be 80°C.

This example illustrates the important case of the failure of the water cooling system and the time available for recovery. Single time constants for the transient curves have been assumed for simplicity. The following relationships apply:

$$\theta_2 - \theta_1 = (\theta_{2m} - \theta_1)(1 - e^{-t_2/\tau_2})$$

(see Figure 11.29)

$$\theta_3 - \theta_{3m} = (\theta_2 - \theta_{3m})\, e^{-t_3/\tau_3}$$

Hence,

$$\theta_2 - 70 = (120 - 70)(1 - e^{-7/7})$$

$$\therefore\ \theta_2 = 101.6°C$$

With water cooling on:

$$\theta_3 = 80°C$$

and

$$(80 - 70) = (101.6 - 70)\, e^{-t_3/0.65}$$

$$\therefore\ \frac{t_3}{0.65} = 1.15$$

$$t_3 = 0.75\ h$$

Time to fall to 80°C

$$= 0.75\ h$$

11.13 Underground Distribution Systems

Although the emphasis in this book tends to be on technological problems concerned with transmission at high voltages, the distribution of energy to individual consumers should not be thought of as of lesser importance. The capital cost of installed distribution plant is vast and improvements in technique are constantly being made. Traditionally the main transmission circuits feed 11 kV (15 kV in the U.S.A.) substations from which a network of feeders, usually underground in urban areas, in turn supply local substations where 415 V, three-phase and 240 V, single-phase (208 V three-phase and 120 V single-phase in the U.S.A.) feeders form a network supplying consumers. Arrangements at this level vary somewhat from one supply company to another, but where radial feeders are used effective switching to ensure continuity of supplies in case of faults is made. Considerable developments in distribution in the U.S.A. have been brought about by the growing public pressures to underground circuits in suburban areas.

Improved cables and multiple earthing

Although oil-impregnated paper insulation is still widely used in Britain for medium-voltage cables, elsewhere synthetic insulations dominate. PVC insulated and sheathed split-concentric neutral-earth cable for house services is being used by some authorities.

The use of the neutral for the dual purpose of neutral connexion and consumers' earth connexion requires the use of a cable with only four conductors, the neutral being in the form of a concentric conductor as shown in Figure 11.34. The neutral conductor is earthed at the supply transformer, at the

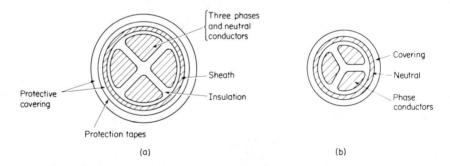

Figure 11.34   Relative size of a conventional cable with the concentric neutral type. (a) Conventional. (b) Concentric neutral type.

remote end, and at intermediate points such that the resistance to earth is always less than 10 Ω. This system is known as *protective multiple earthing* (*p.m.e.*). The conductors are normally of aluminium and one possible hazard is the neutral becoming discontinuous because of corrosion.

High-voltage single-phase networks

There is a growing tendency in the U.S.A. to use high voltage for single-phase. A schematic diagram of such a system is shown in Figure 11.35. The distribution point separates the three-phase supply into three single-phase to neutral supplies, each of up to 1 MVA. These cables radiate from the distribution point supplying 12 kV/120 V transformers of 50–100 kVA capacity which in turn supply the customers. These transformers are frequently situated underground. The economics of the system depends on the housing density, at 10 houses per hectare (typical of the U.S.A.) the h.v. single-phase system is considerably cheaper than traditional schemes; at 30 houses per hectare (typical of Britain) the system is marginal.

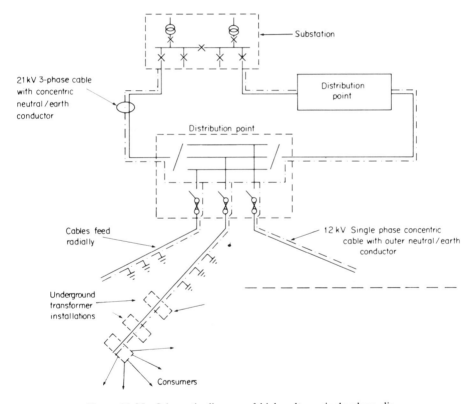

Figure 11.35   Schematic diagram of high-voltage single-phase distribution scheme.

11.14   Cryogenic Cables

With the magnitudes of power to be transmitted steadily increasing attention is constantly being given to new materials and techniques. Ideas which theoretically are attractive frequently cannot be translated into an engineering

468

entity because of lack of suitable materials of the fact that an existing process is cheaper. The continued use of paper/oil insulation up to the highest voltages over most of this century is a sobering lesson on the availability of materials and the importance of economics on engineering processes. The application of superconductivity to electric transmission was for many years a dream; the superconductors available lost their remarkable properties at relatively low magnetic fields, being of little use in engineering applications. In 1961 niobium and a compound of niobium and tin (Nb_3Sn) became available and remained superconducting at 8.8 Wb/m^2 (tesla) with a current density of 10^5 A/cm^2 and their application to electric transmission and machinery became the object of intense research activity.

A possible alternative to the superconducting cable is the resistive operation of a conductor of conventional material (probably aluminium) at cryogenic temperatures. This utilizes the decrease in resistivity with decrease in temperature (see Figure 11.36) and uses a cryogenic fluid in the same manner as a conventional coolant; such a cable is known as cryoresistive.

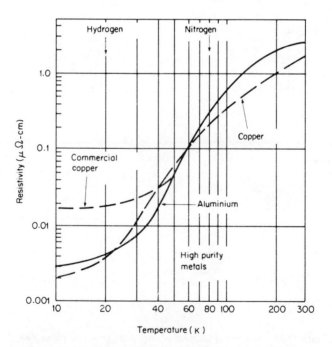

Figure 11.36   Relation between the electrical resistivity of metals with temperature.

A major cost in any cryogenic scheme will be the refrigerators, the size of which will be considerable owing to their low efficiencies, especially at very low temperatures. The basic parameters of coolants and refrigerators related to a conventional cable are summarized in Table 11.7.

Table 11.7 Properties of cryogenic liquids

Cable coolant	Temperature (K)	Relative resistivity copper and aluminium	Refrigerator ratio watts in/watts load	
			Theoretical	Practical
Conventional oil	293	1	—	—
Liquid N_2	77	$\frac{1}{8}$	3	6–10
Liquid H_2	20	$\frac{1}{500}$	14	50–100
Liquid He	4.2	Super-conducting	75	500–1000

For resistive cables the optimum coolant theoretically would appear to be liquid hydrogen, but the cheapness and complete safety of nitrogen make it more attractive.

Although it would seem attractive to operate superconducting cables at low voltages there are practical difficulties such as conductor a.c. loss and the transformer requirements at the ends. Also high operating voltages appear to be economically advantageous. Some design aspects are common to both superconducting and resistive cables, e.g. electrical insulation, cooling systems and the cryogenic envelopes to minimize heat entering the cable from the ambient (inleak); these requirements will be discussed first and then possible designs described.

It is necessary to keep the heat inleak to a very low value, e.g. in liquid helium systems a few microwatts per square centimetre. The envelope which encloses the cable cores and coolant system comprises superinsulation which is made up of multiple reflective metallic surfaces spaced in an angular region under vacuum.

Cooling systems and refrigeration

The cost of refrigerators approximates to $(a + bq^{\frac{1}{2}})$, where a and b are constants and q the heat to be removed, hence with larger systems refrigeration costs are relatively less. Two factors control the maximum length of cable between refrigerator and pumping stations, these are the maximum temperature rise and maximum pressure drop of the coolant. Two-phase flow, i.e. a mixture of liquid and gas sets up high-pressure drops in the cooling channels and should be avoided. Coolants may be pumped through conductor ducts and returned through pipes separate from the cable cores or returned through the conductor ducts of a second cable.

Dielectrics

A number of possibilities exist for the electrical insulation which, because of the use of high voltages, is of paramount importance. As vacuum is needed for thermal insulation it could be used as the dielectric as well. The presence of spacers to separate the annular surfaces reduces the breakdown voltage compared with that of plain vacuum gaps. The low-temperature environment will enhance the dielectric properties of vacuum which include very low dielectric loss. The use of the cryogenic fluids themselves as impregnants with taped insulation using plastic, e.g. polyethylene tapes is attractive. Both liquid nitrogen and hydrogen have breakdown and loss characteristics superior to those of conventional paper/oil, while those of liquid helium are slightly worse although still usable. A further advantage is the low permittivity compared with paper/oil giving lower capacitance. If tapes are not used then the problems associated with spacers are again encountered. Typical dielectric loss angles (tan δ) are 2×10^{-7} for hydrogen at 14 K, 2×10^{-6} for helium at 4.2 K, and 2×10^{-3} for paper/oil at 293 K. Breakdown properties are summarized in Figure 11.37

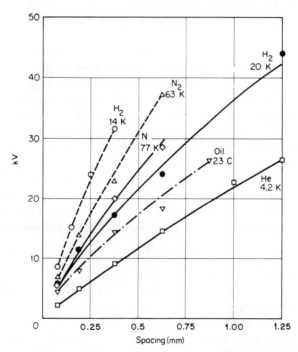

Figure 11.37   Electrical breakdown properties of cryogenic fluids. (*Permission of the Institute of Electrical and Electronics Engineers.*)

Superconducting cables

Below their critical temperature (T_c) superconductors exhibit perfect diamagnetism and the current flows on the conductor surface; above T_c they become resistive. Metals which switch state in this manner are called Type I and include most pure metals (e.g. lead); another variety known as Type II superconductors switch to a mixed state, partly superconducting and partly resistive, above the original critical field and do not become completely resistive until a much higher magnetic field strength. When the Type II conductor is in the mixed state the current and flux penetrate the conductor, and on alternating current a loss occurs which at 50 Hz in a niobium strip of thickness 0.0025 cm and carrying 400 A/cm of circumference amounts to 10μ W/cm^2; for a thickness of 0.016 cm with 800 A/cm the loss is 40 μW/cm^2. It is usual to express currents in terms of distance around the circumference as the current largely flows on the surface. Type II conductors comprise mainly brittle intermetallic compounds such as niobium tin (Nb$_3$Sn) niobium–zirconium (NbZr).

In the practical cable the conductor must withstand several times the rated current on faults, and to meet this condition the superconductor is formed on to a strip of high-purity aluminium or copper of thickness 0.25 cm.

The thermal contraction of polymer tapes is about 10 times the value for the conductor material. To avoid undue stresses in the insulation at low temperatures (the tapes are of course applied at ambient) in superconducting cables the conductor is arranged to contract radially, thus compensating for the contraction. As well as the more obvious electric parameters such as breakdown voltage on impulse and a.c., crucial quantities at low temperatures are low values of ε_r and tan δ. Mechanical properties at low temperatures are also vital and from the large number of synthetic tape materials available few are worth consideration, and of these polyethylene at the moment appears to be predominant. The impregnated taped insulation has now emerged as the preferred insulation system because of its electrical integrity (no spacer or particle problems) and facility for bending, i.e. a flexible core is possible. It will be necessary to operate such insulation below the partial (corona) discharge level.

Of the various a.c. versions which have been developed one type appears to be nearest final development. This embodies a superconductor in the form of helical strips on a former, insulated by polyethylene tapes. The three insulated conductors are situated in a common cryogenic envelope. A British design, shown in Figure 11.38, uses niobium as the superconductor and has a

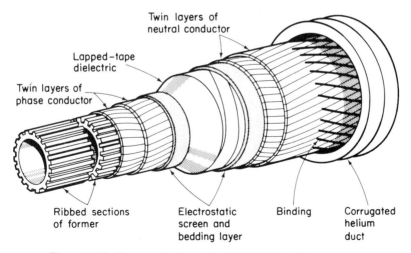

Figure 11.38   Superconducting cable—British design. (*Permission of Central Electricity Generating Board.*)

temperature range up to 6 K. An American design developed by the Brookhaven Laboratories is similar in concept but uses niobium tin as the superconductor. It uses supercritical helium as the coolant, with inlet and outlet temperatures of 6 and 8 K and an inner-bore pressure of about 15 atm.

Such cables are expected to become less expensive than conventional cables at around 5 GVA capacity but because of the reduced load growth the use of superconducting cables in Europe is unlikely this century and the development

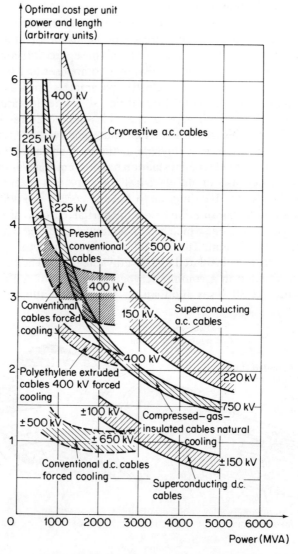

Figure 11.39 Relative costs of various types of cable system. (*Permission of the Institute of Electrical and Electronics Engineers.*)

effort has been reduced. In the U.S.A., Japan, and Russia, however, considerable development is continuing and the first full-size test cable was commissioned in 1982.

With operation on d.c. similar designs may be used although the situation is much eased by the absence of the conductor, dielectric, and screen losses. Hence, much greater powers are possible (e.g. 10 GW), but as with conventional cables the cost of the conversion equipment is a serious drawback except for long links. A review of costs for the entire range of available and projected cables is given in Figure 11.39.

References

BOOKS
 1. Peek, F. W. Jr., *Dielectric Phenomena in High Voltage Engineering*, McGraw-Hill, New York, 1929.
 2. *EHV Transmission Line Reference Book*, Edison Electric Institute, New York, 1968.
 3. *Transmission Line Reference Book, 345 kV and Above*, Electric Power Research Institute, Palo Alto, U.S.A., 1975.
 4. McCombe, J., and F. J. Haigh, *Overhead-line Practice*, Macdonald, London, 1966.
 5. *Underground Systems Reference Book*. Edison Electric Institute, 1957.
 6. Weedy, B. M., *The Underground Transmission of Electric Power*, Wiley, Chichester, 1980.
 7. Emanueli, L., *High Voltage Cables*, Chapman and Hall, London, 1926.
 8. McAdams, W. H., *Heat Transmission*, 3rd edn., McGraw-Hill, New York, 1961.
 9. *Electrical Transmission and Distribution Reference Book*. Westinghouse Electric Corp., Pennsylvania, 1964.
10. Endrenyi, J., *Reliability Modeling in Electric Power Systems*, Wiley, Chichester, 1978.

PAPERS
11. Cox, E. H., 'Overhead line practice', *I.E.E. Review*, **122** (1973), 1009–1017.
12. Leclerc, M. *et al.*, 'Wetting processes on a contaminated insulator surface', *I.E.E.E. Trans.*, **PAS-101** (1982), 1005.
13. Allan, R. N., and J. E. J. Cottrill, 'Design and parameters affecting the surface stress of overhead-power-line conductor systems', *Proc. I.E.E.*, **118** (1971), 1486–1492.
14. I.E.E.E. Committee Report, 'Survey on extra high voltage line noise', *Trans. I.E.E.E.*, **PAS-92** (1973), 1019–1028.
15. I.E.E.E. Working Group, 'Electrostatic effects of overhead transmission lines', *Trans. I.E.E.E.*, **PAS-91** (1972), 422–444.
16. Shah, K. R. *et al.* 'More than appearance to EHV line design', *Energy International (U.S.A.)*, **July 1975**.
17. I.E.E.E. Committee Report, 'A survey of the problem of insulator contamination in the U.S.A. and Canada', *Trans. I.E.E.E.*, **PAS-91** (1972), 1948–1954.
18. Avila, C. G., and A. F. Corry, 'Underground transmission in the United States', *I.E.E.E. Spectrum*, **March 1970**.
19. 'Study of environmental impact of underground electric transmission systems', *E.P.R.I. Report 7826*, E.R.D.A. E(49-18)-1608, **May 1975** (U.S.A.).
20. I.E.E.E. Committee, 'A comparison of methods for calculating audible noise of high voltage lines', *I.E.E.E. Trans.*, **PAS-101** (1982), 2694.
21. Gazzana Priaroggia, P. *et al.*, 'The Long Island Sound submarine cable interconnection', *Trans. I.E.E.E.*, **PAS-90** (1971), 1863–1873.

22. Miranda, F. J., and P. Gazzana Priaroggia, 'Recent advances in self-contained oil-filled cable systems', *I.E.E. Electronics and Power*, **Feb. 1977**, 134–140.
23. Balet, W. J., and R. L. Webb, 'Reactive power and its control in a large metropolitan electric supply system', *I.E.E.E. Trans.*, **PAS-87** (1968), 49–52.
24. 'Calculation of the current rating of cables (100% load factor)', *I.E.C. Report 287* (1969).
25. Allen, P. H. G., 'The thermal properties of high voltage insulants', *Proc. I.E.E., C,* Monograph, No. 250 M (1957).
26. Ball, E. H., and G. Maschio, 'The a.c. resistance of segmented conductors used in power cables', *I.E.E.E. Trans., P.A. & S.*, **87** (1968), 1143.
27. Ball, E. H., E. Occhini, and G. Luoni, 'Sheath overvoltages in H.V cables resulting from special sheath bonding connexions', *Trans. I.E.E.E., P.A.S.*, **84** (1965), 974.
28. Beale, H. K., 'Some economic aspects of extra high-voltage underground cables', *I.E.E., Proc.*, **112** (1965), 109.
29. Ralston, P., and G. H. West, 'The artificial cooling by water of underground cables', *C.I.G.R.É.*, Paris, Paper 215 (1960).
30. Weedy, B. M., 'Thermal transients in a high voltage cable system with natural and artificial cooling', *Proc. I.E.E.*, **109A** (1962).
31. Wormer, I., 'Calculating cable temperature transients', *Trans. A.I.E.E.*, **74** (1955).
32. Neher, J. H., 'The transient temperature rise of buried cable systems', *Trans. A.I.E.E.*, **83** (1964).
33. Weedy, B. M., and J. P. Perkins, 'Steady state thermal analysis of a 400 kV cable through joint', *Proc. I.E.E.*, **114** (1967), 109.
34. Kitagawa, K., 'Forced cooling of power cables in Japan. Its studies and performance', *C.I.G.R.E.*, Paper 213 (1964).
35. Wilkinson, K. J. R., 'Prospect of employing conductors at low temperature in power cables and in power transformers', *Proc. I.E.E.*, **113** (1966), 1509.
36. Mathes, K. N., 'Cryogenic cable dielectrics', *Trans. I.E.E.E. Electrical Insulation*, **EL4**, No. 1 (1969).
37. Graneau, P., 'Economics of underground transmission with cryogenic cables', *I.E.E.E., P.A.S.*, **89** (1970), 1.
38. McAnulla, R. J., 'Sodium conductor for power cables', *I.E.E. Electronics and Power*, **Nov. 1968**, 434.
39. Paul, H., 'Power transmission of the future', *I.E.E. Electronics and Power*, **May 1970**, 171.
40. Weedy, B. M., and S. J. Rigby, 'Thermal and electrical assessment of flexible cryoresistive cables', *Cryogenics* **87** (1977), 453–459.
41. Forsyth, E. B. *et al.*, 'The technical and economic flexibility of superconducting power transmission', *I.E.E.E. Trans.*, **94** (1975), 161–167.
42. Swift, D. A., 'Dielectric design for a superconducting cable with solid insulation', Revue Générale d'Electricité, No. 10 (1975), 741–746.
43. Deschamps, L. *et al.*, 'Prospects of development in France of electric power transmission by insulated EHV cables', *Proc. I.E.E.E. Underground Transmission and distribution Conference*, 1976, pp. 459–464.

Problems

11.1 A transmission line consists of three conductors of radius 0.585 cm (0.23 in) equilaterally spaced 306 cm (10 ft) apart. The line voltage is 139 kV at 60 Hz and the line conductor surfaces are smooth giving a value for m of 0.96. The barometric pressure is 72.2 cm of mercury, the temperature 20°C, and the weather fair. Calculate the corona loss per km.
(Answer: 2.24 kW/km)

11.2 A three-phase overhead line consists of three conductors in equilateral formation spaced 244 cm apart. The conductor diameter is 1.04 cm and the conductor surface factor (m) is 0.85. The air temperature and pressure are 21.1°C and 74 cm of mercury. Calculate the critical visual voltage for corona.

(Answer: 81 kV)

11.3 The span between two towers is 300 m. If the conductor weight per metre is 0.87 kg and the conductor tension (assumed equal to the horizontal tension) is 14.3 kN, calculate the length of the conductors.

(Answer: 300.4m)

11.4. A steel-cored aluminium overhead line has a total area of 850 mm² and diameter of 20 mm. The a.c. resistance is 0.045 Ω/km. at 65°C. If the working temperature is 65°C calculate the percentage increase in conductor length. The temperature coefficient of expansion is 11.5×10^{-6} degC⁻¹. Calculate the current rating for an ambient of 25°C and a wind velocity of 0.5 m/s. Ignore the radiated loss and solar heating.

(Answer: 955 A)

11.5. An overhead line of span 250 m and effective cross-sectional area of 300 mm² has a tension of 37 kN at 25°C. The conductor weight per metre is 0.82 kg. The modulus of elasticity is 16.2×10^{10} N/m² and the temperature coefficient 11.5×10^{-6} degC⁻¹. Calculate the tension at 60°C.

(Answer: 15 kN)

11.6. An overhead line has steel-core aluminium conductors of allowable tension 35 kN. The conductor weight per metre is 0.82 kg and the ice thickness is 3.75 mm over the conductor diameter of 19.5 mm ($\rho_{ice} = 915$ kg/m³). A wind pressure of 37.5 N/m² is assumed. Calculate the sag, assuming towers of equal height, for a span of 260 m.

(Answer: 3.34 m)

11.7. Calculate the percentage of the line-to-ground voltage across each disc of a four-unit insulator string. The capacitance of each unit to the tower is one-quarter of the self-capacitance of the unit (see equivalent circuit in Figure 3.20). Leakage resistance may be ignored.

(Answer: 14.5 (top); 18.1; 26.3; 41.1)

11.8. The insulator string in problem 11.7 is to be more accurately represented by the inclusion of the capacitances between the connectors or link pins and conductor. These are assumed equal and of magnitude one-eighth of the self-capacitance of each unit. Calculate the perceentage of conductor–ground voltage across each unit.

(Answer: 23.2 (top); 19.3; 22.8; 34.7)

11.9. An overhead test line is energized at 865 kV, 60 Hz (line to ground). At the position of greatest sag the conductor is 12 m above the fround surface. Calculate the magnitude and direction of the electric field at a point directly below the lowest conductor position and 1 m above the ground. Comment on the environmental aspects of this situation.

(Answer: 20.5 kV/m)

11.10. Given a BIL of 1425 kV determine the maximum impulse stress in a cable insulation such that the overall diameter is not greater than 140 mm. The diameter over the conductor is 50 mm.

(Answer: 27.7 kV/mm)

11.11. Each core of a self-contained oil-filled cable consists of a conductor of physical area 2500 mm² with a central duct. The impulse design strength of the insulation is 110 kV/mm and the outer diameter of the insulation 150 mm. The system voltage is 765 kV and the BIL 2100 kV. Determine the diameter over the conductor and conductor duct. Calculate the maximum working stress. (Guess a diameter and then iterate.)

(Answer: 51.9 mm; 43.5 mm)

11.12. The insulation round a conductor of radius 20 mm is to be capacitance-graded by the use of two types of tape of relative permittivity 3 and 5. The voltage between the conductor and outer screen is 200 kV and the maximum working a.c. stress is to be 15 kV/mm. The design impulse stress is 90 kV/mm and the BIL 1350 kV. Design the cable, noting that for minimum thickness of insulation the maximum stress in each wall of tape must be the same. Calculate the radius over the insulation if only one type of tape is used.

(Answer: $\varepsilon_r = 5$ tape next to conductor and radius to interface is 33.3 mm. Radius to outer screen 32.5 mm. Using one type of tape, outer radius 42.4 mm)

11.13. A 66 kV single-core lead covered cable is laid direct in soil of thermal resistivity 1.2 degC-m/W at a depth of 1 m. The cable specification is as follows.

Cross section of copper conductor	260 mm²
Insulation: thickness	7.62 mm
thermal resistivity	550°C/W per cm
Lead sheath: thickness	2.03 mm
diameter over sheath	42 mm
diameter over serving	47 mm
thermal resistivity of serving	400°C/W per cm
Conductor alternating-current resistance at 20 °C	0.0658 Ω/km
Copper temperature-resistance coefficient	0.004/°C

Calculate the maximum current to be taken by the cable if the conductor temperature rise is not to exceed 50°C above an ambient of 15°C.

(Answer: 670 A)

11.14. Investigate the variation of permissible steady-current in the cable of problem 11.13 with different values of soil thermal resistivity in the range 0.7 to 1.5 degC m/W.

11.15. Determine the equivalent thermal circuit for transient studies for the 66 kV cable in problem 11.13. The following specific heats (in J/°C per gm) and densities (gm/cm³) obtain:

Copper	0.39 and	8.93
Dielectric	1.6 and	1.05
Lead	0.13 and	11.37

11.16. A length of three-core, three-phase, metal-sheathed cable gave the following results on test for capacitance:

(a) capacitance between shorted conductors and sheath, 1.0 μF;
(b) capacitance between two conductors shorted with the sheath and the third conductor, 0.6 μF.

With the sheath insulated, determine the capacitance

(1) between any two conductors; and
(2) between any two shorted conductors and the third conductor.
(3) Calculate the capacitance current per conductor per km when connexion is made to 100 kV, 60 Hz busbars.

(Answer: (1) 0.367 μF; (2) 0.489 μF; (3) 10 A)

11.17. A three-phase underground circuit consists of three single-core cables each of effective conductor radius of 0.75 cm and spaced 5.1 cm apart (axis to axis) in equilateral formation. The diameter of the lead sheath is 2.3 cm and the sheath thickness 0.15 cm. The specific resistance of lead is 22.0×10^{-6} Ω-cm at the working temperature and the conductor resistance per 1.6 km (1 mile) is 0.26 Ω at 65°C. For a cable length of 1.6 km (1 mile) and a load of 300 A, determine, (a) the induced sheath voltage without bonding (47 V per sheath); (b) the ratio of sheath loss to conductor loss (0.029).

11.18. An underground tunnel has a cross-section of 5.4 m^2 and is 250 m in length. A group of cables laid along the length of the tunnel dissipates a total of 10 W/cm. The effective thermal resistance from the air to ambient (air to tunnel wall heat-transfer and transfer through wall and soil) is 6°C cm/W. If the maximum steady-state temperature rise of the air blown through the tunnel is to be not greater than 32°C determine the air velocity required.
(Answer: 1.3 m/s)

11.19. Determine the summer rating (normal and emergency) of a 230 kV HPOF pipe-type cable. The relevant data are as follows:

  pipe depth 0.91 m;
  pipe diameter 254 mm;
  Conductor temperature $T_c = 80$°C (normal), 95°C (300 h emergency)
  T_{amb} (summer) = 25°C;
  load factor 100 per cent;
  1000 mm^2 (1970 kc mil) copper conductor (diameter 41.5 mm),
  conductor a.c. resistance (in pipe) 0.044 $\mu\Omega$/ph per km;
  trench backfill; thermal resistivity 0.4 degC per m/W;
  insulation thickness 19.3 mm
  insulation thermal resistivity 5 degC per m/W;
  ε_r (insulation) 3.5.

(Answer: 400 MVA normal)

12

Protection

12.1 Introduction

In Chapter 7 attention is confined to the analysis of various types of faults which may occur in a power system. Although the design of electrical plant is influenced by a knowledge of fault conditions, the major use of fault analysis is in the specification of switchgear and protective gear. Circuit-breaker ratings are determined by the fault MVAs at their particular locations. At the time of writing the maximum circuit-breaker rating is of the order of 50,000 MVA, and this is achieved by the use of several interrupter heads in series per phase in an air-blast system. Not only has the circuit breaker to extinguish the fault current arc, it has also to withstand the considerable forces set up by short-circuit currents which, as indicated in Chapter 7, can be very high.

A knowledge of the currents resulting from various types of fault at a location is essential for the effective operation of what is known as system protection. If faults occur on the system the control engineers, sensing the presence of the fault, can operate the appropriate circuit breakers to remove the faulty line or plant from the network. This, however, takes considerable time and experience. Faults on a power system resulting in high currents and also possible loss of synchronism must be removed in the minimum of time. Automatic means, therefore, are required to detect abnormal currents and voltages and, when detected, to open the appropriate circuit breakers. It is the object of protection to accomplish this. In a large interconnected network considerable knowledge and skill is required to remove the faulty part from the network and leave the healthy remainder working intact.

There are many varieties of automatic protective systems, ranging from simple overcurrent electromechanical relays to sophisticated electronic systems transmitting high-frequency signals along the power lines. The simplest but extremely effective form of protection is the electromechanical relay which closes contacts and hence energizes the circuit-breaker opening mechanisms when currents larger than specified pass through the equipment.

The protection used in a network can be looked upon as a form of insurance in which a percentage of the total capital cost (about 5 per cent) is used to safeguard apparatus and ensure continued operation when faults occur. In a highly industrialized community the maintenance of an uninterrupted supply to consumers is of paramount importance and the adequate provision of protection is essential.

Summarizing, protection and the automatic tripping (opening) of associated circuit breakers has two main functions: (a) to isolate faulty equipment so that the remainder of the system can continue to operate successfully; and (b) to limit damage to equipment due to overheating, and mechanical forces, etc.

12.2 Switchgear

Some of the functions of the switches or circuit breakers, as they are usually called, are obvious and apply to any type of circuit, others are peculiar to high-voltage equipment. For maintenance to be carried out on plant, it must be isolated from the rest of the network and hence switches must be provided on each side. If these switches are not required to open under working conditions, i.e. with fault or load current and normal voltage, a cheaper form of switch known as an *isolator* can be used; this can close a live circuit but not open one. Owing to the high cost of circuit breakers much thought is given in practice to obtaining the largest degree of flexibility in connecting circuits with the minimum number of switches. Popular arrangements of switches are shown in Figure 12.1.

High-voltage circuit breakers take four basic forms; oil immersed, small oil volume, air-blast, and SF_6. These will be briefly described below.

(a) The bulk oil circuit breaker. A cross-section of an oil circuit breaker with all three phases in one tank is shown in Figure 12.2. There are two sets of contacts per phase. The lower and moving contacts are usually cylindrical copper rods and make contact with the upper fixed contacts. The fixed contacts consist of spring-loaded copper segments which exert pressure on the lower contact rod when closed to form a good electrical contact. On opening, the lower contacts move rapidly downwards and draw an arc. When the circuit breaker opens under fault conditions many thousands of amperes pass through the contacts and the extinction of the arc and hence the effective open circuiting of the switch are major engineering problems. Effective opening is only possible because the instantaneous voltage and current per phase reduces to zero during each alternating current cycle. The arc heat causes the evolution of a hydrogen bubble in the oil and this high-pressure gas pushes the arc against special vents in a device surrounding the contacts called a turbulator (Figure 12.3). As the lowest contact moves downwards the arc stretches and is cooled and distorted by the gas and so eventually breaks. The gas also sweeps the arc products from the gap so that the arc does not re-ignite when the voltage rises to its full open-circuit value.

(b) The air-blast circuit breaker. For voltages above 120 kV the air-blast breaker is popular because of the feasibility of having several contact gaps in series per phase. Schematic diagrams of two types of air-blast head are shown in Figure 12.4. Air normally stored at 1.38 MN/m^2 is released and directed at the arc at high velocities, thus extinguishing it. The air also actuates the

480

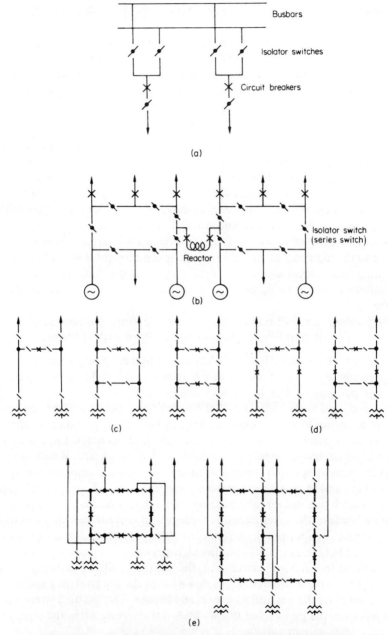

Figure 12.1 Possible switchgear arrangements. (a) Double-busbar selection arrangements. (b) Double-ring busbars with connecting reactor busbars can be isolated for maintenance, but circuits cannot be transferred from one side of the reactor to the other. (c) Open-mesh switching stations, transformers not switched. (d) Open-mesh switching stations, transformers switched. (e) Closed-mesh switching stations. (Isolators are sometimes called series switches.)

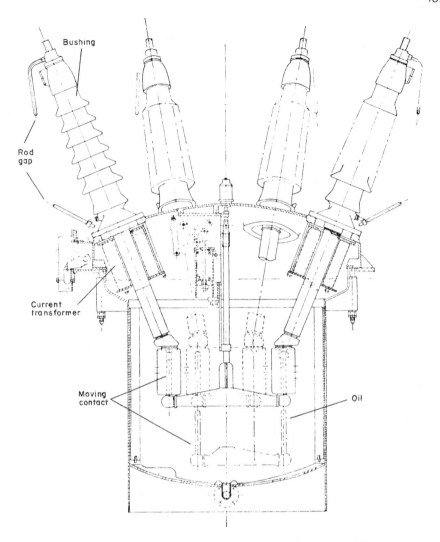

Figure 12.2   Cross-section of a 66 kV bulk oil circuit breaker. Three
phases in one tank. (*Permission of English Electric Co. Ltd.*)

mechanism of the moveable contact. Figure 12.5 shows a 132 kV air-blast
breaker with two breaks (interrupters) per phase and its associated isolator or
series switch. A 245 kV circuit breaker is shown in Figure 3.50 in Chapter 3. A
similar circuit breaker has been developed using sulphur hexafluoride (SF_6)
instead of air. When the main contacts have opened, an isolator (or series)
switch, in series with the main contacts, opens and then the main contacts
reclose when the air pressure is cut off.

(c) Small or low oil-volume circuit breakers.   The quenching mechanisms
are enclosed in vertical porcelain insulation compartments and the arc is

482

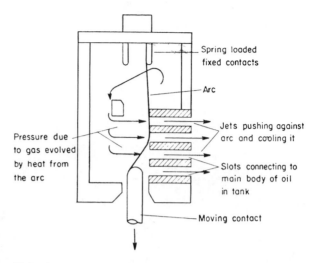

Figure 12.3   Cross-jet explosion pot for arc extinction in bulk oil circuit breakers.

extinguished by a jet of oil issuing from the moving (lower contact) as it opens. The volume of oil is much smaller than in the bulk-oil type.

A further arc-quenching medium is *vacuum*. An arc is possible only if contact material is vaporized, thus forming a conducting path. The maximum acceptable switching voltage at the present time is about 150 kV (r.m.s.) per vacuum chamber. This type of circuit breaker is under consideration for possible use in high-voltage direct-current schemes.

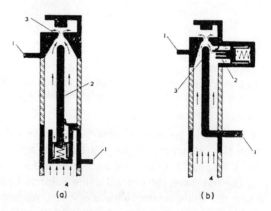

Figure 12.4   Schematic diagrams of two types of air-blast head. (a) Axial flow with axially moving contact. (b) Axial flow with side moving contact. 1. Terminal, 2. moving contact, 3. fixed contact, and 4. blast pipe. (*Permission of A. Reyrolle & Company Ltd.*)

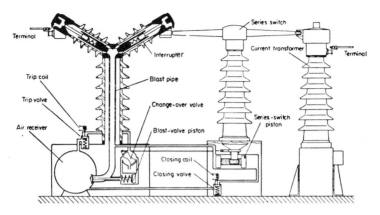

Figure 12.5 Schematic arrangement for typical air blast circuit breaker.

A circuit breaker must fulfil the following conditions:

(a) open and close in the shortest possible time under any network condition;
(b) conduct rated current;
(c) withstand thermally and mechanically any short-circuit currents;
(d) maintain its voltage to earth and across the open contacts under both clean and polluted conditions;
(e) not create any large overvoltage during opening and closing;
(f) be easily maintained.

Air is very convenient to use and its performance as an interrupting medium improves linearly with pressure over a good range. Circuit breakers of 400 kV have to clear short-circuit currents up to 60 kA (rating of 40 GVA) within 40 ms after the receipt of a tripping signal. In order to equalize the voltages across the various series-units of the breaker, a 500 pF shunt capacitor is installed across each unit. In the future a shorter clearance time than 40 ms will be necessary. Also resistive switching will be adopted, the resistors being inserted into the circuit in a time of 7–8 ms. The noise caused by operation will have to be limited to less than 95 dB, and particular attention will have to be paid to corona discharges, and especially the radio interference and audible noise so caused. An 1100 kV breaker will have to be able to clear a current of 50–60 kA and must be able to withstand power-frequency alternating-voltage tests for 1 min at 1900 kV (dry) and 1500 kV (wet), and a lightning surge test at 2800 kV.

The advantages of using sulphur hexaflouride (SF_6) as an insulating and interrupting medium in circuit breakers arise from its high electric strength and outstanding arc-quenching characteristics. SF_6 circuit breakers are much smaller than air breakers of the same rating, the electric strength of SF_6 at atmospheric pressure being roughly equal to that of air at the pressure of

484

10 atm. Temperatures of the order of 30,000 K are likely to be experienced in arcs in SF_6 and these are, of course, well above the dissociation temperature of the gas (about 2000 K); however, nearly all the decomposition products are electronegative so that the electric strength of the gas recovers quickly after the arc has been extinguished. Filters are provided to render the decomposition products harmless and only a small amount of fluorine reacts with metallic parts of the breaker.

The arrangement of equipment in a 380/110 kV substation SF_6 is illustrated in Figure 12.6. The switchgear bay also contains all necessary measuring and

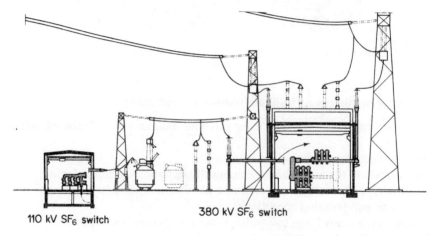

110 kV SF_6 switch   380 kV SF_6 switch

Figure 12.6   Arrangement of 380 kV/110 kV substation using SF_6 switchgear. (*Permission of Brown Boveri.*)

other facilities as follows: SF_6 insulated toroidal current transformers and voltage transformers, cable terminations, gas storage cylinders, cable isolator, grounding switches, bus isolator, and bus system. Such substations are of immense value in urban areas because of their greatly reduced size compared with air blast. SF_6 circuit breakers rated at 45 GVA are available and designs for 1300 kV have been produced.

The main advantages of a vacuum interrupter are: (a) the very small damage normally caused to the contacts on operation, so that a life of 30 years can be expected without maintenance; (b) the small mechanical energy required for tripping; and (c) the low noise caused on operation. The nature of the vacuum arc depends on the current; at low currents the arc is diffuse and can readily be interrupted, but at high currents the arc tends to be constricted. Electrode contour geometries have been produced to give diffuse arcs with current densities of 10^6–10^8 A/cm^2; electron velocities of 10^8 cm/s are experienced in the arc and ion velocities of 10^6 cm/s.

Considerable progress has been made in increasing the current-breaking capacity of vacuum interrupters, but not their operating voltage. Today they can be produced with ratings up to 100 MVA and 5000 MVA interrupters are being designed. Vacuum interrupters are cheaper than other types of circuit breaker at comparatively low voltages; an 11 kV interrupter, rated at 500 MVA, is available and it is expected that it will be used in distribution systems. The maximum voltage of each interruption unit is about 150 kV so that more interruption units would be needed in a vacuum interrupter than in an equivalent SF_6 breaker.

12.3 Qualities Required of Protection

A few terms often used to describe the effectiveness of protective gear will now be described.

Selectivity or discrimination—its effectiveness in isolating only the faulty part of the system.

Stability—the property of remaining inoperative with faults occurring outside the protected zone (called external faults).

Speed of operation—this property is more obvious. The longer the fault current continues to flow the greater the damage to equipment. Of great importance is the necessity to open faulty sections before the connected synchronous generators lose synchronism with the rest of the system. This aspect is dealt with in detail in Chapter 8. A typical fault clearance time in h.v. systems is 140 ms, the future requirement is 80 ms, and this requires very high-speed relaying.

Sensitivity—this is the level of magnitude of fault current at which operation occurs, which may be expressed in current in the actual network (primary current) or as a percentage of the current transformer secondary current.

Economic consideration—in distribution systems the economic aspect almost overrides the technical one owing to the large number of feeders, transformers, etc. provided that basic safety requirements are met. In transmission systems the technical aspects are more important. The protection is relatively expensive, but so is the system or equipment protected and security of supply is vital. Two separate protective systems are used, one main (or primary) and one back-up.

Reliability—This property is self-evident. A major cause of circuit 'outages' is mal-operation of the protection itself. On average, in the British system (not including faults on generators), nearly 15 per cent of outages are due to this cause.

Back-up protection

Back-up protection as the name implies is a completely separate arrangement which operates to remove the faulty part should be main protection fail to operate. The back-up system should be as independent of the main protection as possible, possessing its own current transformers and relays. Often only the circuit-breaker tripping and voltage transformers are common.

Each main protective scheme protects a defined area or *zone* of the power system. It is possible that between adjacent zones a small region, e.g. between the current transformers and circuit breakers, may be unprotected, in which case the back-up scheme (known as remote back-up), will afford protection because it overlaps the main zones, as shown in Figure 12.7. In distribution the application of back-up is not as widespread as in transmission systems, it often being sufficient to apply it at strategic points only. Remote back-up is slow and usually disconnects more of the supply system than is necessary to remove the faulty part.

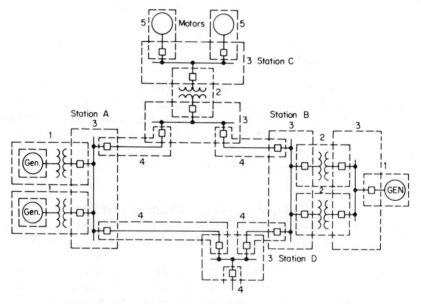

Figure 12.7   Line diagram of a typical system and the overlapping zones of protection. (*Permission of Westinghouse Electrical Corporation.*)

12.4   Components of Protective Schemes

Current transformers. (CTs)

In order to obtain currents which are proportional to the system (primary) currents and which can be used in control circuits, current transformers are used. Often the primary conductor itself, e.g. an overhead line, forms a single primary turn (bar primary). Whereas instrument current transformers have to remain accurate only up to slight overcurrents, protection current transformers must retain proportionality up to, say, 20 times normal full load.

A major problem exists when two current transformers are used which should retain identical characteristics up to the highest fault current, e.g. in pilot wire schemes. Because of saturation in the silicon steel used and the possible existence of a direct component in the fault current the exact matching of such current transformers is difficult. The nominal secondary current rating of current transformers is now usually 1 A, but 5 A has been used in the past.

Linear couplers   The problems associated with current transformers have resulted in the development of devices called linear couplers, which serve the same purpose but, having air cores, remain linear at the highest currents.

Voltage (or potential) transformers. (VTs or PTs)

These provide a voltage which is much lower than the system voltage, the nominal secondary voltage being 110 V. There are two basic types, the wound (electromagnetic), virtually a small power transformer, and the capacitor type. In the latter a tapping is made on a capacitor bushing (usually of the order of 12 kV) and the voltage from this tapping stepped down by a small voltage transformer. The arrangement is shown in Figure 12.8, the reactor (X) and the

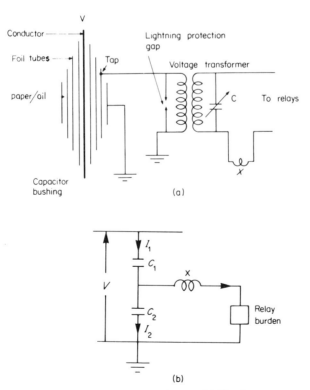

Figure 12.8   Capacitor voltage transformer. (a) Circuit arrangement.
(b) Equivalent circuit—burden = impedance of transformer and load referred to primary winding.

capacitor (C) constitute a tuned circuit which corrects the phase angle error of the secondary voltage.

Relays

A relay is a device which when energized by appropriate system quantities indicates an abnormal condition. When the relay contacts close the associated circuit-breaker trip-circuits are energized and the breaker contacts open, isolating the faulty part from the system. There are two main forms of relay, electromagnetic and semiconductor. For some purposes (e.g. overload protection) a bimetallic-strip thermal action is used.

The basic forms of the electromagnetic type comprise induction disc, induction cup, hinged armature, and plunger action. The hinged armature and plunger-type devices (Figure 12.9) are the simplest and rely on the attraction of an armature or plunger due to an electromagnet which may be energized by a.c. or d.c.

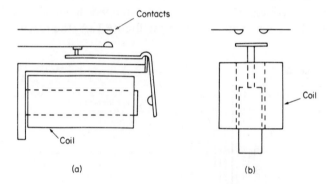

Figure 12.9   (a) Hinged armature relay. (b) Plunger-type relay.

Induction disc relay

A copper disc is free to rotate between the poles of an electromagnet which produces two alternating magnetic fields displaced in phase and space. The eddy currents due to one flux and the remaining flux interact to produce a torque on the disc. In early relays the flux displacement was produced by a copper band around part of the magnet pole (shading ring) which displaced the flux contained by it. Modern relays employ a wattmetric principle in which two electromagnets are employed, as shown in Figure 12.10. The current in the lower electromagnet is induced by transformer action from the upper winding and sufficient displacement between the two fluxes results. This, however, may be adjusted by means of a reactor in parallel with the secondary winding.

The basic mode of operation of the induction disc is indicated in the phasor diagram of Figure 12.11. The torques produced are proportional to $\Phi_2 i_1 \sin \alpha$ and $\Phi_1 i_2 \sin \alpha$ so that the total torque is porportional to $\Phi_1 \Phi_2 \sin \alpha$ as Φ_1 is proportional to i_1 and Φ_2 to i_2.

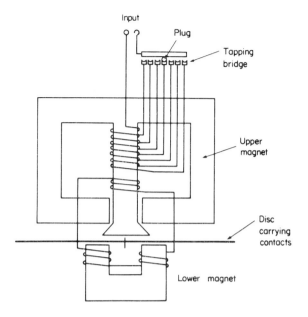

Figure 12.10 Induction disc relay.

This type of relay is fed from a CT and the sensitivity may be varied by the plug arrangement shown in Figure 12.10. The operating characteristics are shown in Figure 12.12. To enable a single characteristic curve to be used for all the relay sensitivities (plug settings) a quantity known as the current (or plug) setting multiplier is used as the abscissa instead of current magnitude, as shown in Figure 12.12. To illustrate the use of this curve (usually shown on the relay casing) the following example is given.

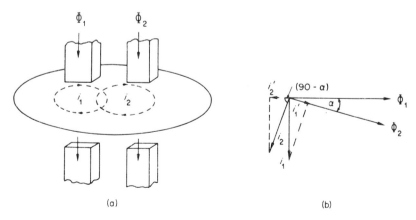

Figure 12.11 Operation of disc-type electromagnetic relay (a) Fluxes. (b) Phasor diagram. i_1 and i_2 induced currents in disc.

490

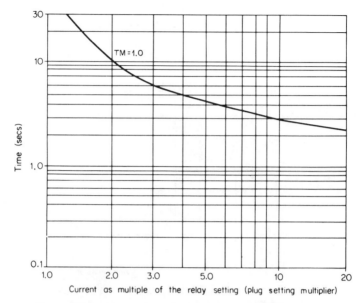

Figure 12.12   Time–current characteristics of a typical induction disc
in terms of the plug-setting multiplier. TM = time multiplier.

Example 12.1   Determine the time of operation of a 1 A, 3 s over-current
relay having a plug setting of 125 per cent and a time multiplier of 0.6. The
supplying CT is rated 400 : 1 A and the fault current is 4000 A.

Solution   The relay coil current = $(4000/400) \times 1 = 10$ A. The normal relay
coil current is $1 \times (125/100) = 1.25$ A. Therefore the relay fault current as a
multiple of the plug setting = $(10/1.25) = 8$. From the relay curve (Figure
12.12) the time of operation is 3.3 for a time setting of 1. The time multiplier
(TM) controls the time of operation by changing the angle through which the
disc moves to close the contacts. The actual operating time = $3.3 \times 0.6 = 2.0$ s.

Induction disc relays may be made responsive to power flow by feeding the
upper magnet winding in Figure 12.10 from a voltage via a potential transformer and the lower winding from the corresponding current. As the upper
coil will consist of a large number of turns the current in it lags the applied
voltage by 90°, whereas in the lower (small number of turns) coil they are
almost in phase. Hence, Φ_1 is proportional to $\mathbf{V}$ and Φ_2 is proportional to $\mathbf{I}$, and
torque is proportional to $\Phi_1\Phi_2 \sin \alpha$, i.e. to $\Phi_1\Phi_2 \sin (90-\phi)$, or $VI \cos \phi$
(where ϕ is the angle between $\mathbf{V}$ and $\mathbf{I}$).

The direction of the torque depends on the power direction and hence the
relay is directional. A power relay may be used in conjunction with a current-
operated relay to provide a directional property.

Induction cup relay   The operation is similar to the induction disc, here
two fluxes at right angles induce eddy currents in a bell-shaped cup which

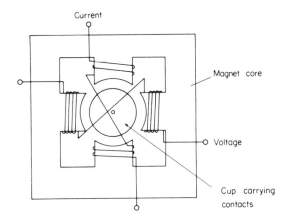

Figure 12.13   Four-pole induction cup relay.

rotates and carries the moving contacts. A four-pole relay is shown in Figure 12.13.

Permanent-magnet moving coil   The action in one type is similar to a moving-coil indicating instrument with the moving-coil assembly carrying the contacts. In a second type the action is basically that of the loudspeaker in which the coil moves axially in the gap of a permanent magnet. The time–current characteristic is inverse with a definite minimum time.

Balanced beam

The basic form of this relay is shown in Figure 12.14. The armatures at the ends of the beam are attracted by electromagnets which are operated by the

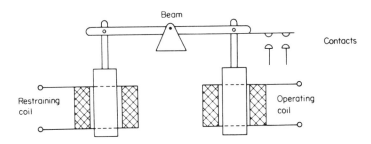

Figure 12.14   Schematic diagram of balanced beam relay.

appropriate parameters, usually voltage and current. A slight mechanical bias is incorporated to keep the contacts open except when operation is required.

The pulls on the armatures by the electromagnets are equal to $K_1 V^2$ and $K_2 I^2$, where K_1 and K_2 are constants, and for operation (i.e. contacts to close), i.e.

$$K_1 V^2 > K_2 I^2$$

$$\frac{V}{I} < \sqrt{\frac{K_2}{K_1}} \quad \text{or} \quad Z < \sqrt{\frac{K_2}{K_1}}$$

This shows that the relay operates when the impedance it 'sees' is less than a predetermined value. The characteristic of this relay when drawn on R and jX axes is a circle as shown in Figure 12.15.

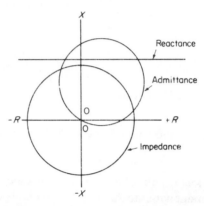

Figure 12.15   Characteristics of impedance, reactance, and admittance (mho) relays shown on the R–X diagram.

Distance relays   The balanced-beam relay, because it measures the impedance of the protected line, effectively measures distance. Two other relay forms also may be used for this purpose:

(a) the reactance relay which operates when $(V/I) \sin \phi \leqslant$ constant, having the characteristic shown in Figure 12.15;

(b) The mho or admittance relay the characteristic of which is also shown in Figure 12.15.

The above relays operate with any impedance phasor lying inside the characteristic circle or below the reactance line. The mho characteristic may be obtained by devices balancing two torque-producing elements and an induction cup could be used. A rectifier bridge supplying a moving-coil relay is often used.

Negative-phase sequence.   This is used in generator protection and is sensitive to the presence of negative-sequence currents. The protection comprises a bridge circuit supplying a current-operated relay.

Solid-state devices

These relays are extremely fast in operation, having no moving parts and are very reliable. Detection involving phase angles and current and voltage magni-

tudes are made with appropriate circuits. Most required current–time characteristics may be readily obtained and solid-state devices are now firmly
established. Inverse-characteristic, overcurrent, and earth-fault relays have a
minimum time lag and the operating time is inversely related to some power of
the input (e.g. current). In practical static relays it is advantageous to choose a
circuit which can accommodate a wide range of alternative inverse time
characteristics, precise minimum operating levels, and definite minimum times.

Electromechanical relays are vulnerable to corrosion, function, shock vibra-
tion, and contact bounce and welding. They require regular maintenance by
skilled personnel. Solid-state relays have been mainly used in areas where the
application of conventional methods is difficult or impossible. Transmission-
line protection by phase comparison is a typical example and in which vacuum
tubes were used prior to solid-state devices. The various advantages of solid
state have to be weighed against their higher cost. Also an advantage of
electromechanical relays is their ability to control a large number of circuits by
banks of contacts. By continuous development and cooperation between
manufacturer and user, solid-state reliability has greatly increased over the
years.

Various functions which have to be performed are: initiation of breaker trip
coil (basically an amplifying function), input and output isolator and buffer
stages, magnitude and phase angle directional comparators, timing, and
indication of trip. For detailed information readers are referred to the literature, especially reference 6. The basic outline of solid-state circuits for some
of these functions will now be given. It should be remembered that these
circuits illustrate only the physical process and are not complete in the full
operational sense, i.e. buffer stages, power supplies, amplifiers, etc. are
omitted.

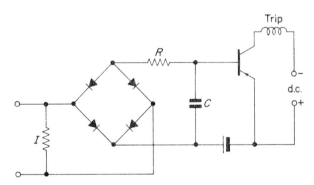

Figure 12.16   Solid-state overcurrent relay-basic circuit.

The basic overcurrent relay is shown in Figure 12.16 and the directional
comparator phase relay (equivalent to the induction disc or four-pole cup
types) in Figure 12.17. An impedance unit (for distance relays) equivalent to
the electromechanical balanced beam or inductor cap relays is shown in Figure

494

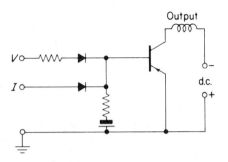

Figure 12.17 Solid-state phase comparator (directional) relay—
basic circuit.

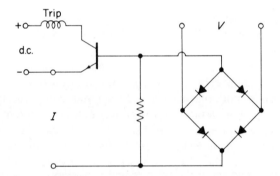

Figure 12.18 Solid-state amplitude comparator (impedance)—basic
circuit.

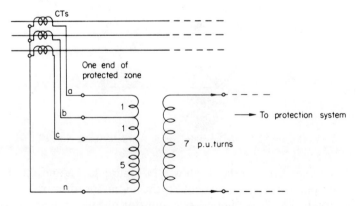

Figure 12.19 Summation transformer.

12.18. This is an amplitude comparator. High-speed operation may be obtained by the use of silicon-controlled rectifiers.

Summation transformer   In some relaying schemes it is necessary to transmit the secondary currents of the current transformers considerable distances in order to compare them with currents elsewhere. To avoid the use of wires from each of the three CTs in a three-phase system, a summation transformer is used which gives a single-phase output, the magnitude of which depends on the nature of the fault. The arrangement is shown in Figure 12.19 in which the ratios of the turns are indicated. On balanced through-faults there is no current in the winding between c and n. The phase (a) current energizes the 1 p.u. turns between a and b and the phasor sum of I_a and I_b flows in the 1 p.u. turns between b and c.

The arrangement gives a much greater sensitivity to earth faults than to phase faults. When used in phase-comparison systems, however, the actual value of output current is not important and the transformer usually saturates on high-fault currents so protecting the secondary circuits against high voltages.

12.5 Protection Systems

The application of the various relays and other equipment to form adequate schemes of protection forms a large and complex subject. Also the various schemes are largely dependent on the methods of individual manufacturers. The main intention here is to present a survey of general practice and outline the principles of the methods used. Some schemes are discriminative to fault location and involve several parameters, e.g. time, direction, current, distance, current balance, phase comparison. Others discriminate according to the type of fault, e.g. negative-sequence relays and some use a combination of location and type of fault.

A convenient classification is the division of the systems into *unit* and *non-unit* types. Unit protection signifies that an item of equipment or zone is being uniquely protected independently of the adjoining parts of the system. Non-unit schemes are those in which several relays and associated equipment are used to provide protection covering more than one zone. Examples of both types are classified in Figure 12.20. Non-unit schemes represent the most widely used and cheapest forms of protection and these will be discussed first.

Overcurrent protection

Theis basic method is widely used in distribution networks and as a back-up in transmission systems. It is applied to generators, transformers, and feeders. The arrangement of the components is shown in Figure 12.21. The relay normally employed is the induction-disc type with two electromagnets as shown in Figure 12.10.

The application to feeders is illustrated in Figure 12.22. Along the radial feeder the relaying points and circuit breakers are shown. The operating times

496

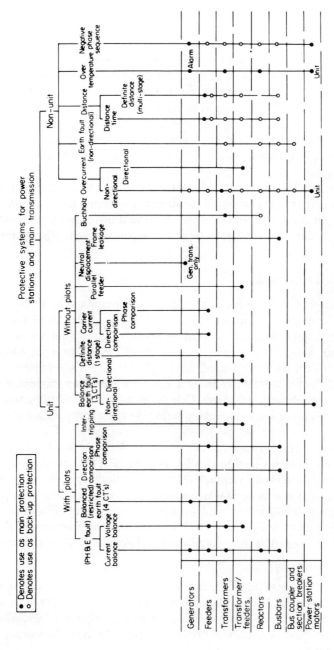

Figure 12.20 Classification of protection schemes.

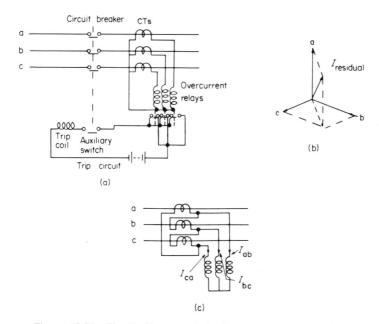

Figure 12.21 Circuit diagram of simple overcurrent protection scheme. (a) CTs in star. (b) Phasor diagram of relay currents, star connexion. (c) CTs in delta.

are graded to ensure that only that portion of the feeder remote from the infeed side of a fault is disconnected. When determining selectivity, allowance must be made for the operating time of the circuit breakers. Assume Figure 12.22 to be a distribution network with slow-acting breakers operating in 0.3 s and that the relays have true inverse-law characteristics. Selectivity is obtained with a through-fault of 200 per cent full load with the fault between D and E as illustrated because the time difference between relay operations is greater than 0.3 s. Relay D operates in 0.5 s and its circuit breaker trips the feeder to the right of the fault in 0.8 s. The fault current ceases to flow (normal load current is ignored for simplicity) and the remaining relays do not close their contacts. Consider, however, the situation when the fault current is 800 per cent of full load. The relay operating times are now: A 0.5 s [i.e. $2 \times (200/800)$], B 0.375 s, C 0.25 s, D 0.125 s, and the time for the breaker at D to open is $0.125 + 0.3 = 0.325$ s. By this time relays B and C will have operated and selectivity is not

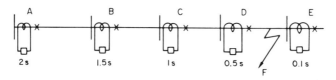

Figure 12.22 Application of overcurrent induction disc relays to feeder protection.

498

obtained. This illustrates the fundamental drawback of this system, i.e. that for correct discrimination to be obtained the times of operation close to the supply point become large.

Overcurrent and directional

In a loop system, to obtain discrimination, relays with an added directional property are required. For the system shown in Figure 12.23 directional and non-directional overcurrent relays have time lags for a given fault current as

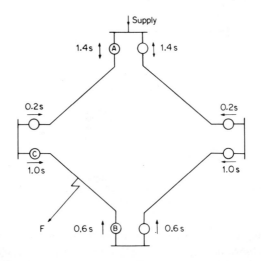

Figure 12.23 Application of directional overcurrent relays to a loop network. ↔ Relay responsive to current flow in both directions. → Relay responsive to current flow in direction of arrow.

shown. Current feeds into fault at the location indicated from both directions and the first relay to operate is at B (0.6 s). The fault is now fed along route ACB only, and next the relay at C (1 s) operates and completely isolates the fault from the system. Assuming a circuit-breaker clearance time of 0.3 s complete selectivity is obtained at any fault position.

12.6 Distance Protection

The shortcomings of graded overcurrent relays have led to the widespread use of distance protection. The distance between any point in the feeder and the fault is proportional to the ratio (voltage/current) at that point and relays responsive to impedance, admittance (mho), or reactance may be used. Although a variety of time–distance characteristics are available for providing correct selectivity the most popular one is the stepped characteristic shown in Figure 12.24. A, B, C, and D are distance relays with directional properties and A and C only measure distance when the fault current flows in the indicated direction. Relay A trips its associated breaker if a fault occurs within the first 80

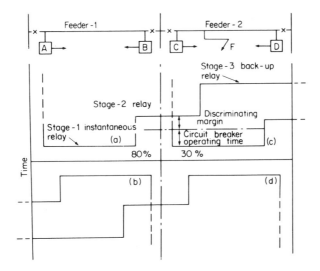

Figure 12.24 Characteristic of three-stage distance protection.

per cent of the length of feeder (1). For faults in the remaining 20 per cent of feeder (1) and the initial 30 per cent of (2) (called the stage 2 zone) relay A initiates tripping after a short time delay. A further delay in (A) is introduced for faults further along feeder (2) (stage 3 zone). Relays B and D have similar characteristics when the fault current flows in the opposite direction.

The selective properties of this scheme can be understood by considering a fault such as at F in feeder 2 when fault current flows from A to the fault. For this fault relay A starts to operate, but before the tripping circuit can be completed relay C trips its circuit breaker and the fault is cleared. Relay A then resets and feeder 1 remains in service. The margin of selectivity provided is indicated by the vertical intercept between the two characteristics for relays A and C at the position F less the circuit-breaker operating time.

It will be noted that the stage 1 zones are arranged to extend over only 80 per cent of a feeder from each end. The main reason for this is because practical distance relays and their associated equipment have errors and a margin of safety has to be allowed if incorrect tripping for faults which occur just inside the next feeder is to be avoided. Similarly the stage 2 zone is extended well into the next feeder to ensure definite protection for that part of the feeder not covered by stage 1. The object of the stage 3 zone is to provide general back-up protection for the rest of the adjacent feeders.

The characteristics shown in Figure 12.24 require three basic features, viz. response to direction, response to impedance, and timing. These features need not necessarily be provided by three separate relay elements, but they are fundamental to all distance protective systems. As far as the directional and measuring relays are concerned, the number required in any scheme is governed by the consideration that three-phase, phase-to-phase, phase-to-earth,

500

and two-phase to-earth faults must be catered for. For the relays to measure the same distance for all types of faults the applied voltages and currents must be different. It is common practice, therefore, to provide two separate sets of relays, one set for phase faults and the other for earth faults, and either of these caters for three-phase faults and double-earth faults. Each set of relays is in practice usually further divided into three, since phase faults may concern any pair of phases, and similarly any phase can be faulted to earth.

12.7 Unit Protection Schemes

With the ever-increasing complexity of modern power systems the methods of protection so far described may not be adequate to afford proper discrimination, especially when the fault current flows in parallel paths. In unit schemes protection is limited to one distinct part or element of the system which is disconnected if an internal fault occurs. On the other hand the protected part should remain connected with the passage of current flowing into an external fault.

Differential relaying At the extremities of the zone to be protected the currents are continuously compared and balanced by suitable relays. Provided

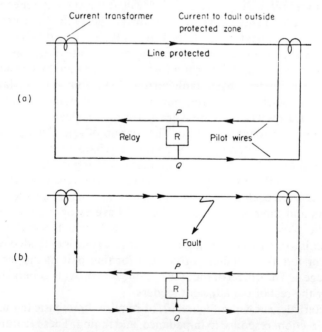

Figure 12.25   Circulating current, differential protection (one phase only shown). (a) Current distribution with through fault—no current in relay. (b) Fault on line, unequal currents from current transformers and current flows in relay coil. Relay contacts close and make operating circuits of circuit breakers at each end of the line.

the currents are equal in magnitude and phase, no relay operation will occur. If, however, an internal fault (inside the protected zone) occurs this balance will be disturbed (see Figure 12.25) and the relay will operate. The current transformers at the ends of each phase should have identical characteristics to ensure perfect balance on through-faults. Unfortunately, this is difficult to achieve and a restraining or bias coil is connected (see Figure 12.26) which carries a current proportional to the full system current and restrains the relay operation on large through fault currents. The corresponding characteristic is shown in Figure 12.26. This principle (circulating current) may be applied to generators, feeders, transformers, and busbars and provides excellent selectivity.

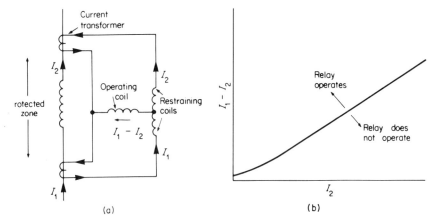

Figure 12.26(a) Differential protection—circuit connexions (one phase only)—relay with bias coil. (b) Characteristic of bias relay in differential protection. Operating current plotted against circulating or restraint current.

12.8 Generator Protection

Large generators are invariably connected to their own step-up transformer and the protective scheme usually covers both items. A typical scheme is shown in Figure 12.27 in which separate differential circulating-current protections are used to cover the generator alone and the generator plus transformer. When differential protection is applied to a transformer the current transformer on each side of a winding must have ratios which give identical secondary currents. In the U.S.A. the generator neutral is often grounded through a distribution transformer. This energizes a relay which operates the generator main and field breakers when a ground fault occurs in the generator or transformer. The ground fault is usually limited to about 10 A by the distribution transformer. In Britain the neutral is usually grounded through a resistor. The field circuit of the generator must be opened when the differential protection operates to avoid the machine feeding the fault.

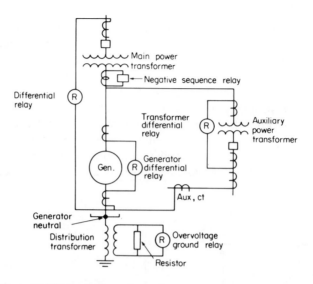

Figure 12.27 Protection scheme for a generator and unit transformer.

The relays of the differential protection on the stator windings (see Figure
12.28) are set to operate at about 10–15 per cent of the circulating current
produced by full-load current to avoid current transformer errors. If the phase
e.m.f. generated by the winding is E, the minimum current for a ground fault at
the star-point end and hence with the whole winding in circuit is E/R, where R
is the neutral effective resistance. For a fault at a fraction x along the winding

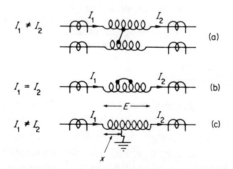

Figure 12.28 Generator winding faults and differential protection.
(a) Phase-to-phase fault. (b) interturn fault. (c) Phase-to-earth fault.

from the neutral, the fault current is xE/R and 10–15 per cent of the winding is
unprotected. With the neutral grounded via the transformer, R is high and
earth faults are detected by a sensitive relay across the transformer secondary.
With an interim fault (turn-to-turn short circuit) on a phase of the stator
winding current balance at the ends is retained and no operation of the

differential relay takes place. The relays operate only with phase-to-phase and ground faults.

On unbalanced loads or faults the negative-sequence currents in the generator produce excessive heating on the rotor surface and generally $(I_2^2 t)$ must be limited to a certain value for a given machine (between 3 and 4 for 500 MW machines), where t is the duration of the fault in seconds. To ensure this a relay is installed which detects negative-sequence current and trips the generator main breakers after a prescribed time. When loss of excitation occurs, reactive power (Q) flows into the machine, and if the system is able to supply this the machine will operate as an induction generator still supplying power to the network. The generator output will oscillate slightly as it attempts to lock into synchronism. Relays are connected to isolate the machine when a loss of field occurs.

12.9 Transformer Protection

A typical protection scheme is shown in Figure 12.29(a) in which the differential circulating-current arrangement is used. The specification and arrangement of the current transformers is complicated by the main transformer connexions and ratio. Current-magnitude differences are corrected by adjusting the turns ratio of the current transformers to account for the voltage ratio at the transformer terminals. In a differential scheme the phase of the secondary currents in the pilot wires must also be accounted for with star–delta transformers. In Figure 12.29(a) the primary side current transformers are connected in delta and the secondary in star. The corresponding currents are shown in Figure 12.29(a) and it is seen that the final currents entering the connexions between the current transformers are in phase for balanced-load conditions and hence there is no relay operation. The delta current–transformer connexion on the main transformer star winding also ensures stability with through earth-fault conditions which would not be obtained with both sets of current transformers in the star connexion. The distribution of currents in a Y–Δ transformer is shown in Figure 12.30.

Troubles may arise due to the magnetizing current inrush on switching operating the relays, and often restraining coils sensitive to third harmonic components of the current are incorporated in the relays. As the inrush current has a relatively high third harmonic content the relay is restrained from operating.

Faults occurring inside the transformer tank due to various causes give rise to the generation of gas from the insulating oil. This may be used as a means of fault detection by the installation in the pipe between the tank and conservator of a gas/oil-operated relay. The relay normally comprises hinged floats and is known as the *Buchholz* relay (see Figure 12.31). With a small fault bubbles rising to flow into the conservator are trapped in the relay chamber, disturbing the float which closes contacts and operates an alarm. On the other hand, a serious fault causes a violent movement of oil which moves the floats, making other contacts which trip the main circuit breakers.

504

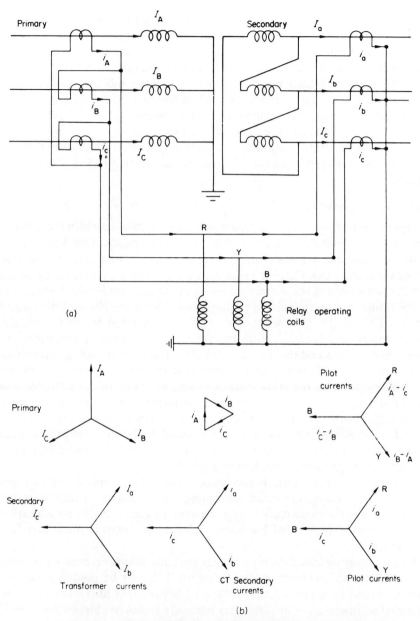

(a)

(b)

Figure 12.29 Differential protection applied to Y–Δ transformers. (a) Current transformer connexions. (b) Phasor diagrams of currents in current transformers.

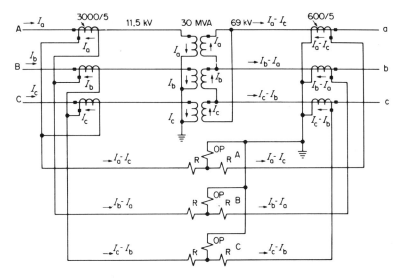

Figure 12.30 Currents in Y–Δ transformer differential protection.
OP = operating coil; R = restraining coil.

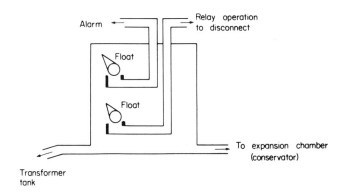

Figure 12.31 Schematic diagram of Buchholz relay arrangement.

12.10 Feeder Protection

Differential pilot wire

The differential system already described can be applied to feeder protection.
The current transformers situated at the ends of the feeder are connected by
insulated wires known as pilot wires. In Figure 12.25, *P* and *Q* must be at the
electrical mid-points of the pilots and often resistors are added to obtain a
geographically convenient midpoint. By reversing the current transformer
connexions (Figure 12.32) the current transformer e.m.f.s oppose and no
current flows in the wires on normal or through-fault conditions. This is known

506

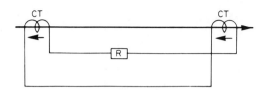

Figure 12.32 Pilot wire differential feeder protection—opposed
voltage connexions.

as the *opposed voltage* method. As under these conditions there are no back ampere-turns in the current transformer secondaries, on heavy through-faults the flux is high and saturation occurs. Also the voltages across the pilots may be high under this condition and unbalance may occur due to capacitance currents between the pilots. To avoid this sheathed pilots are used.

Pilots may be installed underground or strung on towers. In the latter method care must be taken to cater for the induced voltages from the power-line conductors. Sometimes it is more economical to rent wires from the telephone companies, although special precautions to limit pilot voltages are then required. A typical scheme using circulating current is shown in Figure 12.33 in which a mixing or summation device is used. With an internal fault the

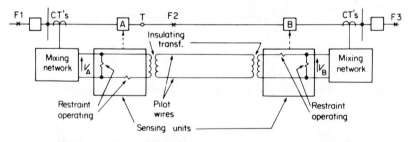

Figure 12.33 Differential pilot wire practical scheme using mixing network (or summation transformer) and biased relays. $V_A = V_B$ for external faults, e.g. at F1 and F3; $V_A \neq V_B$ for internal faults, e.g. at F2.

current entering end A will be in phase with the current entering end B, as in h.v. networks the feeder will inevitably be part of a loop network and an internal fault will be fed from both ends. V_A and V_B become additive, causing a circulating current to flow causing relay operation. Thus, this scheme could be looked on as a phase-comparison method. If the pilot wires become short-circuited, current will flow and the relays can trip on through-faults or heavy overloads. In view of this the state of the wires is constantly monitored by the passage of a small d.c. current.

Carrier-current protection

Because of pilot capacitance the pilot-wire of relaying is limited to line lengths below 48 km (30 miles). Above this, distance protection may be used, although for discrimination of the same order as that obtained with pilots, carrier-current equipment may be used. In carrier-current schemes a high-frequency signal in the band 80–500 kHz and of low power level (1 or 2 W) is transmitted via the power-line conductors from each end of the line to the other. It is not convenient to superimpose signals proportioned to the magnitude of the line primary current, and usually the phases of the currents entering and leaving the protected zone are compared. Alternatively, directional and distance relays are used to start the transmission of a carrier signal to prevent the tripping of circuit breakers at the line ends on through or external faults. On internal faults other directional and distance relays stop the transmission of the carrier signal, the protection operates, and the breakers trip.

A further application known as transferred tripping uses the carrier signal to transmit tripping commands from one end of the line to the other. The tripping command signal may take account of, say, the operation or non-operation of a relay at the other end (permissive intertripping) or the signal may give a direct positive instruction to trip alone (intertripping).

Carrier-current equipment is complex and expensive. The high-frequency signal is injected on to the power line by coupling capacitors and may either be coupled to one phase conductor (phase-to-earth) or between two conductors (phase-to-phase), the latter being technically better but more expensive. A schematic diagram of a phase-comparison carrier-current system is shown in Figure 12.34. The wave or line trap is tuned to the carrier frequency and

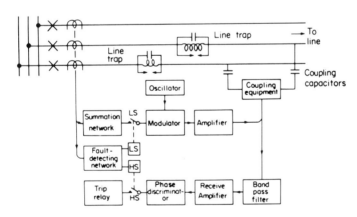

Figure 12.34 Block diagram measuring and control equipment—for carrier-current phase comparison scheme. LS = low set relay; HS = high set relay.

presents a high impedance to it but a low impedance to power-frequency currents; it thus confines the carrier to the protected line. Information regarding the phase angles of the currents entering and leaving the line is transmitted from the ends by modulation of the carrier by the power current, i.e. by blocks of carrier signal corresponding to half-cycles of power current (Figure 12.35).

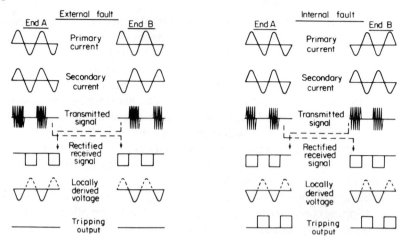

Figure 12.35   Waveforms of transmitted signals in carrier-type line
protection.

With through or external faults the currents at the line ends are equal in
magnitude but 180° phase displaced (i.e. relative to the busbars, it leaves one
bus and enters the other). The blocks of carrier occur on alternate half-cycles of
power current and hence add to form a continuous signal which is the condition
for no relay operation. With internal faults the blocks occur in the same
half-cycles and the signal comprises non-continuous blocks; this is processed to
cause relay operation (Figure 12.34).

The currents from the current transformers are fed into a summation device
which produces a single-phase output fed into a modulator (Figure 12.34). This
combines the power frequency with the carrier to form a chopped 100 per cent
modulated carrier signal which is then amplified and passed to the line-
coupling capacitors. The carrier signal is received via the coupling equipment,
passed through a narrow-bandpass filter to remove any other carrier signals,
amplified, and then fed to the phase discriminator which determines the
relative phase between the local and remote signals and operates relays
accordingly. The equipment is controlled by low-set and high-set relays that
start the transmission of the carrier only when a relevant fault occurs. These
relays are controlled from a starting network. Although expensive, this form of
protection is very popular on overhead transmission lines.

12.11   Busbar Protection

The protection of the various connexions in a substation is of vital importance.
It is essential to obtain correct operation and good discrimination on external
faults, as to isolate busbar faults all source connexions to the buses must be cut
off. Usually differential relaying is used because of its great selectivity. A simple
scheme using circulating current is shown (for one phase only) in Figure

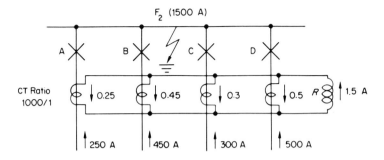

Relay current $= 0.25 + 0.45 + 0.3 + 0.5 = 1.5$ A

(a)

Relay current $= 1.0 - 0.25 - 0.45 - 0.3 = 0$ A

(b)

Figure 12.36 Busbar protection—circulating-current differential scheme. (a) Currents on internal fault. (b) Currents on external fault.

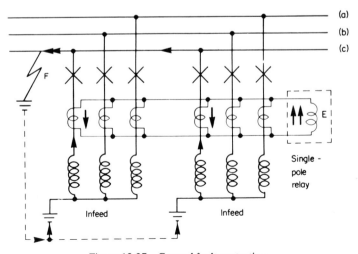

Figure 12.37 Ground fault protection.

12.36(a) and (b). The distribution of secondary currents for both internal and external faults is shown.

If the switchgear is phase-segregated, all faults are initially earth faults. A typical arrangement for earth or ground fault protection is shown in Figure 12.37.

12.12 Protection using Digital Computers

There is a constant requirement as systems become more complex for the speed and reliability of conventional protection to be improved, an increasingly difficult process. Also the protection is in operation for very brief periods at very infrequent intervals, which to some extent decreases the reliability of the equipment and the confidence which can be placed in it. This could be avoided if local substation computers are used. To avoid instability occurring with a severe short circuit close to generator terminals, the machine must be isolated very quickly. Whereas this in itself is attainable, correct discrimination is a major problem.

Most measurement and comparison techniques at the moment performed by relays and associated gear could be carried out by digital computers fed by inputs in digital form derived from the system. The instantaneous system quantities would be obtained via current transformers and voltage transformers and converted to digital form using analogue to digital conversion techniques. The system parameters are sampled at intervals and then processed according to the particular method in use. A proposed 4 ms tripping of severe faults requires a fault-current sampling interval of 0.5 ms and a minimum of six samples is needed to trip, thus requiring a total of 3 ms.

Data storage

The number of quantities to be stored is substantial when the protection of a large complex substation is required. Ten samples of all phase currents and phase-to-ground voltages in the various substation connexions are stored together with two peak magnitudes of these quantities. Also stored are the sample number, i.e. position in time, of the current peaks and other quantities such as differential currents. In all, for an average (500 kV–230 kV) substation, storage of the order of 2000 words has been estimated.[13]

Several subroutines are stored and perform the logical and comparison functions previously carried out by relays. These routines include: analogue-to-digital conversion, busbar differential, line-current peak determination, current comparison, station overall differential and transformer differential, waveshape analysis, and various distance schemes.

Proposed bus-zone protection[16]

A method for bus-zone protection using current balance has been described. Samples of the currents are simultaneously read in to the digital processor and for each set of samples if

$$\sum_{n=1}^{n=f} i_n \geqslant M$$

a trip signal is produced to all the circuit breakers in the protected zone, where i_n is the current in the nth feeder and f the number of feeders. If M is not exceeded no internal fault exists. A block diagram for fault detection in one zone of the protection is shown in Figure 12.38. The flow diagram of a possible program for this protection is shown in Figure 12.39.

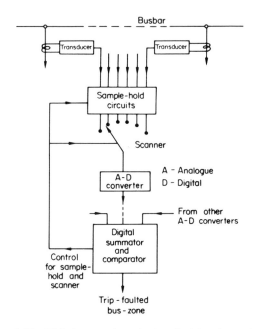

Figure 12.38 Digital protection—basic principle of sample-scan-summate operation. (*Permission of the Institution of Electrical Engineers.*)

Digital protection is at the moment in the initial development stage. It appears to offer several technical advantages including fast operating times. As in all new developments both economic as well as technical factors must be considered.

12.13 System Security and Emergency Situations

The various methods of relaying have now been reviewed and the underlying assumption has been that everything will work as planned. However, in practice owing to human error, malfunction of equipment, lack of complete information, and natural disasters the integrity of the system is occasionally at

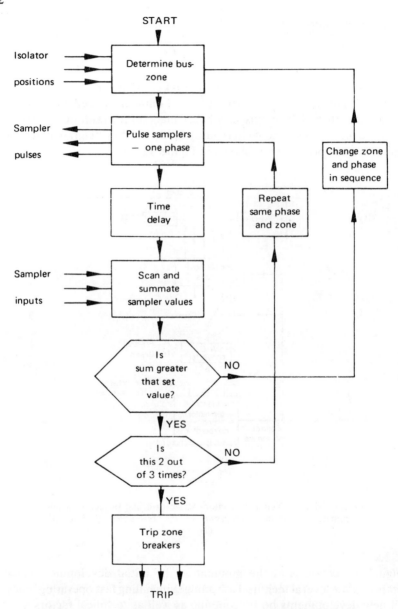

START

Isolator ──▶ → Determine bus-zone

positions ──▶

Sampler ◀── Pulse samplers — one phase

pulses ◀──

Time delay

Sampler ──▶ Scan and summate sampler values

inputs ──▶

Is sum greater that set value? — NO →

YES

Is this 2 out of 3 times? — NO →

YES

Trip zone breakers

Repeat same phase and zone

Change zone and phase in sequence

TRIP

Figure 12.39   Flow diagram for direct digital protection—applied to busbars. (*Permission of the Institution of Electrical Engineers.*)

risk. Usually this is at a local level without major repercussions, but occasionally involves millions of consumers, e.g. the famous north-eastern blackout and the New York City blackout in 1977.

This section discusses in general terms the idea of system security and the various events which could result in major loss of supply. Often the cause of trouble is the progressive tripping out of feeders in parallel. If one line is tripped out for some reason the load is thrown into the others resulting in their overload settings being exceeded and they too trip out one by one. This is known as cascading.

Security level

This is an instantaneous time-varying condition which indicates the ability of the system to withstand imminent disturbances. It is therefore an operating problem and is determined by the margin between actual power flows and the line transfer capabilities, and the probability of a disturbance. The latter is a function of factors such as storm activity which require a higher margin than normal conditions. Major system disruptions usually result from the inability of the system already operating at a reduced security level to withstand the effects of a series of relatively minor events, e.g. loss of a line, smaller generator, etc. The security level is also reduced under normal conditions with plant out for maintenance, etc. The system operator must be provided with up-to-date knowledge of the reserve margins of his equipment and of the probabilities of a disturbance.

A particularly difficult location from a security viewpoint is New York City where a considerable amount of power is imported from the north through very restricted right of ways. On 13 July 1977 lightning struck two 345 kV lines to the north of the city resulting in a considerably lower security level although the system remained intact. Later a further lightning strike initiated switching and a further removal of transmission plant which under normal conditions would have been easily coped with. With the already reduced security, critical overloads on lines resulted exceeding the short time (emergency) ratings. Hence these lines tripped out, throwing extra load on the remainder which then systematically tripped out. New York City was effectively cut off from external support and the 1700 MW deficiency between the load and in-city generation could be provided only by kinetic energy of the machines resulting in a fall in speed and frequency (see Chapter 4). With the fall in frequency automatic load shedding commenced after 2 s (frequency 58.3 Hz). The generation then exceeded the load and the frequency began to rise. With the now decreased load the generators were faced with a light-load situation which in conjunction with the capacitive underground cable system produced a rise in voltage and consequent decreased excitation on the generators. The heavy flow of reactive power into the 1000 MW Ravenswood No. 3 generator and consequent field current reduction resulted in the operation of the 'loss of field' relay and the unit (carrying 850 MW) was tripped. Again the load exceeded generation and the frequency continued its decline. Although further load was shed this proved insufficient to stem the fall in frequency. One by one generators were tripped either automatically or by personnel to avoid damage. At 21.36 hours the city was without electricity supply, i.e. blacked out.

Apart from the more obvious causes of the failure the situation with possible voltage rises and resulting generator problems on load shedding is very interesting. It would seem desirable that automatic load shedding be introduced before the city is completely isolated and not activated by falls in frequency.

Emergency control

The following system operating states and transition between states may be identified.

(a) Normal to alert—reductions of security level. This could be caused by unexpected load increases, loss of generating units, derating of plant due to environmental constraints, and rescheduled maintenance.
(b) Alert to emergency—inability of parts of system to meet requirements, e.g. lines (emergency ratings), voltage levels, frequency, machine and bus voltage angles. Caused by malfunction or loss of major items of equipment due to internal fault, malfunction of protection, lightning, etc.
(c) Emergency to extreme condition (collapse)—loss of integrity. Caused by loss of ties resulting in isolated generation islands which are unable to carry their internal loads. This is triggered by prolonged overloading of critical ties, malfunction of protection, and successive disturbances.

Corresponding control measures to meet the above situations are as follows.

Alert Restoration of reserve margins, increased generation reserves, rescheduling of tie-lines, and voltage reduction.

Emergency Fast valving on steam-turbines, dynamic breaking of generators, load control, capacitor switching, and immediate action to clear equipment overloads.

Extreme All of the above plus load shedding and controlled operation of isolated power groups.

Following these emergencies action is taken to re-establish a viable system. This involves restarting and synchronization of generation units, load restoration, and resynchronization of all areas. Although action to prevent transient instability of individual generators has been a major factor in emergency control, such instabilities are not necessarily a major factor in system extreme emergencies. Local immediate (reflexive) action may prevent damage to the equipment involved, but the resulting system security may be reduced to a dangerous level.

12.14 Security Analysis and State Estimation Security Analysis

The system must be operated such that should a fault or maloperation occur, the system will not be in a dangerous situation. To achieve this three major functions are performed in the control centre.

(a) System monitoring.

(b) Contingency analysis.
(c) Corrective action analysis.

In (a) measurements of voltages, currents, power flows, etc. are transmitted to the centre. This is a vast quantity of information and can only be handled by computers. These systems are also used for supervisory control, i.e. control of circuit breakers, transformer taps, etc.

(b) Comprises an analysis of the system using models of the problems as they arise. This analysis provides the necessary constraints on the normal operation of the system. The methods for achieving this are given in references 11 and 12. The initial conditions for the analysis are based on measured system data and static estimation.

The final action (c) is self-evident but must be achieved quickly to be effective. Most power systems are designed to have sufficient redundancy to withstand major problems; however, the operators endeavour to maximize the reliability of the system at any given time. Two major events are transmission line and generator outages. Both these cause changes in the flows and voltages in the system.

State estimation

This involves the determination of unknown variables based on measurements in the system. The state variables are voltage magnitudes and relative phase angles. The system performance is estimated in real time from the measurements. Errors may occur in the measurements which if small can go undetected. Large errors are easily detected and hence ignored. State estimation techniques smooth out random measurement errors and use samples to calculate the value of unknown parameters. Various criteria exist for state estimation as follows:

(a) Maximum likelihood—this maximizes the probability that the estimate of a variable is the true value.
(b) Weighted least-squares—the sum of the squares of the weighted deviations of the estimated from the actual measurements is minimized.
(c) Minimum variance—minimize the sum of the squares of deviations of the estimated values from the true values.

The detection of bad data is a complex statistical procedure and not appropriate to this book. The network analysis is usually performed using the fast decoupled low flow method discussed in Chapter 6. For a detailed account of security and state estimation see references **11** and **12**.

References

BOOKS
1. Lythall, R. T., *The J. & P. Switchgear Book*, Johnson & Phillips, London, 1969.
2. Cook, V., *Analysis of Distance Protection*, Research Studies Press Ltd, 1985.

3. Klewe, H. R. J., *Interference between Power Systems and Telecommunication Lines*, Edward Arnold, London, 1958.
4. Willheim, R., and M. Waters, *Neutral Grounding*, Elsevier, New York, 1956.
5. Mason, C. M., *The Art and Science of Protective Relaying*, Wiley, New York, 1956.
6. Warrington, A. R. C. Van, *Protective Relays*, Vol. 1, 2nd edn. 1968, Vol. 2, 3rd edn. 1978, Chapman and Hall, London.
7. Westinghouse Electric Corporation, *Applied Protective Relaying*, Newark, New Jersey, U.S.A., 1968.
8. General Electric, *The Art of Protective Relaying*: Series of booklets: *Introduction, Transmission and Subtransmission Lines and E.H.V. Systems*, Switchgear Dept., Philadelphia, U.S.A.
9. Electricity Council (edited), *Power System Protection*, Vols, 1–3, Macdonald, London, 1969.
10. Mathews, P., *Protective Current Transformers and Circuits*, Macmillan, London, 1955.
11. Wood, A. J., and B. F. Wollenburg, *Power Generation, Operation and Control*, Wiley, New York, 1984.
12. Heydt, G. T., *Computer Analysis Methods for Power Systems*, Collier-McMillan, New York, 1986.

PAPERS

Each year many papers are published, most of which are very specialist in nature. The reader is advised to consult the I.E.E.E. committee reports which appear from time to time giving a bibliography of relays and systems of protection. These are published in the *I.E.E.E., P.A.S. Trans.*

11. Bibliography of Relay Literature for the Period 1965–1966. I.E.E.E. Committee Report, *Trans. I.E.E.E., P.A.S.*, **PAS-88** (1969), 244.
12. I.E.E.E. Committee Report, 'Bibliography of gas insulated substations', *Trans. I.E.E.E.*, **PAS-101** (1982), 4289.
13. Rockefeller, G. D., 'Fault protection with a digital computer', *Trans. I.E.E.E.*, **PAS-88** (1969), 437.
14. Bornarde, P., and Bastide, J. C., 'A prototype of multiprocessor based distance relay', *Trans. I.E.E.E.*, **PAS-101** (1982), 491.
15. Dy Liacco, T. E., and T. J. Kroynak, 'Processing by logic programming of circuit breaker and protective relaying information', *Trans. I.E.E.E.*, **PAS-88** (1969).
16. Cory, B. J., and J. F. Mount, 'Application of digital computers to busbar protection', *Proc. of I.E.E. Conference, The application of Computers to Power System Protection and Metering* (May 1970).
17. I.E.E.E. Power System Relaying Committee, 'Transient response of current transformers', *Trans. I.E.E.E.*, **PAS-96** (1977), 1809–1814.
18. Miki, Y., 'Distance relay using a microcomputer, *Trans. I.E.E.E.*, **PAS-96** (1977), 602–613.
19. Kotheimer W. C., 'Electromagnetic interference and solid-state protective relays', *Trans. I.E.E.E.*, **PAS-96** (1977), 1311–1317.
20. Aoki, F., and M. Noda, 'Distribution line carrier system for switch telecontrol', *Trans. I.E.E.E.*, **PWRD-1** (1986), 35–42.
21. Bowe, T. R., *et al.*, 'A decision analysis model to determine the appropriate level of protection of the small producer/utility interconnection', *Trans. I.E.E.E.*, **PWRD-1** (1986), 78–90.
22. Brand, K. P., *et al.*, 'Topology based interlocking of electrical substations', *Trans. I.E.E.E.*, **PWRD-1** (1986), 118–127.
23. Balser, S. J., *et al.*, 'A microprocessor-based technique for detection of high impedance faults', *Trans. I.E.E.E.*, **PWRD-1** (1986), 252–259.

24. Schweppe, F. C., and E. Handschin, 'Static state estimation in power systems', *Proc. I.E.E.E.*, **62**, (1974).
25. Dopazo, J. F., O. A. Klitin, G. W. Stagg, and L. S. VanSlyck, 'State calculation of power systems from line flow measurement', *Trans. I.E.E.E.*, **PAS-89** (1970), 1698–1708.
26. Dopazo, J. F., O. A. Klitin, and L. S. VanSlyck, 'State calculation of power systems from line flow measurements, Part II', *Trans. I.E.E.E.*, **PAS-91** (1972), 145–151.
27. Allemong, J. J., L. Radu, and A. M. Sasson, 'A fast and reliable state estimation algorithm for AEP's new control center', *Trans. I.E.E.E.*, **PAS-101** (1982), 933–944.
28. Dopazo, J. F., O. A. Klitin, and A. M. Sasson, 'State estimation for power systems: detection and identification of gross measurement errors', *Proceedings 8th PICA Conference, Minneapolis* (June 1973).
29. Debs, A. S., 'Estimation of steady-state power system model parameters', *Trans. I.E.E.E.*, **PAS-93**, No 5 (1974).
30. Handschin, E., F. C. Schweppe, J. Kohlar and A. Fiechter, 'Bad data analysis for power system state estimation', *Trans. I.E.E.E.*, **PAS-94** (1975), 329–337.
31. Krumpholz, G. R., K. A. Clements, and P. W. Davis, 'Power system observability: a practical algorithm using network topology', *Trans. I.E.E.E.*, **PAS-99** (1980), 1534–1542.
32. Dy Liacco, T. E., K. Ramarao, and A. Weiner, 'Network status analysis for real time systems', *Proceedings 8th PICA Conference, Minneapolis* (June 1973).

Appendix I

Optimal System Operation

The limitations resulting from the 'B coefficient' method for economic load dispatching have been referred to in Chapter 4. A more general approach involving the use of non-linear programming has been reported by Carpentier[1] and Sirioux[2].

Non-linear programming classifies the method used to optimize non-linear problems. Normally it is required to minimize a function $f(x)$ subject to certain constraints, e.g.

and
$$\left.\begin{array}{ll} g_i(x) \geqslant 0 & \text{when } i = 1 \ldots, m \\ h_j(x) = 0 & \text{when } j = 1 \ldots, p \end{array}\right\} \tag{1}$$

The simplest case resolves to the solution of the set of simultaneous equations

$$\frac{\partial f(x)}{\partial x} = 0$$

to obtain the minimum but the solution is often very impracticable. Several numerical methods are available for minimization without constraints, especially gradient methods, e.g. steepest descents,[3] parallel tangents, conjugate gradients,[4,5] and the generalized Newton–Raphson.[3] With constrained minimization the classical approach is that of the Lagrange multipliers.[6] $f(x)$ is minimized with $m = 0$ and $p > 0$.

The solution of the constraints, $h_j(x) = 0$ together with

$$\frac{\partial f}{\partial x} + \sum_{j=1} \lambda_j \frac{\partial h_j(x)}{\partial x} = 0$$

is required and this is often difficult. An extension of the Lagrange method by Kuhn and Tucker[7] deals with equation (1) with $p > 0$ and $m > 0$. This method requires the solution of the constraints, $h_j(x) = 0$, together with

$$\frac{\partial f}{\partial x} + \sum_{j=1}^{p} \lambda_j \frac{\partial h_j(x)}{\partial x} - \sum_{i=1}^{m} u_i \frac{\partial g_i(x)}{\partial x} = 0$$

and,

$$u_i g_i(x) = 0, \qquad i = 1 \ldots, m$$

$$u_i \geqslant 0 \qquad i = 1 \ldots, m$$

Again it may be difficult to solve these equations but there are non-linear programming numerical methods which can cope with the solution.

The solution of the power-system dispatching problem describes in fact an operating point for the power system. This may be considered the result of a standard load flow given optimal boundary conditions. The following equations describe the nature of the problem to be solved. No attempt to give a solution will be made as this is outside the scope of this text. It is hoped, however, that the reader will gain some appreciation of this important topic and with the references given be able to commence a detailed study if required.

The more general approach involving the use of non-linear programming reported by Carpentier[1] and Sirioux[2] is formulated as follows. Let P_i and Q_i be the active and reactive powers generated at the ith node of an n node system, and C_i and D_i be the active and reactive powers consumed at the ith node. The total cost, $F_t = f(P_i, Q_i)$ for $i = 1 \ldots, n$.

As for modern systems

$$\frac{\partial F_i}{\partial P_i} \gg \frac{\partial F_i}{\partial Q_i}, \qquad F_t = f(P_i) \qquad i = 1 \text{ to } n$$

The steady-state equations for a transmission link (see section 2.7) between nodes i and k are,

$$P_i - C_i = \sum_k \frac{V_i V_k}{Z_{ik}} \sin(\theta_i - \theta_k - \delta_{ik}) + \sum_k \frac{V_i^2 \sin \delta_{ik}}{Z_{ik}} \tag{2}$$

and

$$Q_i - D_i = -\sum_k \frac{V_i V_k}{Z_{ik}} \cos(\theta_i - \theta_k - \delta_{ik}) + \sum_k \frac{V_i^2 \cos \delta_{ik}}{Z_{ik}} - V_i V_k^2 \tag{3}$$

where, $i = 1, \ldots, n, i \neq k, \delta_{ik}$ = angle across link = $\cos^{-1}(R_{ik}/Z_{ik})$ and θ_i and θ_k are the angles at nodes i and k respectively. Along with these two equations the following inequality constraints apply,

$$P_i^2 + Q_i^2 - {}_M S_i^2 \leqslant 0 \tag{4}$$

$$_m P_i - P_i \leqslant 0 \tag{5}$$

$$Q_i - {}_M Q_i \leqslant 0 \tag{6}$$

$$_m Q_i - Q_i \leqslant 0 \tag{7}$$

$$V_i - {}_M V_i \leqslant 0 \tag{8}$$

$$_m V_i - V_i \leqslant 0 \tag{9}$$

where the prefixes m and M denote lower and upper limits, respectively,

520

determined by MVA ratings, generator limits, transformer tap ranges, and boiler limits. Also synchronous stability requires that

$$\theta_i - \theta_k - {}_M\theta_{ik} \leqslant 0 \tag{10}$$

The cost F_t has to be minimized with respect to P_i, Q_i, V_i, and θ_i while satisfying equations (2)–(10). To obtain the solution non-linear programming is used. When the system contains appreciable hydroelectric generation the problems become more complex.

References

1. Carpentier, J., 'Contribution a l'etude du dispatching economique', *Bulletin de la Société Française des Electriciens*, Ser. 8, **3** (1962).
2. Carpentier, J., and J. Sirioux, 'L'optimization de la production a l'electricité de France', *Bulletin de la Société Française des Electriciens*, **March 1963**.
3. Householder, A. S., *Principles of Numerical Analysis*, McGraw-Hill, New York, 1953.
4. Powell, M. J. D., 'An efficient method for finding the minimum of a function of several variables without calculating derivatives', *Computer Journal*, **7** (1964), 155.
5. Fletcher, R., and C. M. Reeves, 'Function minimization by conjugate gradients', *Computer Journal*, **7** (1964), 149.
6. Hancock, H., *Theory of Maxima and Minima*, p. 100, Dover, New York, 1960.
7. Kuhn, H. W., and A. W. Tucker, 'Nonlinear programming', *Proc. 2nd Berkeley Symposium on Mathematical Statistics and Probability*, 1951, p. 481.
8. Sasson, A. M., and Merrill, H. M., 'Some applications of optimization techniques to power system problems', *Proc. I.E.E.E.*, **62** (1974), 949.
9. Happ, H. H., and Wirgau, K. A., 'A review of the optimal power flow', *J. Franklin Inst.*, **311** (1981), 231.

Appendix II

Table A2.1   Typical percentage reactances of synchronous machines at 50 Hz—British

Type and rating of machine	Positive sequence			Negative sequence X_2	Zero sequence X_0	Short-circuit ratio
	X_{st}	X_t	X_s			
11 kV Salient-pole alternator without dampers	22.0	33.0	110	22.0	6.0	—
11.8 kV 60 MW 75 MVA Turbo-alternator	12.5	17.5	201	13.5	6.7	0.55
11.8 kV 56 MW 70 MVA Gas-turbine turbo-alternator	10.0	14.0	175	13.0	5.0	0.68
11.8 kV 70 MW 87.5 MVA Gas-turbine turbo-alternator	14.0	19.0	195	16.0	7.5	0.55
13.8 kV 100 MW 125 MVA Turbo-alternator	20.0	28.0	206	22.4	9.4	0.58
16.0 kV 275 MW 324 MVA Turbo-alternator	16.0	21.5	260	18.0	6.0	0.40
18.5 kV 300 MW 353 MVa Turbo-alternator	19.0	25.5	265	19.0	11.0	0.40
22 kV 500 MW 588 MVA Turbo-alternator	20.5	28.0	255	20.0	6.0–12.0	0.40
23 kV 600 MW 776 MVA Turbo-alternator	23.0	28.0	207	26.0	15.0	0.50

Table A.2.2 United States. Approximate reactance values of three-phase 60-HZ generating equipment
(*Values in per unit on rated kVA base*)

Apparatus	Positive sequence						Negative sequence X_2		Zero sequence X_0	
	Synchronous X_d		Transient X_d'		Subtransient X_d''					
	Average	Range	Average	Range	Average	Range	Average	Range	Average	Range
2-pole turbine gen. (45 psig inner-cooled H_2)	1.65	1.22–1.91	0.27	0.20–0.35	0.21	0.17–0.25	0.21	0.17–0.25	0.093	0.04–0.14
2-pole turbine gen. (30 psig H_2 cooled)	1.72	1.61–1.86	0.23	0.188–0.303	0.14	0.116–0.17	0.14	0.116–0.17	0.042	0.03–0.073
4-pole turbine gen. (30 psig H_2 cooled)	1.49	1.36–1.67	0.281	0.265–0.30	0.19	0.169–0.208	0.19	0.169–0.208	0.106	0.041–0.1825
Salient-pole gen. and motors—with dampers	1.25	0.6–1.5	0.3	0.2–0.5	0.2	0.13–0.32	0.2	0.13–0.32	0.18	0.03–0.23
Salient-pole gen.—without dampers	1.25	0.6–1.5	0.3	0.2–0.5	0.3	0.2–0.5	0.48	0.35–0.65	0.19	0.03–0.24
Synchronous condensers air-cooled	1.85	1.25–2.20	0.4	0.3–0.5	0.27	0.19–0.3	0.26	0.18–0.4	0.12	0.025–0.15
Syn. cond.—H_2 cooled at $\frac{1}{2}$ psig rating	2.2	1.5–2.65	0.48	0.36–0.6	0.32	0.23–0.36	0.31	0.22–0.48	0.14	0.03–0.18

Permission: Westinghouse Corp.

Table A.2.3   Principal data of 200–500 MW turbogenerators—Russian

	Units	Values of parameters of turbogenerators of various types					
		1	2	3	4	5	6
Power	MW/MVA	200/235	200/235	300/353	300/353	500/588	500/588
Cooling of winding (stator)		Water	Hydrogen	Water	Hydrogen	Water	Water
(rotor)		Hydrogen‡	Hydrogen	Hydrogen	Hydrogen	Hydrogen	Water
Rotor diameter	m	1.075	1.075	1.075	1.120	1.125	1.120
Rotor length	m	4.35	5.10	6.1	5.80	6.35	6.20
Total weight	kg/VA (N/kVA)	0.93 (9.1)	1.3 (12.7)	0.98 (9.6)	1.05 (10.3)	0.64 (6.26)	0.63 (6.17)
Rotor weight	kg/kVA (N/kVA)	0.18 (1.76)	0.205 (2.01)	0.16 (1.57)	0.158 (1.55)	0.11 (1.08)	0.1045 (1.025)
X_d*	%	188.0	184.0	169.8	219.5	248.8	241.3
X_d'	%	27.5	29.5	25.8	30.0	36.8	37.3
X_d''	%	19.1	19.0	17.3	19.5	24.3	24.3
X_q	%	188.0	184.0	169.8	219.5	248.8	241.3
X_q''	%	28.6	28.5	26.0	29.2	36.0	36.0
X_0	%	8.5	8.37	8.8	9.63	15.0	14.6
τ_1†	s	2.3	3.09	2.1	2.55	1.7	1.63

* For all reactances unsaturated values are given.
† Without the turbine.
‡ Hydrogen pressure for columns 1–4 is equal to 3 atm (304,000 N/m²) and for column 5 4 atm (405,000 N/m²).
Source: Glebov, I. A. *C.I.G.R.E*, **1968**, Paper 11-07.

Table A.2.4 Standard impedance limits for power transformers above 10,000 kVA (60 Hz)

Highest voltage winding: BIL kV	Low-voltage winding: BIL kV. For intermediate BIL use value for next higher BIL listed	At kVA base equal to 55C rating of largest capacity winding							
		Self-cooled (OA), self-cooled rating of self-cooled/forced-air cooled (OA/FA) self-cooled rating of self-cooled/forced-air, forced-oil cooled (OA/FOA) Standard impedance (%)				Forced-oil cooled (FOA and FOW) Standard impedance (%)			
		Ungrounded neutral operation		Grounded neutral operation		Ungrounded neutral operation		Grounded neutral operation	
		Min.	Max.	Min.	Max.	Min.	Max.	Min.	Max.
110 and below	110 and below	5.0	6.25			8.25	10.5		
150	110	5.0	6.25			8.25	10.5		
200	110	5.5	7.0			9.0	12.0		
	150	5.75	7.5			9.75	12.75		
250	150	5.75	7.5			9.5	12.75		
	200	6.25	8.5			10.5	14.25		
350	200	6.25	8.5			10.25	14.25		
	250	6.75	9.5			11.25	15.75		
450	200	6.75	9.5	6.0	8.75	11.25	15.75	10.5	14.5
	250	7.25	10.75	6.75	9.5	12.0	17.25	11.25	16.0
	350	7.75	11.75	7.0	10.25	12.75	18.0	12.0	17.25

Table A.24 (contd.)

550	200	7.25	10.75	6.5	9.75	12.0	18.0	10.75	16.5
	350	8.25	13.0	7.25	10.75	13.25	21.0	12.0	18.0
	450	8.5	13.5	7.75	11.75	14.0	22.5	12.75	19.5
650	200	7.75	11.75	7.0	10.75	12.75	19.5	11.75	18.0
	350	8.5	13.5	7.75	12.0	14.0	22.5	12.75	19.5
	450	9.25	14.0	8.5	13.5	15.25	24.5	14.0	22.5
750	250	8.0	12.75	7.5	11.5	13.5	21.25	12.5	19.25
	450	9.0	13.75	8.25	13.0	15.0	24.0	13.75	21.5
	650	10.25	15.0	9.25	14.0	16.5	25.0	15.0	24.0
825	250	8.5	13.5	7.75	12.0	14.25	22.5	13.0	20.0
	450	9.5	14.25	8.75	13.5	15.75	24.0	14.5	22.25
	650	10.75	15.75	9.75	15.0	17.25	26.25	15.75	24.0
900	250			8.25	12.5			13.75	21.0
	450			9.25	14.0			15.25	23.5
	750			10.25	15.0			16.5	25.5
1050	250			8.75	13.5			14.75	22.0
	550			10.0	15.0			16.75	25.0
	825			11.0	16.5			18.25	27.5
1175	250			9.25	14.0			15.5	23.0
	550			10.5	15.75			17.5	25.5
	900			12.0	17.5			19.5	29.0
1300	250			9.75	14.5			16.25	24.0
	550			11.25	17.0			18.75	27.0
	1050			12.5	18.25			20.75	30.5

Table A2.5   Overhead line parameters—50 Hz (British)

Parameter	275 kV		400 kV	
	2×113 mm^2	2×258 mm^2	2×258 mm^2	4×258 mm^2
Z_1 Ω/km	$0.09 + j0.317$	$0.04 + j0.319$	$0.04 + j0.33$	$0.02 + j0.28$
Z_0 Ω/km	$0.2 + j0.87$	$0.14 + j0.862$	$0.146 + j0.862$	$0.104 + j0.793$
Z_{mo} Ω/km	$0.114 + j0.487$	$0.108 + j0.462$	$0.108 + j0.45$	$0.085 + j0.425$
Z_p Ω/km	$0.127 + j0.5$	$0.072 + j0.5$	$0.075 + j0.507$	$0.048 + j0.45$
Z_{pp} Ω/km	$0.038 + j0.183$	$0.033 + j0.182$	$0.035 + j0.177$	$0.028 + j0.172$
B_1 μmho/km	3.60	3.65	3.53	4.10
B_0 μmho/km	2.00	2.00	2.00	2.32
B_{mo} μmho/km	5.94	7.00	7.75	8.50

Z_1 Positive-sequence impedance.
Z_0 Zero-sequence impedance of a DC 1 line.
Z_{mo} Zero-sequence mutual impedance between circuits.
Z_p Self-impedance of one phase with earth return.
Z_{pp} Mutual impedance between phases with earth return.
B_1 Positive-sequence susceptance.
B_0 Zero-sequence susceptance of a DC 1 line.
B_{mo} Zero-sequence mutual susceptance between circuits.

Note: A DC 1 line refers to one circuit of a double circuit line in which the other circuit is open at both ends.

The areas quoted are copper equivalent values based on 0.4 in^2 and 0.175 in^2.

Table A.2.6 Overhead line data—a.c. (60 Hz) and d.c. lines

	500–550 kV—a.c.		700–750 kV	HVDC	
Region	Pacific	Canada	Canada	Mountain	Pacific
Utility	So. California Edison Co.	Ontario Hydro	Quebec Hydroelectric Com.	U.S. Bureau of Reclamation	Los Angeles Dept. of W. & P.
Line name or number	Lugo–Eldorado	Pinard–Hanmer	Manics.–Levis	Oregon–Mead	Dalles–Sylmar
Voltage (nominal), kV; a.c. or d.c.	500; a.c.	500; a.c.	735; a.c.	750(±375); d.c.	750(±375); d.c.
Length of line, miles	176	228	235.92	560	560
Originates at	Lugo sub	Pinard TS	Manicouagan	Oregon border	Oregon border near, Los Angeles
Terminates at	Eldorado sub	Hanmer TS	Levis sub	Mead substa.	Los Angeles
Year of construction	1967–69	1961–63	1964–65	1966–70	1969
Normal rating/cct, MVA	1000	—	1700	1350 mW	1350 mW
STRUCTURES					
(S)teel, (W)ood, (A)lum	S	S, A	S	S	S or A
Average number/mile	4.21	3.7	3.8	4.5	4.5
Type: (S)q, (H), guy (V) or (Y)	S (S-56)	V (A-51, S-51)	—; (S-71)	S, T (S-72, 73)	S, T (S-72, 73)
Min 60-cps flashover, kV	850	1110	—	—	915
Number of circuits: Initial; Ultimate	1;1	1;1	1;1	1;1	1;1
Crossarms (S)teel, (W)ood, (A)lum	—	S, A	S	S	S or A
Bracing (S)teel, (W)ood, (A)lum	—	—	S	S	S or A
Average weight/structure, lb.	19,515	10,800; 4730	67,3000	11,000	—
Insulation in guys (W)ood, (P)orc; kV	No guys	Nil	—	None	—
CONDUCTORS					
Al, ACSR, ACAR, AAAC, 5005	ACSR 84/19	ACSR 18/7	ACSR 42/7	ACSR 96/19	ACSR 84/19
Diameter, in.; MCM	1.762; 2156	0.9; 583	—; 1361	—; 1857	1.82, 2300
Weight 1 cond./ft, lb	2.511	0.615	1.468	2.957	2.68
Number/phase; Bundle spacing, in	2;18	4;18	4;18	2;16	2;18
Spacer (R)igid, (S)pring, (P)reform	S	R	R	S or P	—
Designed for—amp/phase	2400	2500	—	1800	3380
Ph. config. (V)ertical, (H)orizontal, (T)riangular	H	H	H	H	H
Cond. offset from vertical, ft	—	—	—	—	—
Phase separation, ft	32	40	50	38–41	39–40

Table A.2.6 (contd.)

Region	500–550 kV—a.c.		700–750 kV	HVDC	
	Pacific	Canada	Canada	Mountain	Pacific
Utility	So. California Edison Co.	Ontario Hydro	Quebec Hydroelectric Com.	U.S. Bureau of Reclamation	Los Angeles Dept. of W. & P.
Span: Normal, ft; Maximum, ft	1500; 2764	1400; 2550	1400; 2800	1150; 1800	1175; —
Final sag, ft; at F; Tension, 10^3 lb	46 at 130F; 19.1	45.0 at 60F; 3.4	226 at 120F; 6.2	40 at 120F; 12.4	35 at 60F; 13.0
Minimum clearance: Ground, ft; Structure, ft	40; 12.5	33; 10.5	45; 18.3	35; 7.75	35; 7.8
Type armour at clamps	Performed	Nil	None	—	None
Type vibration dampers	Stockbridge	Spacer damper	Tuned spacer	—	Stockbridge
Line altitude range, ft	1105–4896	800–1300	50–2500	2000–7000	400–8000
Max corona loss, kW/three-ph. mile	5	—	74 ($\frac{1}{2}$in. snow/h)	—	—
INSULATION					
Basic impulse level, kV	2080	1800	—	—	—
Tangent: (D)isk or (L)ine-post	D	D	D	D	D
Number strings in (V) or (P)arallel	2V	1 or 2P	4V	Single	Single
Number units; Size; Strength, 10^3 lb	27; $5\frac{3}{4} \times 10$; 30	23; $5\frac{3}{4} \times 10$; 25	35; 5×10; 15	—; —; —	—; —; —
Angle: (D)isk or (L)ine-post	D	D	D	D	D
Number strings in (V) or (P)arallel	2V	3P	4P	Single	—
Number units; Size; Strength, 10^3 lb	25; $6\frac{1}{4} \times 10\frac{3}{8}$; 40	26; $5\frac{3}{4} \times 10$; 25	35; $6\frac{1}{4} \times 10$; 36	—; —; —	—; —; —
Insulators for struts	None	None	—	—	None
Terminations: (D)isk or (L)ine-post	D	D	—	D	D
Number strings in (P)arallel	2P	3P	—	—	4P
Number units; Size; Strength, 10^3 lb	25; $6\frac{1}{4} \times 10\frac{3}{8}$; 40	26; $5\frac{3}{4} \times 10$; 25	—; —; —	—	—; —; —
BIL reduction, kV or steps	3.5 steps	—	—	—	—
BIL of term, apparatus, kV	1150–1425	1675–1900	2050–2200	—	1300

Table A.26 (contd.)

PROTECTION					
Number shield wires; Metal and Size	2; 7 No. 6 AW	2; $\frac{5}{16}$ in. galv.	2; $\frac{7}{16}$ in. galv.	1;—	2;—
Shield angle, deg; Span clear, ft	24.5; 28	20; 48	20; 50	30;—	20;—
Ground resistance range, ohms	—	15			30 max
Counterpoise: (L)inear, (C)rowfoot	C	L	L		
Neutral gdg: (S)ol, (T)sf; Eq. ohms	S; 0	S;—	S;—	S;—	S;—
Arrester rating, kV; Horn gap, in	420; None	480/432; None	636;—	—;—	—;—
Arc ring diameter, in.: Top; Bottom	None	None	—;—	—;—	—;—
Expected outages/100 miles/year	0.5	1	—	—	0.2
Type relaying	Ph. compar., carr.	Dir'l compar.		Variable	Grid control
Breaker time, cyc; Reclos, cyc	2; None	3; None	6; 24	—;—	—;—
VOLTAGE REGULATION					
Nominal, %	±10	—	—	Variable	Grid control
Synch. cond. MVA; Spacing, miles	None	None	—;—	—;—	None
Shunt capac. MVA; Spacing, miles	None	None	330; 236	—;—	On a.c. terminals
Series capac. MVA; Compens., %	None	None	None	—;—	—;—
Shunt reactor, MVA; Spacing, miles	—;—	—	—	—;—	—
COMMUNICATION					
(T)elephone circuits on structures	None	None	None	None	None
(C)arrier; (F)requency, kc	C;—	C; 50-98	C; 42.0 and 50.0	None	None
(P)ilot wire	None	None	P	None	None
(M)icrowave; Frequency, Mc	M; 6700	—;—	M; 7777.35	M;—	M;—

Permission Edison Electric Institute

Appendix III

Transmission Study—190

(Information reproduced by permission of the United States Department of the Interior)

This is a reconnaissance study conducted by the United States Bureau of Reclamation, Bonneville Power Administration, and Southwestern Power Administration to investigate e.h.v. interconnexions between the utilities of the north-western and central areas of the United States of America. In particular these interconnexions would carry *seasonal* peak diversity inter-changes between the north-west and the central states. Also, a limited con-sideration is made of the magnitude of the time-zone diversity between these regions. In Figure A3.1 the seasonal load diversity expected by 1980 is shown for the areas under consideration which are designated 1, N, C, and S, respectively.

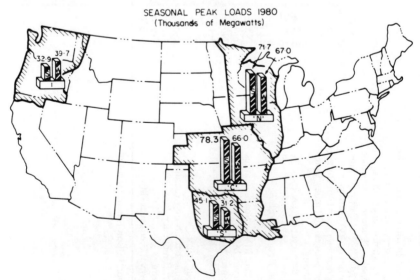

SEASONAL PEAK LOADS 1980
(Thousands of Megawatts)

Figure A3.1 Loads in areas under study (predicted 1980 peak values). S = summer; W = winter.

Apportioning non-committed resource additions in two areas

The supply and need for the two annual peak-load periods in an area are balanced with the lesser seasonal load diversity and also the need of the combined areas at the season of greater load is considered. Consider areas 1 and C for the summer and winter of 1980.

Let

$$R_1 = \text{resource addition required in area 1 (MW)}$$
$$R_C = \text{resource addition required in area C (MW)}$$
$$P_S = \text{sending-end value of interchange power}$$

(applies to '(1) end' for summer and '(C) end' for winter),

$$P_r = \text{receiving end value of interchange power}$$

('(1) end' in winter and '(C) end' in summer).

Summer 1980	Area 1	Area C	Total
Resource (MW)	27,571	48,538	76,109
Need (MW)	32,947	78,311	111,258
Surplus or deficit	−5376	−29,773	−35,149

Balancing the supply and need for area (1) in the summer of 1980,

$$27,571 + R_1 = 32,947 + P_s$$

For the same area in the winter,

$$(\text{Resource}) \ 28,086 + R_1 + P_R = 39,707 \ (\text{Need})$$

Balancing supply and need for combined areas for the summer,

$$R_1 + R_C = 35,149 + P_s - P_r$$

Also assuming a 10 per cent loss in interchange power at maximum seasonal exchanges,

$$P_s - P_r = 0.1(P_s + P_r)/2$$

Solving these four equations yields:

$$R_1 = 8655 \text{ MW} \qquad R_C = 26,806 \text{ MW}$$

$$P_s = 3278 \text{ MW} \qquad P_r = 2966 \text{ MW}$$

$$\left(\frac{P_s + P_r}{2}\right) = \text{average interchange power}$$

$$= 3122 \text{ MW}$$

and the transmission loss $= P_s - P_r = 312 \text{ MW}$. This indicates a 3000 MW seasonal interchange from east to west in winter and west to east in summer.

Similar analysis indicates the following equal and opposite seasonal interchanges:

Areas (1) and (N), 2500 MW N $\rightarrow$ 1 in winter
and 1 $\rightarrow$ N in summer

Areas (1) and (C), 3100 MW C $\rightarrow$ 1 in winter
and 1 $\rightarrow$ C in summer

Areas (1) and (S), 3100 MW S $\rightarrow$ 1 in winter
and 1 $\rightarrow$ S in summer

The minimum requirements for transmission capacity between areas will result from equal and opposite seasonal exchanges.

Transmission system (a.c.)

The above data indicate a firm power transfer of the order of 3000 to 4000 MW with a distance between end terminals of over 1500 miles (2400 km). In view of this, a.c. systems operating at 700 kV (nominal), 765 kV (maximum) are proposed. All lines have 70 per cent series capacitor compensation with the capacitors rated to cover an outage of one circuit. The quad-bundle conductors proposed reduce the line inductive reactance by 35 per cent compared with the single conductor line. Connexions are made every 480 to 800 km and the angle across a single line section is 8–12°. The total angle across the whole system is 40° for 3000 MW.

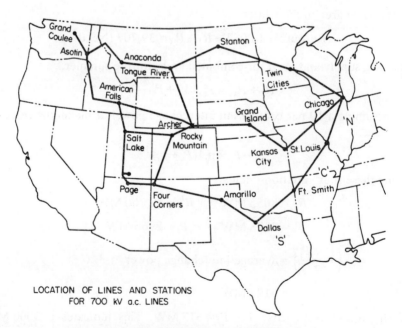

LOCATION OF LINES AND STATIONS
FOR 700 kV a.c. LINES

Figure A3.2 Possible 700 kV network for interchanges.

The general requirements would be:

4300 circuit miles, 20,000 towers;
80,000 insulation strings (35 units in each);
43 series capacitor stations (22,540 MVAr);
45 shunt reactors at 700 kV (13,500 MVAr);
80 shunt reactors at 15 kV (4000 MVAr);
83,700 kV circuit breakers and finally transformers 700/500 kV 8, 700/345 kV 5, 700/230 kV 1.

A geographical outline of the a.c. scheme is shown in Figure A3.2.

Transmission system (d.c.)
A system operating at 1200 kV (±600 kV) is proposed. Firm capability is ensured by appropriate switching to transfer full power on to the remaining lines in the event of an outage. The general routing of the d.c. system (untapped) is similar to that for a.c. shown in Figure A3.2. An economic analysis shows the d.c. scheme with no tappings to be cheaper than a tapped d.c. one, both d.c. schemes are cheaper than the a.c. proposals.

Index